The Paper Road

The publisher gratefully acknowledges the generous support
of the Philip E. Lilienthal Asian Studies Endowment Fund
of the University of California Press Foundation, which
was established by a major gift from Sally Lilienthal.

The publisher also gratefully acknowledges
the generous contribution to this book provided
by the University of Michigan.

The Paper Road

Archive and Experience in the Botanical
Exploration of West China and Tibet

ERIK MUEGGLER

University of California Press

BERKELEY   LOS ANGELES   LONDON

University of California Press, one of the most distinguished university presses in the United States, enriches lives around the world by advancing scholarship in the humanities, social sciences, and natural sciences. Its activities are supported by the UC Press Foundation and by philanthropic contributions from individuals and institutions. For more information, visit www.ucpress.edu.

University of California Press
Berkeley and Los Angeles, California

University of California Press, Ltd.
London, England

Library of Congress Cataloging-in-Publication Data

Mueggler, Erik, 1962-
    The paper road : archive and experience in the botanical exploration of West China and Tibet / Erik Mueggler.
        p.   cm.
    Includes bibliographical references and index.
    ISBN 978-0-520-26902-6 (cloth : alk. paper)
    ISBN 978-0-520-26903-3 (paper : alk. paper)
    1. Forrest, George, 1873-1932—Travel.   2. Rock, Joseph Francis Charles, 1884-1962—Travel.   3. Botany—Fieldwork—China—Yunnan Sheng—History—20th century.   4. Botany—Fieldwork—China—Gansu Sheng—History—20th century.   5. Botany—Fieldwork—China—Tibet Autonomous Region—History—20th century.   6. Yunnan Sheng (China)—Description and travel. 7. Gansu Sheng (China)—Description and travel.   8. Tibet Autonomous Region (China)—Description and travel.   9. Botanists—Scotland—Biography.   10. Botanists—United States—Biography. I. Title.
    QK355.M83   2011
    580.92'2—dc23   2011017740

20   19   18   17   16   15   14   13   12   11
10   9   8   7   6   5   4   3   2   1

For Sean McCabe, 1962–2009

Contents

Illustrations

Acknowledgments

So many people contributed ideas to this project that I have lost count. I can name only a few: Steven Feierman for a crucial early suggestion; Michael Watts, George Marcus, and Gunnar Olsson, for stimulating discussions at meals and evening seminars; participants in colloquia and conferences at the Kent University, the School for American Research, Cornell University, the University of Michigan, the University of Texas at Austin, Brandeis University, Harvard University, New York University, the CUNY Graduate School, Syracuse University, Brown University, Yale University, and the London School of Economics, for extraordinarily rich discussions of several chapters. Arthur Kleinman, Robert Weller, P. Stephen Sangren, Terrence Turner, Lucien Castaing-Taylor, Angela Zito, Faye Ginsburg, Stevan Harrell, Richard Handler, Elizabeth Ferry, Paul Eiss, Mandana Limbert, Bruce Grant, Nancy Jacobs, Michael Lambek, Stephan Feuchtwang, and Charles Stafford all provided unforgettable critiques, few of which I was able to do adequate justice. Sydney White and Charles McKhann were generous consultants on Naxi matters. David Porter, Webb Keane, and Gillian Feeley-Harnik provided sustained moral support and intellectual inspiration. Readers for the University of California Press and the University of Chicago Press gave indispensable advice during the last stages. And I thank Max Mueggler for his unlimited kindness, patience, and interest during the late stages of this project, which has lasted nearly his entire short life so far.

I have benefited from the generosity of many institutions while writing this book. The MacArthur Foundation, the British Academy, and the University of Michigan Office of the Vice President for Research provided funding for research. The latter institution also contributed a publication subvention that paid for the illustrations. A fellowship at the Center for Advanced Study in the Behavioral Sciences, two brief residencies at the Mesa Writer's

Refuge, and a writing fellowship at Deep Springs College gave me an embarrassment of peace-filled days for writing. The Royal Botanic Gardens at Edinburgh and Kew, the Arnold Arboretum, the Hunt Institute for Botanical Documentation, and the Royal Geographical Society were generous with their collections, their photographs, and the patient aid of their librarians and archivists.

Note on Transliteration

The Western botanists who provide much of this book's source material were self taught in Chinese, and they did not consistently adhere to any system of romanization when writing Chinese proper names. For this reason, I have chosen to use the contemporary *hanyu pinyin* system of romanization for the names of places and geographical features usually written in Chinese. Where the botanists use a widely accepted English equivalent I often give it side by side with the Chinese (or sometimes Tibetan or Naxi) name.

For the names of people, I also use the *hanyu pinyin* system, except where I have been unable to find or make a precise guess at the Chinese characters. In those cases, I have written the names as my sources did, choosing a single version for any one person when they were inconsistent.

There are several systems of romanization for the Naxi language. Since Joseph Rock is one of this book's main characters, however, and since he provides some of the Naxi source material, I have chosen to use a simplified version of the complex and awkward system of transcription that he invented, given in Joseph Rock, *A Na-Khi–English Encyclopedic Dictionary*, vol. 1 (Rome: Instituto Italiano per il Medio ed Estremo Oriente, 1963), xxxi–xxxvii. My simplifications are as follows. I have eliminated all of Rock's diacriticals except for his umlauts over the o, a, and u. I have replaced with diacriticals the numbers that Rock uses to indicate tones: I use a grave accent over a vowel to indicate a low tone, which Rock marks with the number 1; I place no accent mark over the vowel to indicate a mid-level tone, which Rock marks with the number 2; and I use an acute accent over a vowel to indicate a high tone, which Rock marks with the number 3. In addition, I eliminate the hyphens that Rock places between syllables. In verse, I separate all syllables with spaces; when quoting an individual word I place its syllables together without spaces or hyphens.

For Tibetan terms and proper names, I have used pronounceable equivalents, supplementing them with the Wyle system in parentheses where it seems appropriate.

Introduction

Zhao Chengzhang unrolled a sheet of paper. It was special paper, large and nearly transparent, purchased in Burma. This enterprise was all about paper. Each time he walked out the city gate, one of his mules carried a full load of paper, textured and absorbent, made of a dwarf bamboo that grew in thickets on the lower mountainsides. Piles were sold in every market town in the province. When he reentered the city after weeks or months of rough travel, he led a string of mules carrying stacks of paper neatly bundled and pressed between boards. Folded into each sheet was a plant specimen: *Meconopsis, Rubia, Primula, Gentiana, Potentilla, Rhododendron.* The plant presses were stacked in the courtyard now, under rocks. Over the next few days he would unfold each rough sheet, rearrange the specimen to accord with his exacting sense of space and proportion, and refold it into smooth writing paper. This paper could not be found in local markets: it came by rail, barge, and ox cart from Rangoon to the Burmese frontier town of Bhamo, then by mule over the long border road to this little trading city, Tengyue. He would attach a paper tag printed with a number and his patron Fu Laoye's name in English to each specimen. The tags came by parcel post from Edinburgh where Fu's own patrons lived. He would divide the specimens into bundles of one hundred, tie each bundle with tape made locally for binding sandals, then wrap it with yellow oil paper, manufactured in England and imported through Rangoon. He would place the packages in wooden crates made for this purpose by one of his own men, a skilled carpenter. Then he would send the crates off to Bhamo, two per mule.[1]

Many other kinds of paper were also involved. Before packing the specimens, he discussed each with Old Fu, who took notes, filling reams and reams of writing paper from Rangoon. In earlier years, he had hauled around

Figure 1.   Zhao Chengzhang. Courtesy of the Royal Botanic Garden, Edinburgh.

bound notebooks from Edinburgh in which Old Fu wrote as he traveled, though these days Fu mostly sat in Tengyue waiting for the plants to come to him. For his own daily accounts, Zhao used ordinary, unlined writing paper, which he found in the markets in Tengyue or the city of Lijiang, a month's walk away. Old Fu also drafted many letters on this paper before making fair copies on imported paper. And he consumed great quantities of photographic paper, which always seemed scarce: he wrote many letters frantically begging his employers to send more. Even that paper, slick and sharp-smelling, was made of plants—bits of the earth's flesh ground, soaked, screened, and pressed. Back home, Zhao had seen the old ritualists boil, pound, and screen tree bark to make the narrow chunks of thick, durable paper onto which they copied their ceremonial books. Many of these books, he knew, described journeys in long lists of place names that stretched all the way to this distance place and to the gigantic mountains that lay weeks to the north—mountains he and his companions had explored, enduring immense difficulty. In a funeral ceremony, which he had seen many times, the ritualists moved clay figurines along a long, painted paper road, a narrow map that led north into those ranges.

It was 1925. He had been at this since 1906. He walked; he gathered plants; he memorized attributes—numbers of petals, shapes of leaves, types of hairs and scales on brackets and leaves—and he thought about where he might find species he did not yet know. He learned strings of place names from travelers, listened to their stories, puzzled out new routes. He slept in inns, in village courtyards, on goatskins laid on the ground. He hired many parties of others from his village to make repeated excursions of days or months. And all this walking, searching, and gathering found its way into piles of paper: names on paper, lists, notes, maps, diaries, letters, accounts, and photographs on paper and, in particular, specimens and seeds folded into paper. For him, this region was made of earth and his experience of the earth. But it was also a thing made of paper, a thing just as real.

He dipped his pen into the ink bottle. Taking care not to tear the sheet with the crude steel nib, he drew eight wavy lines, roughly parallel, from the top edge to the bottom. The four longest were sharply jagged near the top, echoing the great limestone crags of the north. They rounded off near the bottom with the gentler mountains of the south. Between the ranges, he drew parallel, sinuous lines, indicating the great rivers: the Jinsha, or Yangtze; the Lancang, or Mekong; the Nu, or Salween. On the left, the Xiao snaked off into Burma; below that, he traced two branches of the Long, or Shweli. Along the rivers, he wrote the names of villages and mountains in neat Chinese characters. Then he drew another map, and a third, describing a vast, river-shorn region: north up the Nu towards Tibet, west to the Enmaikai, or Nmai Kha, and east to the great, rugged bend in the Jinsha, within which lay the city of Lijiang and his own village.

Rolled and packed in trunks stacked around the room were other maps of the same region. In the eighteenth century, scholars in Lijiang had com-

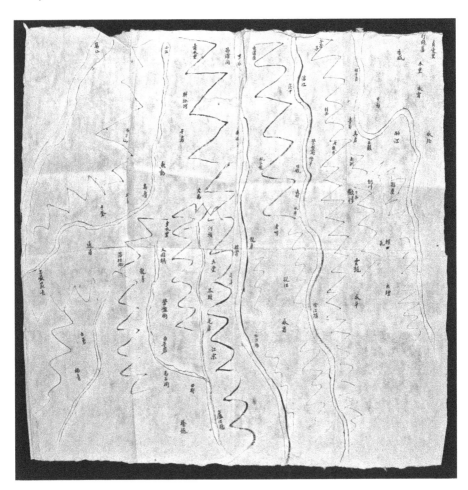

Figure 2.   Northwest Yunnan, by Zhao Chengzhang, 1925. Courtesy of the Royal Botanic Garden, Edinburgh.

piled the reports of traders and soldiers to make a map for an official gazetteer.[2] It centered on Lijiang, shown nestled within protective canals, but this center was dislocated to the far right-hand side. The map's left half described the city's vast northwest hinterland, once its far-flung empire, with the Nu river as its western edge. The far side of the Nu was blank, with a dotted line to mark the "boundary of the Nu barbarians." Here the great military machine that had swept out of Lijiang during the Ming dynasty had ground to a halt, unable to subdue the Lisu peoples who had taken refuge in the Nu gorge.

In 1900, in the service of another empire, the Great Trigonometric Survey of India had published two sheets showing Burma's Northeast Frontier and including most of Yunnan Province. They were compiled from route surveys drawn by British army officers traveling in secret. Large portions were white space, particularly between the Mekong (Lancang) and Salween (Nu) rivers. Parts of the Salween and the entire Taron (Qiu) were shown as speculative dotted lines. The Survey issued an improved map in 1909, taking into account the travels of Major H. R. Davis, who had marched through Yunnan with fifty men of the Nineteenth Yorkshire Regiment of Light Infantry.[3] Botanists had also drawn maps. Heinrich Handel-Mazzetti, professor of botany at the Natural History Museum in Vienna, had wandered Yunnan for five years during the Great War, making a map of his journeys. It included much information not on the India Survey maps, but the place names were written in his own idiosyncratic, sometimes indecipherable, system of romanization.[4] Zhao Chengzhang had studied all of these maps, tracing his journeys and pointing out to Old Fu where he had found each new specimen. He was also familiar with the most detailed maps of all—map-like texts, really—the long lists of place names drawn in elegant pictographic script in and around his home village of Nvlvk'ö, near Lijiang.

The maps Zhao now drew were different from any of these. He drafted them freehand from memory, rather than from route surveys, with no attempt to mark latitude or longitude, and little concern for accuracy of scale. Still, in many respects they were more comprehensive than the other maps, describing in more detail the most remote and difficult parts of the region. In style, they drew from the tradition represented by the eighteenth-century map from Lijiang. But they were not contributions to cartographic knowledge in any tradition. They were not guides to travelers nor aides to administration. They were records of experience. Some lines condensed memories of hundreds of days of arduous travel in the shadows of snowy peaks; others were records of a single glimpse, from a distance, of a range of peaks trending off to the north or west. Some place names were as familiar to him as his own name, part of his world since birth; others were rumors, heard once or twice from the lips of other travelers, tentative transcriptions of sounds in tongues foreign to him. The maps' casual, freehand style, combined with their careful descriptions of relationships between ranges, rivers, villages, and roads, trace, with some precision, a lengthy process of gathering disparate experiences into an imagined abstraction.

.   .   .

Figure 3. Northwest Yunnan, from Guan Xuexuan and Wan Xianyan, *Lijiang Fu Zhi Lue*, 1743.

東北至鏡可度
下瓦塘止距府
城四百五十里
江外係永北府
永盈界

永盈界

北至阿喜渡
口止距府城
五十五里江
外係中甸界

金沙江

塔城

塔城汛

久淺州

汛明目

鐵橋遺址

橋頭汛

石門關

嚴渡東

中甸木

汛鼓石

河九關

老君山

老君河

老志
潭

石鼓江

大中甸

小中甸

刺喔

金江

阿喜
渡口

雲山

玉龍山

象山峻

麗江府

黃山青龍橋

壽奇汛

本子關

閂雷門關

吾汛

吾鳴音

山州

舊五

打古汛

瀘沽宗池

永北府

永北界

永北界

東
東至瀾滄江渡口
距府城一百三十
里江外係永北府

吳烈山

舊通安州

近塘汛

關

馬欄橋

七河

府界

瀘沽江廿渡

係羅

剌字寨

刺鶴慶

府鶴慶界

金沙江

東

夷南至剌字
宗村止距府
城一百三十
五里接鶴慶
慶府黑泥箐
府界

南至千子哨止距
府城四十五里哨
南鶴慶府界

劍川州

千木河

界

界文慶鵝哨圍界

Zhao Chengzhang was likely the most prolific Western botanical explorer of the early twentieth century. The term Western, though never accurate, is appropriate shorthand here, since Zhao's project lay at the tail end of the long Linneaen enterprise in which European botanists and their collaborators cataloged the world's flora and participated in the creation, consolidation, and conceptualization of colonial empires. His home was a village at the foot of the luminous Yulong mountain range, known as Xuecongcun in Chinese and Nvlvk'ö in Naxi, the first language of its inhabitants. His given name reflected the aspirations of many in his village to literacy: *chengzhang* means to write or speak lucidly or logically. It sounds like a *xueming*, a school name, given a child when he or she began to attend school. And indeed he must have had at least a few years of education in one of the several public elementary schools the Qing government had established near his village, for he wrote in a clear, precise hand.[5]

From 1906 to 1932, Zhao worked for the indefatigable Scottish botanical explorer George Forrest, whom he knew as Lao Fu ("Old Fu"), Fu Laoye ("Master Fu"), Fu Lishi (his official Chinese name), or Fuzi (a respectful term for a Confucian teacher). During and between Forrest's seven lengthy expeditions to Yunnan, Zhao operated a network of botanical collectors—mostly kin and friends from his own village. As many as twenty-two worked for him at one time. During Forrest's first two expeditions, these men guided him on excursions in the Yulong range and on longer trips west and north. Later, Forrest set up bases in Nvlvk'ö, in Tengyue far to the west, or in tiny Tibetan hamlets along the upper Mekong in the remote northwest. Zhao and his men fanned out from these bases in parties of three or four on excursions that lasted from a week to two months. Zhao was the consigliere who ran these expeditions, hiring and paying the men and directing them where to go and what to look for. He and his "special gang" of six to eight men made the most difficult journeys, to the mountains of the extreme northwest and beyond. In the 1920s and 1930s, Zhao and his men continued to work between expeditions after Forrest returned to Edinburgh. In 1929, Zhao mounted a major expedition for four wealthy Scottish gardeners, buying supplies, deciding routes, and labeling and shipping the specimens and seeds. For the work of writing him a letter from Edinburgh and sending him the sponsors' money, Forrest collected fifty pounds sterling and credit for all his discoveries. In all, Zhao, Forrest, and the men from Nvlvk'ö sent tens of thousands of plant specimens and hundreds of birds, butterflies, snakes, and mammals to the Royal Botanic Garden, Edinburgh. Thousands of the plants and many of the insects and animals were new to science. More than 150 plant species remain in cultivation, and the rhododendrons alone are the parents of hundreds of hybrids. Species from 192 genera bear Forrest's name. These plants transformed the garden landscapes of the British Isles.

Until now, however, Zhao Chengzhang's full name has never been recorded. It appears nowhere in the many thousands of pages penned by his patron, and indeed it is unlikely that Forrest ever learned it. Forrest referred to him always as Lao Chao ("Old Chao"), and this is the name all Forrest's biographers have also given him.[6] Almost nothing exists in Zhao's own hand. None of the letters he wrote Forrest survive, and his labels were replaced by labels in English once the specimens arrived in Edinburgh. But inserted into the stacks of Forrest's correspondence in the archive of the Royal Botanic Garden, Edinburgh, and unnoticed by any biographer, are two ledger pages in precise Chinese characters.[7] They record the expenses of six men from Nvlvk'ö over a three-month period in 1925, during Forrest's sixth sojourn in Yunnan. Each entry begins with the date in two calendrical systems, then lists locations explored and expenses accrued. The expenses are mainly for food, salaries, and mule hire, but the collectors also purchased some curios: a sword, a quiver, a bow and arrows, some boar tusks, some rat skins.[8] In May, He Nüli and Zhao Chengzhang returned to Tengyue from the Pangdi River area and were paid seven yuan each as salary. Zhao Tangguang and Li Wanyun also returned; they were paid five yuan. Forrest annotated the ledger's margins in blue pencil, transcribing place names into Roman characters, transliterating Chinese numbers into Arabic numerals, and identifying the men with the nicknames by which he knew them: "Lao Ho" for He Nüli, "Lao Lu" for Li Wanyun, "Lao Sheung" for Zhao Tangguang, and "Lao Chao" for Zhao Chengzhang. The ledger's author could have been no one but the latter. Other than three maps, these pages are the only remaining artifacts that bear the mark of his pen.

I like the combination of precision and casual confidence in Zhao's maps, the way the tight, accurate writing balances the flowing lines. These maps were the explorer's attempt to express his sense of this enormously complex region, accumulated over many years of arduous travel. Here is how the river Long lies in relation to the great gorge of the Nu; these are the villages through which one passes as one travels north up the Long's west branch. I have seen farmers in Yunnan take up pens in the same casual way to trace in long bold strokes the boundaries of their villages and the valleys in which they lie—places as familiar as their own courtyards. The maps communicate the sense that Zhao held all of west Yunnan in his memory in the same intimate fashion that I hold the rooms of my house. A photograph Forrest made of Zhao around 1910 echoes this combination of assurance and precision (see figure 1). The explorer stands in a courtyard in the city of Dali,

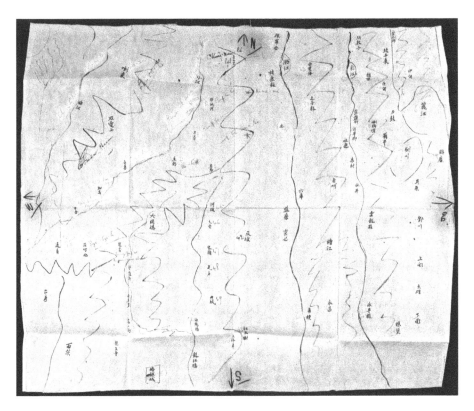

Figure 4.   West Yunnan, by Zhao Chengzhang with additions by George Forrest, 1925. Courtesy of the Royal Botanic Garden, Edinburgh.

upright as the pillars behind him. He is neatly dressed in the tunic, sash, turban, and trousers characteristic of the Lijiang area. His feet rest lightly on the flagstones, but it is also as though they are gripping those stones through the straw of his sandals. (Zhao and his colleagues from Nvlvk'ö wore only straw sandals as they wandered some of the roughest mountain country in the world: photographs taken decades later show them at rest on the road, sprawled amongst their plant presses, still in sandals). Zhao stands comfortably balanced, hands on hips, face sober. He radiates a formidable air of poised self-assurance.

I also like these maps because they record a dialog. The conversation centers on two parallel valleys, north of Tengyue and west of the Salween (Lu and Nuzi on the maps), through which two branches of the river Long flow. Zhao initiated the dialog by giving his sense of these valleys' location and orientation in relation to the region as a whole. Forrest then read and deciphered the maps, working two of them over with pens and pencils. The map in figure 4 bears evidence of at least three stages of translation. On what was probably a first pass, Forrest used a black pen, romanizing the names of a few key places: Ching-mu-li in the north, Imaw Bum in the far west. Could that be a crude attempt at the Chinese character *shan* (mountain) beneath Ching-mu-li? The emphasis on this place is evidence that the maps belong to the same conversation as Zhao's 1925 ledger pages, with which they were preserved. On the latter, Forrest used what seems to be the same black pen to circle Ching-mu (Qingmu), a briefer name for this mountain. His letters home were full of enthusiasm for Zhao's finds there.[9] In a second pass, Forrest went over figure 4 in gray pencil, transliterating a few more place names in the west, sketching in a watershed in the northwest margin, and numbering village and mountain names in and around the two focal valleys. Many names and some numbers in gray pencil appear to have been erased, a sign that this was a stage in a longer process of translation. Finally, Forrest went over the maps with a blue pencil, with more assurance. He marked up the map in figure 5 in detail, writing in all the names for which he gave numbers in figure 4. These contributions reveal his discomfort with Chinese characters even after four lengthy expeditions in Yunnan. One imagines that Zhao sat by his side reading off the names as he transcribed the sounds with the Roman alphabet. That he often wrote the same name differently supports this conjecture: thus "Ching-mu-li" on figure 4 is "Chimili" on figure 5, an inconsistency it is difficult to imagine the botanist committing could he read these very simple characters.

The dialog has another dimension too. Zhao's maps were preserved with a much smaller sketch of the two focal valleys in Forrest's hand. This map is cramped and hesitant, a dramatic contrast to Zhao's exact, fluid drawings. Mountain ranges appear to be represented by diagonal slashes, routes by solid black lines, rivers by light sketchy lines. The drawing is tightly focused on the two valleys, with none of Zhao's sweeping regional coverage. Its uncertainty and limited scope likely derive from its place in this cartographic conversation: it was an attempt to translate Zhao's knowledge into Forrest's vernacular—to understand what he was being told rather than to demonstrate his own knowledge. When one places the sketch beside Zhao's maps, one notices immediately that the latter do not show routes. This is remarkable, for the expertise they demonstrate was all gained by walking, and one would think that the lie of paths through this deeply scored country would be the most important form of knowledge they had to offer. But perhaps the routes were too obvious to note explicitly. They are implicit in the long valleys and strings of village names: drawing them may have been as unnecessary

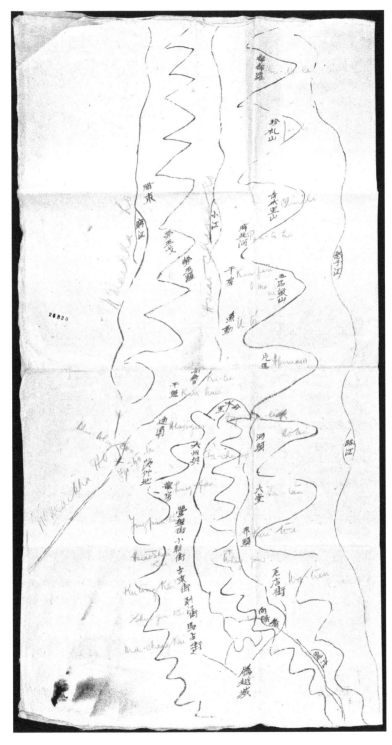

Figure 5. The Minguang valley, by Zhao Chengzhang with additions by George Forrest, 1925. Courtesy of the Royal Botanic Garden, Edinburgh.

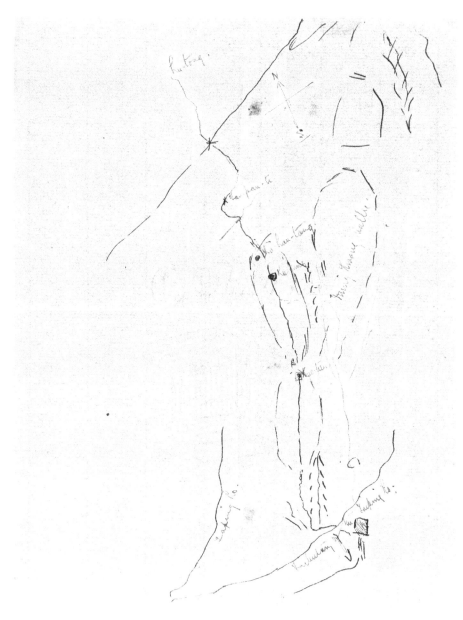

Figure 6.   The Minguang valley, by George Forrest, 1925. Courtesy of the Royal
Botanic Garden, Edinburgh.

as writing "Yunnan" across the top of the maps. On the other hand, Forrest's sketch of the two focal valleys is all about routes. It centers on the path, drawn in deep black, from Tengyue, below the bottom of the map, up through the western valley and over the Burmese border. "Is this the way?" it asks. If Zhao's maps are about the lie of the land, Forrest's sketch is about a particular, speculative experience: it is a crude, hypothetical route map of a possible journey.

This conversation exemplifies the tense relationship between these two men. Zhao's task was to create representations of his experience with specimens, talk about specimens, routes, talk about routes, maps, sketches, and gestures. Forrest's task was to understand representations, translate them, and recast them as bits of taxonomies and potential taxonomies. But Zhao often took the opportunity to create robust taxonomical abstractions, his maps being a clear example. And Forrest borrowed these abstractions, sometimes explicitly, sometimes covertly, as he struggled to create maps of plants and places, to superimpose these maps one on the other, and to force this superimposition to reveal blank areas where the taxonomies were not yet filled in. He would never acknowledge these borrowings. His enterprise depended on resolute commitment to the principle that imperial institutions, particularly the Royal Botanic Gardens, the India Survey and the Royal Geographical Society, were the sources of all taxonomical authority. In any case, this was the defining question of the relationship between the two men: how to take perceptions of the earth and forge them into representations of the earth for the imperial archive; how to take representations of the earth and make them a guide for perceptions of the earth; how to repeatedly traverse this complex, ambiguous, power-laden, inherently social territory between experience and archive.

•   •   •

What is a landscape? Perhaps for Zhao Chengzhang the landscape of West Yunnan was his experience of the various "cultural images" of the earth (to borrow a phrase from Cosgrove and Daniels's influential study of landscape), images that mediated his daily acts of walking, looking, and gathering. Yet his walking, looking, and gathering were mediated not so much by "cultural images" as by varied acts of putting the landscape on or between sheets of paper: if there were images, they were produced by these acts. In material form, his experience persisted only as paper. His maps bear traces of his personality—not merely the dedication, perseverance, and capacious memory that made him such a profoundly successful explorer and collector but also the grounded, commanding presence revealed in his photograph. As he made his paper Yunnan, he also made himself: his habitual attitudes, his habitual affects, even his wire-hard physical form. But this was hardly a solitary project. All his acts of putting the earth onto paper were made in dialog. Most immediately, these acts were in conversation with his patron, who wrote himself into each of them, bringing them to bear on his own sustained project of remaking himself from a rough, working-class wanderer into a bemedalled member of the Royal Horticultural Society. Fashioned in many such projects of encounter and self-making, the landscape of west Yunnan was a social relation.

Any landscape is more a social relation than an image or representation. We imagine social relations to link parties endowed with autonomy and agency, limited or expansive in nature. The landscapes Forrest and Zhao helped produce brought many parties together: wandering botanists and collectors, their scientific and cultural patrons, their domestic circles in Nvlvk'ö and Britain, to mention only some of the most immediate. Some other central participants were the screes, passes, meadows, and gorges through which these explorers and their parties trudged and out of which they fashioned their piles of specimens, descriptions, maps, and photographs. These participants too had a form of autonomy and agency. Their obdurate material existence, which shaped every moment of Zhao's experience, was their autonomy. And nearly everyone who lived in this dramatic landscape for long attributed social agency to its great mountains, finding them to be gods, and to its rivers and springs, treating them as manifestations of lesser animate forces. Here, through its drama, its difficulty, its fragmented nature, the earth created the conditions for an enormous variety of plant and animal life and an exceptional diversity in ways of human living. Through all the attempts to gather it, represent it, or abstract it, the earth retained agency, forcefully shaping the ways it could be apprehended. Here, even more than in most places, to think about the landscape as a social relation is to think about the social being of the earth.

If a landscape is a social relation, it is archival in form. The movement of the earth in time is not evolutionary like that of most living beings: it is accretive. And ever since our ancestors began to write, texts have been among the accretions that have formed its body. A landscape is a part of this body with a particular accretive structure, produce by sedimentation, erosion, upheaval, eruption, metamorphosis—forces that fold into the earth multiple images of itself. Landscapes grow, transform, and destruct as we delve through sediments, select among them, make montages of them, and add to their accretions in other directions. To think about the archival quality of a landscape is to reject the dogmatic version of representation: the world over there, images of the world over here, perceptions the troubled link between. It is to understand representations as folded into the world—as part of its substance.

ARCHIVE AND EXPERIENCE

This book is about the ways some wandering botanists put the earth onto or between sheets of paper: collecting, writing, and photographing. How are paper landscapes made? How does this making create, mobilize, and transform social relations? And how do these activities bring the earth into social being: how do they remake the earth and its inhuman inhabitants as participants in human social relations? At the heart of this book is a metaphor, with ancient and extensive roots: the earth as a book, which can be read, copied, transposed, revised, and written again. In a refinement of this metaphor, which emerges here and there, the act of walking is associated with the movement of a pen on paper; the body mediates between reading the book of the earth and writing it. A landscape might be understood as a social process of layering the earth with paper. We inscribe perceptions of the earth onto paper (or electronic media) to store, manage, interpret, organize, communicate, or create value from them. We fold this paper back into the earth as we name, plan, build, order, and administer. As an analytical term, landscape holds together these elements: earth and paper, experience and archive. This book explores the seam between immanent experience and abstracted archive, along which bodies walk and pens write.

The Paper Road tells the story of a remarkable set of encounters that spanned most of the first half of the twentieth century. They began in 1906 when George Forrest first walked up the main street of the tiny, poor village of Nvlvk'ö and hired a couple of men to guide him into the great range that towered above. They ended in 1950, when the Viennese-American botanist Joseph Francis Charles Rock, confused and frightened by the new rulers of northwest Yunnan, loaded his books into a China National Aviation Corporation airplane on a field just north of Nvlvk'ö, waved a hurried goodbye to the villagers who had packed his belongings at considerable personal risk, and flew off to India. Between 1906 and 1950, two generations of men from Nvlvk'ö explored west China for alpine flora for Western gardens and scientific institutions. The first were some twenty-five to thirty tough, skilled, knowledgeable, and adventurous botanical explorers, led by Zhao Chengzhang, and in George Forrest's employ. They scoured the entire vast, fissured region of northwest Yunnan for alpine flora during and between Forrest's seven lengthy expeditions to Yunnan, until his death in 1932. The second generation were twelve of their sons and nephews, who traveled all of north Yunnan, made a long, dangerous trek to Gansu and the high grasslands of what is now Qinghai, and explored the great mountain ranges of Minya Konka and the Konka Ling in west Sichuan, in the employ of Joseph Rock. Eventually, Rock turned from botany to his own brand of philological ethnography, making his life's work the investigation and translation of a strange, charming pictographic script he found written in piles of old manuscripts stored in homes in Nvlvk'ö and surrounding villages. Many of the young adventurers from Nvlvk'ö collaborated with him as collectors, companions, translators, and coauthors.

This book investigates the river of specimens, notes, diaries, letters, photographs, ritual texts, manuscripts, articles, and books that flowed from these collaborations. It is a work of spatial history, a term I borrow from Paul Carter's innovative study of explorers and settlers in Australia. This is to say it begins with the spatiality of individual experience: of bodies, hands and eyes moving along lines etched into the earth and inscribed on paper. It does not end there, however. The activities explored here transform individual spatial experience into more abstract and comprehensive visions of the earth and the social life in which it is involved. Zhao Chengzhang's maps are effects of such a transformation: an attempt to create out of two decades of arduous experience a comprehensive vision of a region. It is through such efforts at abstraction that the earth emerges most powerfully as a participant in social life. The chapters that follow trace such transformations as they emerge out of the contingent heterogeneity of daily experience.

Carter claims that the subject of his history is "not a physical object but a cultural one . . . the spatial forms and fantasies through which culture declares its presence."[10] Though spatial history pays close attention to space and language, in this formulation its ultimate goal is the recovery of a "cultural object." I could not make such a claim here even were I to wish to: the nature of the landscapes in question precludes it. The border regions between China, Burma, and Tibet were not inscribed or organized by any one set of "spatial forms and fantasies." They were the intersections of multiplicities of cultures, languages, empires, forms of economic activity, forms of political organization. This book investigates a few encounters along the paths that crisscrossed these landscapes. Most of these paths had existed already for hundreds of years, and they had been inscribed many times in maps, texts, taxonomies, or drawings on temple walls. Western botanists walked these paths rooted firmly, at first, in certainties granted by their own cultures of imperialism. Very quickly they became less certain, as the visions of others inflected their experience. Their collaborators walked with them, bending their own notions and intentions to the wills of their employers. But at the same time, they had other things in mind—other ways of organizing experiences of walking, looking and gathering, other ways of building social relations with the earth and its inhuman inhabitants. And some of these made their way into that river of paper flowing to the West.

THE ROAD TO WEST CHINA

Europeans began to search for garden plants in China in the seventeenth century. Many Jesuit missionaries were amateur botanists who shipped specimens and seeds to Paris in great numbers.[11] In 1757, the Qing court confined European trade to Canton, and for nearly a century that city was the sole Chinese center for European natural history. Traders and envoys explored the city's gardens and markets for attractive flowers, its environs for ornamental trees, and its shops and homes for botanical paintings. The Royal Botanic Garden at Kew and the Horticultural Society of London (which became the Royal Horticultural Society in 1861) sent salaried naturalists to Canton to look for plants and learn horticultural techniques. Sir Joseph Banks (1743–1820), president of the Royal Society and adviser to the Royal Botanic Garden, used his influence to recruit members of the Canton factory to search out plants new to science and send specimens, seeds, and live plants back to Kew. Banks appointed naturalists to accompany the Macartney embassy (1792) and the Amherst embassy (1816–17), rare opportunities to collect flora and fauna in the interior.[12] The Opium War (1840–42) and the Treaty of Nanjing (1842) opened up new territories around the ports of Shanghai, Ningbo, Fuzhou, Amoy (Xiamen), and Hong Kong. Almost immediately, the Horticultural Society of London hired an experienced botanist, Robert Fortune (1812–1880), to go to south China to collect seeds, specimens, and live plants, taking special notice of "the plants that yield tea of different qualities," "the plant which furnishes Rice Paper," and "Peonies with blue flowers, the existence of which is, however, doubtful."[13] Fortune's journey was a spectacular success; he made four more, and a host of botanical and zoological collectors followed him into the interior.

The European powers quickly built a new circulatory system for trade and natural history: the network of British consulates and Chinese Maritime Customs stations. By 1880 the British Consular Service had opened consulates in more than twenty Chinese cities and employed over two hundred officers; the Maritime Customs Service employed more than six hundred Europeans, two thirds of them British.[14] Many of these officials made botany a pastime or passion, working nearby territories on their summer breaks, and hiring Chinese peasants and hunters to collect for them when they were confined to office work. Like Banks before him, Sir Joseph Hooker (1817–1911) exercised his considerable political influence as director of Kew Garden in the service of natural history in China, getting the Foreign Office to assign young men interested in botany to the consular service. Among these were Augustus Margary, murdered while collecting orchids in Yunnan, Alexander Hosie, consul in Chongqing, who wrote a treatise on the industry of insects bred to produce white wax, and Augustine Henry, posted to remote Yichang in Sichuan, who collected an enormous herbarium and became a leading authority in Chinese upland flora.[15] Consulates and customs houses, often built in the bungalow style developed for colonial officials in India, housed microcosms of British domestic society. They were comfortable and convenient staging points for those who traveled to China to collect for horticultural institutions and, increasingly, for private firms and syndicates.

Plants from China joined the enormous flow of new floral species that were challenging scientists at the great botanical centers in Paris, Kew, London, Edinburgh, and Glasgow to create new taxonomical categories and relate new problems of geographical distribution to the emerging theories of descent. But Chinese plants were also big business. The Crystal Palace of 1851 had been modeled on a greenhouse; the Great Exhibition it housed resembled an overgrown version of the yearly Chelsea flower show, and garden plants were prototypes for the great commodity spectacle it showcased. "Like domesticated species in Darwin's world," writes Thomas Richards, "the commodities in the Crystal Palace exemplify plentitude and multifariousness and resist being resolved into a straightforward order. Well supervised and carefully tended in their garden environment, the things of the world have become so abundant that they have made it difficult to draw any clear distinction between individual and differences and slight varieties. They have become . . . hyper-productive to the point of riot."[16] Infinite in novelty and variety, evocative of every exotic territory of the empire and the globe beyond, garden plants were exemplary forms of the commodity kitsch that transformed Victorian and Edwardian domestic spaces in the decades following the Exhibition.

Until the late nineteenth century, nearly all trade between the Chinese and British empires traversed the oceans. In 1886, seeking to remove obstacles to commercial expansion posed by the unpredictable Burmese monarchy, the British annexed Upper Burma. A long series of guerilla wars in the border zones where Burma met Yunnan, Assam, and Tibet followed.[17] The border with Yunnan was particularly problematic. Between the Irrawaddy basin and Yunnan were interposed a large number of small, prosperous, and well-organized semi-independent principalities, the Shan and Wa states. Some of these states had confederated to oppose the chaotic last kingdom of Burma and now resisted British domination. By 1897, the British had pacified the region and imposed "indirect rule" on most Shan and Wa states and territories in the Irrawaddy watershed. The Qing court signed a treaty allowing the British a trading port in Tengyue (now Tengchong). This was an old trading city, occupied mainly by Han Chinese in a region where the other prosperous populations were concentrated in Shan states.[18] The merchants of Tengyue worked the trade between Yunnan and Burma. From the Indian empire flowed raw cotton and Bombay yarn, needles, ribbons, matches, umbrellas, cube sugar, tinned milk, metal buckets, enameled ware, and carpenter's tools. From Yunnan came hides, raw yellow silk, and the musk glands from several species of *Moschus,* the musk deer.[19] The new treaty port, with its British consulate and Chinese Maritime Customs office, opened this route to direct British exploitation.

Earlier in the century, French missionaries had established a network of Catholic missions in Yunnan's interior. Many French priests resided in the province for long periods; several became prolific botanical collectors. The route through Tengyue gave the British new opportunities to compete with the French in scientific endeavor, missionary work, and colonial expansion. Cartographers, geographers, mining engineers, and military explorers mounted expeditions from Bhamo through Tengyue and east to Dali and beyond. Botanists soon followed. Ernest Henry Wilson made several very successful journeys to central and southwest China for the British seed firm Veitch and Sons and for the Harvard Arnold Arboretum. The famous alpine collector Reginald Farrer visited the mountains of Gansu in 1914–1916 and the Burma-Yunnan border in 1919, where he died of dysentery.[20] The indefatigable Francis Kingdon Ward made twelve expeditions to northwest Yunnan, southeast Tibet, and northeast Burma between 1911 and his death in 1958.[21] Austrian botanist Heinrich Handel-Mazzetti, trapped in China by the Great War, wandered Yunnan and west Sichuan from 1913 through 1918. And Frank Ludlow, a colonial schoolmaster, and his friend George Sherriff, a retired professional soldier, walked through southeast Tibet in several expeditions from 1934 through 1949.[22] George Forrest and Joseph Francis Charles Rock were among the most prolific participants in this late, exuberant flourish of the great Linnaean project.

GEORGE FORREST

This book is not biographical. It focuses on the relationships and historical contexts in which these botanists found themselves embedded in China, and my attention to their lives in the chapters that follow is episodic and roughly chronological rather than comprehensive. Here I offer brief, introductory biographical sketches of Forrest and Rock to prepare the reader to place these episodes in their context of social relations among these botanists and their indigenous collaborators, scientific patrons, commercial and private sponsors, and domestic circles.

George Forrest grew up in the Scottish industrial and manufacturing towns of Falkirk and Kilmarnock. His parents, George Forrest, son of a blacksmith, and Mary Bain, daughter of a sailor, had moved to Falkirk from Paisley, where their first four infants had died. There George had been employed as a grocer; in Falkirk he apprenticed to a draper and built a draper's shop. Mary bore ten more children; eight survived; George was the youngest. The family was deeply invested in the Evangelical Union, or Morisonians, for which James, George's only brother, became a minister.[23] This branch of the evangelical Free Church of Scotland was founded in Kilmarnock in 1843 by James Morison who preached, against the accepted doctrine of the Church of Scotland, that Christ's atonement saved nonbelievers as well as believers.[24] The Morisonians and the other branches of the Free Church supported numerous missionary efforts overseas. Its members shared the strong sense of a racially differentiated global religious family that Catherine Hall describes for the Baptist Union in England.[25] His early religious education probably shaped the stance of limited openness that Forrest eventually developed towards his Naxi and Tibetan collaborators in China.

When George was twelve, the family followed James to Kilmarnock, where he had his congregation. There, supported by his brother and four unmarried sisters, George attended the Kilmarnock Academy, a secondary school reputable for its science teaching. His father, ill since the move, died when George was sixteen. George graduated at eighteen and went to work in a pharmaceutical chemist's shop. He worked for six years, drying herbs and packing capsules, living with James. A bequest of fifty pounds from a maternal uncle provided an opportunity for escape: he spent it on passage to Australia, where he dug for gold and hired out as a sheep farmer in New South Wales. He had little success, and when he inherited another small bequest from another of his mother's brothers, he returned to Scotland, moving into a cottage with his mother and sisters in Loanhead, a coal mining village south of Edinburgh. Unemployed, he wandered the North Esk Valley, botanizing for a Glasgow natural history society of which James was a member. On a fishing trip he found a stone casket with some human bones. He took it to the National Museum of Antiquities in Edinburgh; the Museum's keeper, John Abercromby, returned with him to the site, eventually publishing a paper on the three ancient stone coffins they found there.[26] Abercromby, well connected in Scotland's robust network of natural history societies, took a liking to Forrest. He wrote Isaac Bailey Balfour (1853–1922), regius keeper of the Royal Botanic Garden, Edinburgh, asking if he knew of any society in need of a young man to search for botanical specimens abroad. Balfour did not. He did, however, offer Forrest employment as a gardener at about half the usual wage.[27] Forrest was thirty-one, with no trade, no profession, and no prospects. For six months, he walked six miles from Loanhead to work in the Garden's herbarium. There, he prepared botanical specimens, courted Clementina Trail, also employed in the herbarium, and pressed "the Professor," Balfour, to help him find a position abroad as a plant collector.[28]

During the nineteenth century, the most prominent horticultural firm in Britain had been James Veitch and Sons. Beginning in 1840 this firm sent horticultural collectors over virtually the entire world: South and Central America, West Africa, Malaysia, Borneo, Australia, New Zealand, Korea, Japan, and, finally, China.[29] The firm's first collector in China was William Maries, who explored the Yangtze river basin in 1878 and 1879 with moderate success.[30] In 1899, the firm sent Ernest Henry Wilson to Yunnan and Sichuan via the new route from Burma through Tengyue. Wilson walked across Yunnan to the Yangtze gorge near Yichang, Sichuan, and sent home seeds of 305 species, hundreds of Wardian cases of live plants, and specimens of 906 species.[31] He made two more journeys to Sichuan for Veitch and Sons before going to work for the Arnold Arboretum at Harvard in 1906. From 1903 until the firm closed its doors in 1914, Veitch's seed catalogs contained special illustrated sections advertising many novelties from southwest China. Unlike most previous finds from China, gathered in the subtropical regions of the southern coast, many could be grown in Britain outside of hothouses. They were immensely popular with the gardening public.

After Wilson's dramatic hauls, Veitch's directors decided that there were likely few new profitable species to be found in China. It was a serious failure of imagination. By 1900, Augustine Henry, employed by the British Consular Service, had sent some 158,000 dried specimens to Kew, mostly from Sichuan. The French Catholic missionaries Jean Marie Delavey, Armand David, and Jean André Soulié had collected specimens and seeds of thousands of species from Sichuan for the Muséum national d'Histoire naturelle, including, spectacularly, thirty new species of *Primula*. Kew and Edinburgh competed for samples of seeds from Paris.[32] To Isaac Bailey Balfour at Edinburgh, these collections were evidence of an extraordinary abundance of mountain flora connecting the flora of central China, then being explored by Wilson, with that of the western Himalaya, described in some detail by Sir Joseph Hooker from 1848 to 1851.[33] One of Balfour's acquaintances, Arthur Bulley, a cotton broker and gardening aficionado, was particularly excited by the idea that Yunnan might be a treasury of hardy alpines. In 1904, inspired by Veitch and Sons's success in marketing Chinese plants, he founded his own commercial nursery, A. Bee and Co. (after his initials).[34] He wrote Balfour to ask if he knew of a botanist whom he might send to Yunnan. Balfour did: "an excellent, industrious, and steady man" who had learned something of the world's plants at the Garden's herbarium.[35] A month later, in May 1904, Forrest contracted with Bulley for a journey of three years at a salary of one hundred pounds a year, betrothed Clementina Trail, and sailed for Bombay.

A liner to Bombay, a train to Madras, a boat to Rangoon, a paddle steamer up the Irrawaddy to Bhamo, a flurry of buying and organizing supplies and hiring mules and drivers, then a mule train to Tengyue: it took almost three months, and it would become the most familiar of routes. In Tengyue, the acting British consul, G. Litton, resident there for three years, took him on his first journey in northwest Yunnan, a long loop through Dali and Zhongdian. He made two more journeys on his own, the goal of the second the French Catholic mission at Cigu on the Upper Mekong. He arrived just before a small army of Tibetans, led by lamas from Batang, reacting to Francis Younghusband's invasion of Tibet, swept down the Mekong, attacking Catholic churches. He fled with two priests, Pères Dubernard and Bourdonnec, and a Tibetan Catholic named Ganton, or Anton, who would become one of the most accomplished botanical explorers of Yunnan's far northwest. The priests were killed on the road. Their martyrdom would found a large, thriving Catholic community along the Upper Mekong. Beijing would eventually pay reparations; a cathedral would be constructed; the priests' gravestones would be hidden during the Cultural Revolution and erected again beside the cathedral; the stories of their deaths would become the community's founding myth in the twenty-first century.[36]

Forrest and Ganton struggled down the Mekong, starving, wounded, and in rags, eventually to be rescued near Yezhi. The consulate had already written Forrest's mother and sisters with the news of his death.[37] Not three weeks later, he made a difficult journey up the Upper Salween with Litton; then it was another excursion to Dali and Lijiang, where he camped for the summer on the Yulong range with men from Nvlvk'ö. Chapter 1, "The Eyes of Others," looks at this first expedition, watching Forrest feel his way through the racial rules of empire and his own violent aversion to Chinese stares, toward a tentative accord with Naxi men from Nvlvk'ö, whose sterling virtue was the way they used their eyes.

Back in Scotland, Forrest married Clementina Trail, rented a house in an Edinburgh suburb, and worked in the Royal Botanic Garden sorting, labeling, and describing his specimens. Soon enough, A. Bee and Co., (usually referred to as Bees Ltd.) employed him for a second expedition, cut short after one year by Forrest's disgust with Bulley, who was late in sending funds. After this, Forrest's expeditions would be funded by syndicates of wealthy gardeners. During his second and third expeditions, Forrest began, in collaboration with Balfour, a search for the generative center of the genus *Rhododendron* that would last nearly a decade. His engagement with Yunnan's landscape was dominated by the problem and promise of difference: the extraordinary proliferation of floral species and varieties, each of which demanded intense empirical engagement to which he could see no end. The generative center of *Rhododendron* promised to focus this great scattering of difference on a discrete place. There, perception would transcend the empirical engagement with detail after detail of species after species, and the genus, as well as his history of engagement with this landscape, might be comprehended in a single vision.

Zhao Chengzhang and the botanists from Nvlvk'ö were very active partners in this endeavor. They knew a great deal about plants before they met Forrest, and they had distinctive understandings of travel, exploration, and adventure. Yet the nature of the evidence makes it impossible to know with any precision how these forms of knowledge may have inflected Forrest's enterprise. Chapter 2, "Farmers and Kings," looks at some of the indirect evidence, describing the history of conquest that scattered communities of Naxi speakers all through northwest Yunnan, and notices that the region of conquest coincides with some precision with the areas Zhao Chengzhang and his kin and neighbors explored, despite Forrest's frequent plans to send them elsewhere. The paper landscape produced in these journeys came into contact with another archive: a huge trove of ritual manuscripts in the charismatic pictographic writing system now known as *dongba* script. Chapter 3, "The Paper Road," ventures into the dongba archive. When people in Nvlvk'ö buried their dead, purified their houses, or healed their kin, they listened to dongba read books from this archive describing detailed routes taken by ancestors, spirits, and demons. The best-known was the route that the dead traveled on their way to the lands of the ancestors. It worked northwest, ascending to the lands of the gods in the region where the search for the generative center of *Rhododendron* was most intensive. Chapter 4, "The Golden Mountain Gate," reads Forrest's letters and this archive side by side, speculating as to how the scientific project of mapping Yunnan's flora might have been shaped by the orientation of the earth in dongba pictographic writing.

Forrest worked closely with Isaac Bailey Balfour on this project. Using the explorer's specimens and descriptions, Balfour created a new classification of the *Rhododendron* genus, dividing it first into *lepidote* and *elepidote* series based upon bud construction and microscopic investigations of the leaf integuem. His work became the foundation for all further classifications of the roughly nine hundred species of the genus.[38] After Balfour's death in 1921, Forrest ended his quest for the generative center of *Rhododendron*. Balfour's successor as regent keeper, William Wright Smith, would eventually take over his work and publish it as *The Species of Rhododendron* (1930).[39] Smith was Edinburgh's expert on another very large genus, *Primula*, on which he worked with Forrest through his seven expeditions. Forrest's parties also made enormous bird collections for a wealthy enthusiast, Stephenson Clarke, for Lord Walter Rothschild's private zoological museum in Tring, and for Reginald Cory, an investor in several of the private syndicates that funded the later expeditions. After Forrest died of a heart attack while hunting pheasants near Tengyue, Zhao Chengzhang stayed on to pack up the specimens of his final expedition. Forrest did not have time to remove the original labels, written in Chinese by his taxidermist, and replace them with labels in English: these labels, still at Tring, are the only handwriting remaining from any of the men who worked for decades for Zhao and Forrest.

JOSEPH FRANCIS CHARLES ROCK

He ran away from home at eight; he left for good at eighteen. His father, Franz Rock, was a steward in the Viennese winter palace of the Polish Count Poticki, son of an amateur orientalist, Count Jan Potocki, who authored the *Saragossa Manuscript*, a tale of mystical wanderings in the East. His mother, Franciska Hofer Rock, died when he was six; thereafter his elder sister Lina cared for him. Perhaps he was allowed to use the count's library, for he began to teach himself Arabic and Chinese. His father hoped, then demanded, that he become a priest; he wanted to join the navy. He left directly after finishing *gymnasium*—though perhaps he never did finish, for no record of his diploma exists in any of the schools he might have attended. He wandered Europe and North Africa for four years, nearly penniless. He returned for his father's funeral in 1904, then went to England where he was diagnosed with tuberculosis. Lina nursed him in Vienna for a time, then he was off again, searching for sunny weather in Italy, Tunis, and Malta. He crewed on a ship to Hamburg; he was in and out of charity hospitals in Hamburg and Belgium for months. He missed a train to Aix-La-Chapelle and sailed for New York instead. There, he borrowed a bed, washed dishes, began to learn English, fell ill again, and sailed for dryer climes. In San Antonio, he attended English and Bible study classes at Baylor University for a few months. Then, in 1907, he sailed for Hawai'i. He had a single gold coin in his pocket.[40]

He fell in love with those islands, their strange forests, their sun-flecked oceans. He instructed Latin and natural history at a private middle school in Honolulu and taught himself systematic botany with extraordinary speed. He quit his job to spend more time outdoors; then he convinced Hawai'i's Division of Forestry to hire him as its sole botanist, tasked with building an herbarium of indigenous plants. A few months later, he published his first scientific paper.[41] More than fifty papers and six monographs on Hawai'ian plants would follow.[42] In 1911, the Division of Forestry transferred his burgeoning herbarium and its creator to the College of Hawaii, soon to become the University of Hawaii. He sailed to every island and rode or tramped up every mountain. On Kauai, he met an elderly rancher named Francis Gay who had created a book of plants, each page a splatter print, made by placing specimens on paper and sprinkling paint over them. It included nearly every indigenous plant on Kauai and recorded their Hawai' ian names. Rock would always find irresistible such great, interleaved volumes of paper and plants, images and texts, names and leaves. In 1913, he published his own first great volume, *The Indigenous Trees of the Hawaiian Islands.* It includes Hawai'ian names for each of the three hundred trees it describes.[43] Some of these names were given him by his guides, some by the priests to whom they introduced him, most by Francis Gay, the first of many scarcely acknowledged collaborators.[44]

In September 1919 he boarded a ship and watched the mountains of Hawai'i dissolve into the sea. He had saved up his salary: this was to be his "trip around the world."[45] He was to collect plants for the U.S. Department of Agriculture and exchange specimens of Hawai'ian flora with herbaria in Asia and Europe. For the first time, he kept a diary. In Manila, he visited the Bilibid Prison, run by the U.S. colonial administration: "Dormitories are radiating like the spokes of a wheel; the prisoners march around with their dinner plates and big hats. Poor devils . . . a few years ago they tried to break out and were shot down with machine guns."[46] In Canton, he wandered the alleys with a guide, looking down: "The streets are paved with big oblong granite rocks of various sizes, often not fitting well or uneven, exposing the sewer or drainage underneath, as well as letting escape the odor peculiar to an open sewer system. . . . a busy bee hive."[47] He sailed on to Singapore and Rangoon. In Darjeeling, he looked up, and the great snowy barrier of Kunchenjanga hurt his naked eyes.[48] In Berlin, he retrieved from the Botanisches Museum Berlin-Dahlem a thousand specimens of Hawai'ian flora collected by Hawai'i's first great Western botanist, William Hillebrand.

Back at the College of Hawaii he wrote five more monographs about trees.[49] He dreamed of a textual herbarium, with full descriptions and photographs of every plant of every family, a complete illustrated flora of the Hawai'ian Islands. Assigned to the college's grounds committee, he planted five hundred trees and shrubs, grown from seed he had collected.[50] Still, he could not stop moving. He sailed to Manila, Singapore, and Java in 1916; he botanized in Southern California in 1917; he went to Siam, Malaya, and Java in 1919. While he was away, and without consulting him, the college made plans to transfer his herbarium, now about twenty-eight thousand specimens, from the cramped room where he worked and taught to the Bishop Museum.[51] The surest way to insult this sensitive soul was to ignore or overlook him: he resolved to leave Hawai'i forever. He and his student assistant spent six months going over every specimen of the herbarium; he created a map of the campus, recording on it a number for every ornamental plant indexed to its scientific name.[52] Then, in May 1920, he sailed for San Francisco, hoping to find work as a botanist on the mainland.

In 1910, searching the island of Molokai for trees of the genus *Kokia*, he had come upon the lepers' colony, established in 1865 by King Kamehameha V. The colony was on a peninsula, walled off from the island by sheer sea cliffs, accessible only by sea. Until the 1940s, it received all persons in Hawai'i diagnosed with leprosy.[53] It was the subject of popular articles on the martyrdom of Father Damien deVeuster, a Belgian priest who worked there from 1873 until his death in 1889, and of sensationalist newspaper accounts describing more recent conditions of poverty and starvation.[54] These themes would later stir Rock's imagination in China: isolated communities, unique or charismatic leaders, human abjection. In 1915, patients from the colony were subjects in experiments at the Kalihi Leprosy Hospital in Honolulu with an oil derived from the seeds of a tree from India and Burma called chaulmoogra (several species of *Hydnocarpus*). The oil was an Ayuvedic remedy for skin diseases introduced to Western medicine in 1854 by a British physician. British chemists had isolated several fatty acids from the oil, which physicians at the Kalihi hospital combined with olive oil to inject hypodermically.[55] Alice Ball, a young African American woman, who, after becoming the first woman to receive her Masters degree in science from the College of Hawaii became the first to work at the college as a chemistry instructor, distilled ethyl esters from the fatty acids. The esters proved much more easily injected and readily absorbed than the unrefined oils. About a year later, Ball died of an unidentified illness at the age of twenty-four. Arthur L. Dean, head of the chemistry department and later president of the college, took up the work and, without crediting Ball, published several papers on methods of distillation of and treatment with the ethyl esters. Known as "Dean derivatives," these ethyl esters were used to treat leprosy until sulfones were developed in the 1940s.[56]

Among Rock's correspondents was David Fairchild, head of the Division of Foreign Seed and Plant Introduction in the U.S. Department of Agriculture. In 1920, he visited Fairchild in Washington, D.C. with a proposal. Chaulmoogra seeds had been available on the world market since the early nineteenth century, but they were expensive. Rock convinced Fairchild to fund an expedition to Burma to collect enough *Hydnocarpus* seed to establish a plantation in Hawai'i. "Tears came to his eyes as he pleaded his case," Fairchild wrote later.[57] He departed Washington for Siam, Burma, Bengal, and China in the fall of 1920. He sailed to Bangkok and took a train to Chiangmai. With his interpreter, cook, and "boy," he descended the Meh Ping River, walked over the mountains to the Burmese port of Moulmein, took the train to Rangoon, and made his way to the Upper Chindwin District. He found groves of chaulmoogra trees, first seedless, then, further up in the hills, full of seeds. He shipped seeds to Washington, and went on to Calcutta and then Assam. He sold an account of his journey to the National Geographic Society and then, fat with cash, made a quick trip to Vienna. He bought Lina a house in the country, lavished gifts on her two boys, and enjoyed a night at the opera. In the meantime, the U.S. Department of Agriculture extended his contract: he was to search Yunnan for "economic plants," particularly blight-resistant trees: oaks, chestnuts, walnuts, pears, plums. He would not really return until thirty years later, after the Chinese Communist Party took over his home in Lijiang.

I hesitate to say that his experience of mountains and towns, companions and strangers, filth and purity, and his own misery, loneliness, and occasional good moods, were mediated by writing, as though that were all writing was for him. Writing did mediate his experience, but it was also a foundation for his experience. Of utmost importance were his field notes:

thousands of precise, technical outlines of the miniature topographies of flowers, ovaries, and leaves. And he wrote many letters: in particular his most comprehensive accounts of his first excursions were in correspondence with David Fairchild.[58] But communication by letter was so unsatisfactory: there was so much that could not be said. He turned increasingly to his diary: volumes and volumes of outpourings on roads, plants, moods, musings, disappointments. Then, after a decade in China, he began writing books again: collections, descriptions and taxonomies no longer of plants but now of texts—of the pictographic books he found in the homes of his collaborators and their neighbors, and of the geographical texts in Chinese about the roads he traveled. His life, as I see it, was a grand experiment in working out ways to entangle texts and landscapes: all his creations—herbaria, diaries, geographies, maps, translations—were about making a great book of the earth, piling text on earth and earth on text. It was an inelegant experiment: he began as a competent artificer of sturdy English prose and laboriously taught himself to write very badly indeed, as he sought ways to make words follow footsteps, to trace the rebus-like puzzles of pictographic writing, to double the dual experiences of walking and reading. But it was a contribution nonetheless: a vision of an earth where every footstep was intertwined with deep natural/textual histories of names, of lives and deaths, of human and inhuman inhabitants.

He walked from Chiangmai north into Yunnan in the winter of 1921. On the Chinese side of the border, his party walked through the Tai Lü state of Chiang Rung, or Sipsong Panna, where he met Dao Cheng'an, the head of state, or *cao penliang*, of the polity since 1885, supreme authority over some 220,000 souls.[59] Later Rock would become infatuated with the princes and nobles of China's borderlands, but for the moment he was resolutely unimpressed: the prince was "an ignorant and cooly-like individual who lives in a barn-like structure with pigs and horses living under the house."[60] They walked up the Black River (Heihe) Valley, past graves and fields of opium poppies, where they met Tibetan caravans, five hundred yaks strong, hauling cakes of tea from Sipsong Panna north to the Tibetan town of Zhongdian and beyond.[61] In bustling Dali they met Forrest, coming in from Tengyue. Rock tried to be collegial, writing Fairchild to send to Edinburgh the *Rhododendrons* he had collected in southern Yunnan, where Forrest had never been. Later, William Wright Smith wrote Forrest from Edinburgh: "I have . . . had a very high-falutin letter from the U.S. Dept of Agriculture, describing in very free terms a gentleman known to you as Rock. It appears he has . . . collected nearly 500 packets of Rhodo seeds, and they are insisting on sending us some of this material."[62] The scientists at Edinburgh would gradually warm to Rock, respecting him for his excellent specimens and extraordinary photographs, if not for the originality of his finds. But Forrest disliked him immediately as a blowhard Yankee upstart. Forrest helped Rock establish himself in Nvlvk'ö, but in later years he would arrange to be absent from that village whenever Rock was present and vice versa.

Rock spent the summer in Lijiang, exploring the Yulong range. In September, he hired a party of men from Nvlvk'ö and headed to the far western corner of Yunnan, to Tengyue and the nearby Tai Lü state of Zhanda, looking for chestnuts. There, he met another Tai Lü prince, Dao Baotu, who had ruled the state since the 1911 revolution, remarking only that he had "the bad manners of the Chinese of this region."[63] He went to Sadon in Burma and then returned to Lijiang on a northerly route crossing the great gorges of the Salween and Mekong. In October, he headed west again, to the valley of the Mekong. Chapter 5, "Bodies Real and Virtual," touches on these journeys as it examines the uses to which Rock put his camera and, eventually, his portable gramophone, to create a rigid social diagram in an effort to endure the filth and misery that he encountered, a diagram that finally rendered him capable of limited acts of compassion.

After two years in Yunnan, tired of war, banditry, and the manifold miseries of the roads, he determined to leave China forever. Yet before he left, he wanted to make a visit to the multiethnic polity of Muli (Mili) in southwest Sichuan of which he had learned in conversations with Nvlvk'ö residents, who had explored it thoroughly for both Forrest and Francis Kingdon Ward. Rock and a party from Nvlvk'ö arrived in Muli in January 1924; they made two further visits in 1928 and 1929. Almost immediately, he became immersed in the little state's ritualized theatrics of power. He entered a photographic collaboration with Xiang Cicheng Zhaba, Muli's head of state and supreme religious authority, which continued and deepened during later visits, the theatrics of state power and Rock's own personalized theater of subjectivity reaching happy accord. Eventually, however, the botanist discovered one of the hidden circumstances of state power in Muli: a dungeon with prisoners locked in heavy neck boards and chains. Chapter 6, "Lost Worlds," examines Rock's collaborations with Muli's monarch and several of his most abject subjects.

The eighty thousand plant specimens and sixteen hundred bird skins that Rock and the Nvlvk'ö men collected for Fairchild found a home in the Museum of Natural History at the Smithsonian. Most of these species had already been collected by Forrest and Francis Kingdon Ward for the Royal Botanic Gardens at Edinburgh and Kew. But so much material had flowed from southwest China in the past ten years that no one knew what was duplicated where. The undisputed center in the United States for the botany of west China was the Arnold Arboretum, for which Ernest Henry Wilson had made three expeditions to China. Rock convinced the arboretum's director, Charles Sargent, to generously fund an expedition to Gansu Province—the best possibility for new botanical finds now that Yunnan had been so thoroughly scoured. In December 1924, Rock and a band of twelve young adventurers from Nvlvk'ö set out from the capital of Yunnan Province on a long, dangerous, and eventful trek through northeast Yunnan and west Sichuan to Gansu, which then included the present-day province of Qinghai: that is, most of the Tibetan province of Amdo. They made their base there the semi-independent Tibetan polity of Chone, spending almost two years waiting out a war between the armies of the Muslim warlord Ma Qi and the nomadic Tibetan protectors of Labrang monastery. Eventually, they managed a brief trek to Rock's goal, the great sacred mountain Amnye Machen. Chapter 7, "The Mountain" describes the events in which they found themselves embroiled in Gansu/Amdo. Chapter 8, "Adventurers," examines the conflicted relationships between Rock and the twelve youths from Nvlvk'ö that properly began with this difficult adventure.

More adventures followed. In 1928 and 1929, the twelve men from Nvlvk'ö followed Rock to the great mountain Minya Konka in west Sichuan and the Konka Ling, a range west of Muli, both journeys funded by the National Geographic Society. By this time, he had lost interest in botany and regarded himself as an explorer and geographer. He took bearings obsessively; he was always trying to make maps. Yet the National Geographic Society would fund no more expeditions: he was years late with articles, and he was, by now, an execrable writer. In late 1929, two girls drowned themselves in a pond near Nvlvk'ö. He was always stirred by such tragedies, the more so now, as the girls' families accused him of robbing the bodies when he pulled them out of the pond. As the girls' souls were guided to their ancestors, he took note for the first time of the rich ritual life that had always been happening there, under his nose. He began to collect dongba manuscripts; he hired a dongba to help him understand them; within a few weeks he had decided that this could be his life work. The old books opened up worlds: the world of the Mu monarchs, who had led their Naxi armies to conquer all of northwest Yunnan, worlds of unseen beings sweeping through these landscapes he knew so well: they too had always been there.

He made this his obsession. He worked with an erudite dongba named He Huating and several of his old traveling companions, camped out in the meadows of the Yulong range, in a house in Lijiang; in Da Lat, Vietnam, as the Japanese bombed Yunnan; in Kalimpong, India, after the Communist

Party took the province. He wrote articles first, then books: a huge two-volume work of translations, a two-volume geographical account of the regions once ruled by the Mu kings, a great dictionary of pictographs, the last volume posthumous. Much of this work was funded by the Harvard Yenching Institute. These projects could not have been farther from ethnography in the Malinowskian mode: he had no interest at all in "culture." And they bore little in common with most translation, for the results were not readable texts. They were extensions into texts of the work he had always done, the work of tracing footsteps, taking bearings, collecting specimens, compiling volumes of paper and earth. Chapter 9, "The Book of the Earth" takes account of this final chapter of his grand lifelong quest to take part in an open, cumulative interleaving of earth and word.

TRANSCENDENCE AND ENGAGEMENT

In many respects, Forrest, Rock, and the botanical explorers from Nvlvk'ö seem to belong to a different world, the eighteenth- and nineteenth-century world of imperial botany, which reached its apogee with Joseph Dalton Hooker's explorations of the western Himalaya in 1848–1851. Indeed, both Rock and Forrest thought of themselves as Hooker's successors, extending his work on Himalayan flora eastward. David Arnold argues that Hooker's seven-volume *Flora of British India* (1875) marked the beginning of the end of the era of expeditionary geographical botany: the end came quickly afterward with the expansion of rail travel.[64]

In the first decades of the twentieth century, railways penetrated nearly everywhere. By 1920, China had built, largely on behalf of the imperial powers, about seven thousand miles of railways linking the cities of the east and extending long tendrils to the west. In 1910, the French completed a narrow-gauge line from French Indochina north to Yunnanfu, capital of Yunnan. By 1918, the line north from Canton had reached Wuhan, from whence the difficult waterways of the Yangtze and its tributaries carried travelers and goods into the populous Sichuan basin. In 1922, the line west from Beijing to Zhangjiakou was extended to Baotou, where the Yellow River fed it the wealth of the Tibetan grasslands in the form of baled wool. On the other side of the imperial divide, a line ran from Rangoon northeast to Myitinka from whence a motor road was built to the frontier outpost of Bhamo. Still, between these termini remained an enormous inland region, with the world's highest mountains and some of its most extensive deserts and plateaus. Railways would not even begin to penetrate this region until the

Second World War. From Yunnan north to the Tibetan regions of Kham in west Sichuan and Amdo in what was then Gansu, north to Mongolia, and west and south across the vast new territories of Xinjiang and the vaster Tibetan plateau, travelers walked and rode, and goods were carried in caravans on the backs of mules, oxen, yaks, camels, and humans well into the twentieth century's second half.

Our botanists wandered only a corner of this region, enormous enough. They walked and rode ponies for weeks and months at a time. The rhythms of walking bodies, human and animal, measured their engagement with the earth. But for the British and American participants in this enterprise, walking was different than it had been before the triumph of rail and motor travel. They shared with their interlocutors at home an imaginary of the moral and experiential qualities of walking shaped by the dramatic acceleration of travel in the previous century. Writing of the investment of early photography in romantic visions of nature, Rebecca Solnit comments:

> It is as though the Victorians were striving to recover the senses of place they had lost when their lives accelerated, when they became disembodied. They craved landscape and nature with an anxious intensity no one has had before or since. . . . The ideal landscape seemed formed of a wholeness that was no longer theirs. They looked for this wholeness, and mostly so do we. These histories suggest that nature was equally a kind of time or a pace, the pace of a person walking, of water flowing in a river of seasons, of time told from the sky rather than electrical signals. Nature meant not where you were but how you moved through it. . . . But the Victorian age had launched a juggernaut, and slowing down was the single hardest thing to do.[65]

Forrest and Rock slowed down gradually, taking liners across the ocean, mounting trains, motor cars, and ox carts for west Burma and north Siam, and only then setting out on foot for the Yunnan border. They approached walking self-consciously, comparing their pace to the pace of the world they had escaped. They suffered day by day; they complained incessantly about the roads; they fantasized about airplanes. But they also believed themselves privileged. They imagined their experience to be fundamentally different from that of their families and patrons. And they gave this difference a moral flavor, remarking on the degraded quality of life at a city pace and the aesthetic grace, bracing or languid, with which their own lives were filled. A line jotted by their competitor, Francis Kingdon Ward, as he followed his caravan across the Tibetan plateau comes to mind: "Astonishing the distance one can hear mule bells in the mountains. Propose to float a company for gathering wool from barberry bushes on the trail of the Tibetan caravans."[66]

They felt enormous pressure to make the most of this fortunate circumstance. How were they to plumb the mysteries they felt it must contain? How were they to communicate it?

European and American imperial endeavors had demanded of their participants particular ideologies of representation. Writing of the colonization of Egypt, Timothy Mitchell notes that these ideologies drew unambiguous distinctions between representations and realities, between observing subjects and observed objects, between persons and things.[67] Yet as Webb Keane argues, such ideologies were never stable; they were often confounded by the practical economies of representation in which people actually engaged. It was rarely easy to maintain such clear-cut distinctions in daily life, and the ideas of others tended to complicate matters.[68] Forrest and Rock both found themselves called upon to find ways to refashion their own perceptual relations to the world—the world of things and the social world—with the aid of the sensory tools at their disposal. They carried cameras and field glasses, sometimes prismatic compasses and theodolites; Rock carried a portable gramophone. These instruments were important tools in their efforts to discipline their sight and hearing. But their principle instruments of perception were pens and notebooks. These tools gave them means to organize problems of representation and perception day by day: they gave them that simplest and most fundamental model of subjective experience: feet walking the earth's surface, a pen scratching across a page.

Though they carried on the Linnnean project from the eighteenth and nineteenth centuries, they were modernists nonetheless. They were plagued by modernist uncertainties about perception and representation, and they had recourse to a distinctively modernist dialectic as they sought to deal with these uncertainties. Many modernist ideas about perception have relied on a contrast between a pure presence, separated from time and the body, on the one hand, and lower, quotidian, embodied forms of seeing or listening on the other. For both, walking over the border into Yunnan put all the certainties of empire into question, including ideas about how they should focus and represent their experience. Each attempted to find places of pure vision where the ambiguities of perception and representation might melt away. For Forrest and the elder generation of men from Nvlvk'ö, these places were to be found up beyond the far northwest corner of Yunnan, in the generative center of *Rhododendron*, where the entire genus, the entire region, and their entire history of botanical exploration, would be focused in an ecstatic vision of a mass of color, infinite in variety. Rock, with the younger generation of Nvlvk'ö men, made his searches with his camera and gramophone. He eventually found a way to create a virtual body for himself, with a pure virtual eye. He repeatedly placed this body on a stage, in prose and photographs: a vast, wild landscape centered on a tiny circle of tents containing his twelve "boys" from Nvlvk'ö, his own empty tent among them.

Such transcendent visions could not be sustained for long, however. Both botanists returned repeatedly to a world of sustained engagement with the earth as a social being amongst social beings on the level of time and the body. For Forrest, this engagement was centered on the patient, careful work of crafting specimens: assemblages of words and things, of bits of the earth's flesh and fragments of taxonomical abstractions. For Rock, this engagement was the book of the earth: an intertwining of text and landscape that ultimately made the earth fully social and fully inhabitable for him. For both, transcendence and immanence were thoroughly collaborative projects, involving large casts of interlocutors in China and at home: scientific and commercial patrons, children and wives, reading and gardening publics, and so on. And they were thoroughly historical projects, emerging from an interleave of text and earth that gave character to each of the places they encountered. One purpose of this book is to chronicle the least visible of these collaborations, with the bands of adventurers they mobilized, and the least regarded of these histories, in the borderland places they loved.

At this moment, as the scientific successors of geographical botany project various disturbing futures for the earth, readers might find many parallels with the processes explored here. Dialectics of transcendent imagining and minutely particular engagement are everywhere in the sciences that investigate the effects of our century-and-a-half-long experiment with fossil fuels, for instance. This book points out that it is such efforts at seeing and engaging that make the earth into a social being for us. They bring it to life and weave it into our relations with others. Despite the imperial overtones of their project, Forrest, Rock, and the explorers from Nvlvk'ö all found ways to cultivate the earth as a vibrant social partner. They searched for worlds of transcendence in this uncertain border zone, but they also came back repeatedly to immanent involvement with the earth, making it the substance of relations of love, obligation, patronage, and companionship. Our histories, even in their colonial and semi-colonial phases, contain many such efforts at care-laden engagement. There is no reason why our futures should not extend and deepen them.

Part I

1.   The Eyes of Others

A man is suspended over the river. Leather straps bind him to a half cylinder of bamboo that slides on a rope of twisted bamboo strands greased with yak butter. Having plummeted to a point over the river's center, he will haul himself to the opposite bank, hand over hand. The edges of the strands are sharp, so the crossing is painful as well as physically demanding. The photograph was included in Acting Consul G. Litton's secret report to the British Foreign Office on his journey to the Upper Salween River.[1] It is filed in the archives of the Royal Geographical Society with many other images of suspension bridges over the great rivers that run in parallel gorges through northwest Yunnan: bridges of cane, ropes and planks, timbers and iron chains. The back is stamped with the photographer's name, George Forrest, and the date, 1905. At that time, in the confident opinion of the Royal Geographical Society president, the Salween remained the last great riverine puzzle on earth, its northern reaches yet unexplored by any European.[2]

The drawing looks almost like a cartoon-strip version of the photograph. The rope is stretched between two trees; the bamboo runner and inverted human figure are clearly depicted; a simple tree stands to the right; the whole is surrounded by a square frame. The drawing is from the title page of a hand-written manuscript called *Lònv*. The composition within the frame is this word in what is now known as dongba script. The left side is the word *lò*, meaning "to cross over," the first syllable of *lòk'ö*, to cross a rope bridge. The tree on the right is the word *nv*, which designates a ritual effigy. The manuscript titled *Lònv* was written to be read at a funeral for a child. Joseph Rock bought it in the early 1930s in the Lijiang valley. It now resides with 1,114 other dongba manuscripts in the Staatsbibliothek in Marburgh, Germany.[3]

G. Litton walked up the Salween, or Nu, intending to draw a line between two empires. He had recently been appointed the British representative to

Figure 7.   Crossings. Reproduced by permission of the Royal Geographical Society.

a joint boundary commission charged with fixing the northern border between China and Burma. His Chinese counterpart was Shi Hongshao, the
new trade commissioner *(daotai)* for west Yunnan. An 1897 treaty, in which
the Qing court granted the British a trading post in the border city of
Tengyue, had left this issue unresolved, specifying only that a fixed boundary would be surveyed in the future. Over the next two years, cartographers
for both sides proposed five different boundaries, synthesized in a "five-

coloured map" for the south, in the Wa states, and another "five coloured map" for the north, west of the Salween.[4] The British Foreign Office had declared that it would be "for the advantage of both countries and of their mutual commerce that British jurisdiction should be established over the whole of the upper basin of the Irrawaddy."[5] This in mind, Litton argued that the huge, unsurveyed mountain range rising from the west bank of the Salween, the Gaoligong range, was a "natural and ethnological divide" between the two empires.[6] In response, Shi Hongshao restated the Qing court's position that the local rulers of the peoples Chinese ethnologists called Nu, Lisu, and Qiu, who occupied that range and many territories west of it within the Irrawaddy basin, were *tusi,* native hereditary officials en-feoffed by Ming and Qing emperors, or *fuyi,* hereditary officials granted titles by the Ministry of War. Formally, these rulers were under the juris-diction of the district government in Weixi, a town on the Mekong, and their polities were Chinese territories.[7] Litton believed to the contrary that no actual relationships existed between the district government and the Nu, Lisu, and Qiu chiefs in the region. It was to prove this that, in October 1905, accompanied by two soldiers from the Indian Army and tens of Lisu porters, each carrying sixty to seventy pounds of rice, he walked up the gorge of the Salween.[8]

His other companion was George Forrest, on an expedition to Yunnan for the British seed firm Bees Ltd. Forrest had already been in Yunnan nearly a year. He had made two extensive tours of Yunnan's northwest, the first with Litton, the second on his own—with a cook, a groom, three servants, and two muleteers ("I cannot travel with less," he wrote his fiancée.)[9] His second journey had been particularly eventful: he had barely survived a Ti-betan attack on the French mission at Cigu on the Upper Mekong River, los-ing all his possessions, including an extensive collection of specimens. Two months later, though barely recovered, he happily accepted Litton's offer to make him the naturalist and photographer of this expedition to a new and promising territory.

The two British men began the expedition with a sense of excitement tem-pered by sober purpose. They were performing a vital task for the empire—even if in this rather unimportant periphery. To them, their collaboration was perfectly natural: naturalists had taken part in the military and ad-ministrative surveys of each new territory acquired by the empire in In-dia.[10] They investigated the gorge's climate, botany, zoology, and geogra-phy, and they made ethnological observations of its human inhabitants. The latter, crucially for Litton, were "wild Lisoos with poisoned arrows" under "no control of any sort . . . by any Chinese or other chief."[11] At the bam-

Figure 8.   George Forrest in Dali, Yunnan with boxes of specimens, 1904.
Courtesy of the Royal Botanic Garden, Edinburgh.

boo rope bridge, Forrest took a photograph, and they discharged their
firearms, subduing into "awestruck silence" the crowds of Lisu villagers who
seemed to be vying to control their crossing. They then climbed up the east
bank of the gorge to the summit of the Mekong-Salween divide, where they
accomplished their journey's main objective. "Here a surprise awaited us,"
Litton wrote the Foreign Office, "for the view to the west was perfectly clear,
and the whole of the great Salwin-Irawadi divide was spread out before us.
From a little below the pass, the range could be followed to the north as far
as the eye could reach, until at a distance of about 100 miles from where we
stood, and in approximate lat. 28° 30' N., it was merged into a huge range of
dazzling snow peaks trending westwards."[12] "Not the least doubt remained
in my mind that here was the true frontier between India and China," he
added: it was a "vast wall."[13] Litton did not live to press his case, dying of
malaria only weeks after his return, but his description was repeatedly
quoted in insistent British proposals that the range dividing the Salween
and Irrawaddy watersheds be the border. Though Qing negotiators vigor-
ously opposed these proposals, the British treated the divide as a "provi-
sional border." It became the de facto border in 1913, when British troops

occupied the village of Pianma (Hpimaw), west of the divide, despite furious nationalist protests in China.[14] The border was not finally established until long after Burmese independence in 1960.[15]

As for George Forrest, exhausted by the journey he let Litton's words take the place of his own in letters home and in a paper to the Royal Geographical Society.[16] He was accustomed to being struck silent by views of glistening snow covered peaks, "beyond conception" or "beyond description." They signaled to him the limits of thought and language. He would forget about that distant, shining range for many years as he explored other parts of Yunnan. But eventually, inspired by his indigenous collaborators, he would remember it: as the sign of something always beyond his grasp; as a solution to the great phytogeographical puzzle of the genus *Rhododendron;* as a transcendent Eden, where the difficulties of engagement and representation that this landscape presented might finally melt away.

When a child died in a Naxi home in the Lijiang valley, her parents sent for a ritualist, a dongba, who helped wash the corpse, dress it in new clothing, and place it in the coffin. A little later, the parents held a funeral in their courtyard, attended by a few kin and friends. The dongba read several manuscripts. Only *Lònv* was written specifically to send a child off on her final journey. It seems to have been rare: many funerals for children probably did without it. It described the child's soul, unable to eat, unable to see and hear, unable to use hands or feet, unable to clothe itself. It spoke of the child's journey across a rope bridge separating the world of the living from the world of the dead. It told of how a dongba cut the bridge with a knife, severing the link between the two worlds in order to prevent parents or siblings from following. The manuscript specified the materials to be used: sacrificial animals, black clothing, a strip of blackened hemp for the rope bridge, a fir branch to which the rope was tied. Other texts described the soul's journey, guiding it north through the mountains of northwest Yunnan, toward the lands from which the ancestors had come, and eventually to the great mountain Ngyùná Shílo Ngyù, origin of all things, where the worlds of mountains and forests met the worlds of the gods.

In the early twentieth century the upper Nu (Salween) was about two months travel from the Lijiang valley. There were many rope bridges in closer places, along the Lancang (Mekong) and Yalong rivers. But all these had two ropes, each with one end higher, so the traveler could slide all the way across without resorting to the painful inverted crawl depicted in the *Lònv* manuscript. The distant upper Nu, with its single-rope bridges, lived in the ritual imagination of Naxi farmers and traders.[17] Many Naxi from the Lijiang valley traveled long distances for trade: a late Qing gazetteer mentions that Naxi came over one thousand *li* to sell items to the Nu, who inhabited the far upper reaches of the river that bears their name.[18] The Nu, whom Naxi called Nun, their land on the far upper Nu (Nundù), and the Lisu (Lissuu), who lived in the part of the gorge Litton and Forrest investigated, were all represented in the written lexicon of Naxi ritualists. In particular, a small ceremony to perform rain was performed frequently around Lijiang in the early twentieth century. In this rite, held by a spring, dongba invited the spirits who brought rain to descend from all the great rivers and lakes in four directions. In the west, they came from the lands of Nundù descended to the town of Weixi on the Mekong, circled the mountains to the north and east, and then entered the Lijiang valley. The origin of this journey, and its highest point, was the great snow mountain Nundù Gkyinvlv, the center of that "huge range of dazzling snow peaks" that Litton and Forrest spotted at the apogee of their expedition, shining from a hundred miles further north.[19]

A photograph and a drawing. Each might be said to participate in a separate archival regime. They are archival in the most obvious sense. They are both fragments of large collections that came together in focused historical periods, in projects that followed quite precisely defined rules. I find the word "regime" appropriate because of its use in physical geography to connote all the factors that create the quality of a river's flow: the amount of rainfall, the type of bedrock, the shape of the channel—everything that makes the river meander, braid, flood, disperse into deltas or join with other streams. In an analogous way, an archival regime includes all the rules, habits, and ideologies that decide how perceptions are transferred to paper, collected, and stored. It includes habits and ideologies of seeing and walking, of inscribing and photographing, of collecting, shipping, storing, organizing, and reading.

The British expedition up the Salween was guided by what Bernard Cohn called a "survey modality" of investigation. This mode of summing up a landscape and its inhabitants in a series of comprehensive reports had been central to the colonization and administration of India. Litton's report was added to the heaps of documents written by colonial explorers and administrators in India that had been piling up in the archives of the British Museum, the Royal Geographical Society and related institutions for over two centuries. The British consul who replaced Litton sent all Forrest's photographs from the expedition to the Royal Geographical Society. His images of bridges were filed in drawers marked "bridges," of people in drawers labeled "types," of crossbows in drawers stamped "weapons." Thomas Richards has argued that this kind of archival activity, more than any process of administration or governance, was the central project of the British Empire during the late nineteenth and early twentieth centuries. After the mid-nineteenth century, British imperial power found unity and coherence in an archival myth: the idea that institutions of knowledge production could forge the millions of facts flowing in from all corners of the world into a coherent whole. In an empire where administrative institutions were thinly spread, this myth of archive took the place of actual civil governance in many places. Litton and Forrest participated wholeheartedly in this myth as they walked up the Salween. Despite the complete absence of any British civil authority in the region, the failure of numerous plans for a railway linking Burma to the Yangtze, and the Foreign Office's stated abhorrence for any further expansionist adventures, they both looked forward to the immanent British annexation of Yunnan. Even so, this would be the last time Forrest would imagine himself directly involved in the project of imperial administration. After Litton's death, his activity would be guided by a more limited regime within the survey modality of investigation.

The *Lònv* text was a fragment of another archival regime. Naxi ritualists probably began writing manuscripts in pictographic script sometime during the Ming dynasty, as armies from Lijiang swept through the province's western regions. During the eighteenth and nineteenth centuries, they wrote and copied these books with great assiduity. In the early twentieth century, tens of thousands of manuscripts were stored in the attic libraries of dongba in and around the Lijiang valley. Most were written to be recited during ritual performances; some were for divination and astrology. Many rituals took the form of driving or guiding inhuman entities over the landscapes of northwest Yunnan, and many texts contained long lists of places, some extant, some once extant, some belonging to other worlds. As a whole, the archive of dongba texts contained an extensive geography of northwest Yunnan, far more detailed, at the opening of the century, than any other representation in maps, reports, or herbaria. This archive too was underlain by rules, habits, and ideologies of seeing and walking, writing and collecting, storing, organizing, and reading. Dongba did not create and organize knowledge for purposes of rule, though perhaps they once might have. They wrote and copied texts to regulate the social well-being of people and communities by manipulating relations between human and nonhuman social entities. Their mode of investigation did not survey the landscape. It auscultated its depths, threw out lines of communication to its hidden presences, and divined traces of a past that had vanished from its surfaces.

One might think that each of these two regimes would be incomprehensible from the point of view of the other. And indeed, there is no evidence that either directly incorporated knowledge from the other. Forrest did not find routes, place names, or plant names in dongba texts: he was barely aware of their existence. And no dongba altered any text to include botanical or geographical knowledge created by Zhao Chengzhang and the twenty-five to thirty others from Nvlvk'ö who worked with Forrest. Nevertheless, the two regimes came into close contact for nearly three decades, and at some points of friction they very likely shaped or deformed each other.

By the late 1920s, Zhao Chengzhang, Zhao Tangguang, Li Wanyun, He Nüli, and the rest of the senior generation of explorers from Nvlvk'ö were masters of the taxonomical botany of northwest Yunnan, in large part their own creation. They were expert at keying out plants; they knew the slight differences that separated some species and varieties; they knew far more than anyone else in the world about the geographical distribution of alpine species in the region. By 1921, Forrest and Zhao were having "very heated arguments" about particular species. In one case Zhao and several others maintained, against Forrest's strong objection, that the rare tree-sized *Rhododendrons fastidium*, *giganteum*, and *protestum* were all distinct species. Despite his pride in his knowledge of the complex genus, his stubbornness in argument, and his tendency to tyrannize his employees, Forrest had learned to have great respect for their taxonomical expertise. In his report to his sponsors, he left the issue open, reporting only the differences of opinion.[20]

But Zhao and his fellows were experts in other realms as well. In many ways, dongba culture had been in decline since the eighteenth century. It had largely moved out of the towns of the Lijiang valley into the surrounding mountainous countryside. Located just above the edge of the valley, their village belonged to this drier, poorer, mountainous hinterland. But it was also in the orbit of the market town of Baisha, long the most important center of dongba learning. Nvlvk'ö and the other villages in the southern foothills of the Yulong range, though small and poor, contained the greatest concentrations of dongba and libraries of dongba manuscripts remaining within the loop of the Yangtze. It was from Nvlvk'ö that the dongba archive moved out larger world: when Joseph Rock began to collect dongba books there, he had a library of hundreds in within days and thousands in a few months.

The money they earned from their quest for flowers made Zhao Chengzhang and his colleagues their village's most prosperous and important citizens. As elders, they officiated at village-wide events like the yearly *muanbpò* ritual to regulate the village's health; they were the central guests at funerals; and they were in a position to sponsor rituals to cleanse their houses, cure their kin, defeat slander, and overcome drought. Dongba books were central to all these activities: from just a few to as many as seventy were read aloud at each. There is no evidence that Zhao or any of his colleagues were trained as dongba, but it is certain that they listened to many key books being recited many times: they all had a general knowledge of the dongba corpus. Despite the slow general decline of the dongba cult, it was still, for these men, a deep source of knowledge, attitudes, and orientations, particularly about its central protagonist, the living, spirit-inhabited landscape of northwest Yunnan.

In Part 1, I place these two archival regimes side by side in the context of Forrest's and Zhao's explorations of northwest Yunnan. For the men from Nvlvk'ö these regimes were very much coexistent. But to understand the specific ways they coexisted and, perhaps, intersected, can only be a speculative exercise. As I mentioned in the Introduction, almost nothing written by the first generation of explorers from Nvlvk'ö remains: there are only Zhao's four maps and two scraps of ledger, a few labels on bird specimens, and one brief, crudely written thank-you note. Any ideas about their relations with the gorges, mountains, and meadows they wandered have to be pieced together from two very different sources. On the one hand, there are Forrest's letters, reports, notes, and photographs, tightly focused on the botanical enterprise. On the other, there is the massive archive of dongba texts, collected within the Yangtze loop, supplemented by a small descriptive and ethnological literature about the Lijiang valley and the Baisha region.

Tacking between these sources, I make several conjectures about how the dongba archive might have rubbed up against, and created new and compelling possibilities for, the phytogeography of northwest Yunnan. In particular, I show how the dongba corpus might have shaped the decade-long quest of Forrest and his mentor Isaac Bailey Balfour for the center of origin of the genus *Rhododendron*. I speculate about how the encounter between the dreams of geographical botany and the words of dongba texts might have tilted that search north and west, focusing it finally on that gigantic range that Forrest saw glowing in the distance during his walk up the Salween river with Litton in 1905.

These arguments are the path I have chosen to trace through the geographies of these two archival regimes. Route maps such as these, even if limited in scope, are perhaps the best way to come to know such a geography. The alternative—maps that pretend to a comprehensive domain view—would fail this vast, varied, complex, and interesting landscape in important ways. Along this path, I open up some of the questions that animate the rest of this book. How do rules about how we must perceive inflect how we walk and see? How do perceptions inscribed on paper become interleaved with the substance of the earth, to inflect other perceptions? How in this process does the earth emerge into social being? In what ways might this social being serve as a resource for experiences that circumvent established ways of thinking and living the divides we make between the social and the natural?

LEARNING RACE

He had never quite found his footing in the British class system. His parents were of the ranks of workers, clerks, shop assistants, displaced tradespeople, and aspiring, mobile entrepreneurs who had wreaked massive transformations in class relations during the late nineteenth century. His own opportunities to move within the class hierarchy—to gain a secondary education, to try immigrating to Australia, to become a colonial explorer—were due to unforeseen gifts or patronage from others of superior class position. In India and Burma, colonial travelers and functionaries reoriented themselves to accord with an immediately apparent cartography of difference. As Partha Chatterjee has put it, colonial states operated through a "rule of difference," and in India the salient differences were racial.[21] Ann Stoler suggests that the work of drawing and redrawing maps of racial difference produced new middle-class sensibilities, organized around images of racial purity, sexual virtue, and proper masculinity.[22] This cartography gave imperial men who moved between metropolitan and colonial societies while occupying underprivileged class positions in Britain new opportunities to recraft their class identities. "White Englishmen were able to use the power of the colonial stage to disrupt the traditional class relations of their country and enjoy new forms of direct power of subject peoples," Catherine Hall writes. "At the same time . . . their own identities were ruptured, changed, and differently articulated by place."[23]

As he began his first expedition in 1904, traveling through India and Burma towards the western entrance to China, he was unsettled, in a way an upper-class Englishman might not have been, by the brutal distinctions of race and class on display. Upon arriving at the Watson Esplanade Hotel in Bombay, he set out to see the sights. He did not walk; "no European ever walks"; he rode in a gharry, pulled by "gharry-wallas." All the servants were servile; they kept salaaming him: "it makes one feel uncomfortable at first." But the ideologies of race were quickly enforced and quickly absorbed; the servants worked for almost nothing, as he had, but "apparently they are quite pleased with such payment, and to give them more, as I was inclined to do, only makes them lose respect for you." From the gharry, he looked down on the streets: men and women carrying away waste on their backs, men naked but for turbans and loincloths, beggars "afflicted with some disabling and generally loathsome disease." "I saw the [British] captain kick and hammer one of the porters until I thought he intended killing him, for a most trivial offense." He watched himself reshape his attitude and comportment toward the lowest classes: "I can swear at them and order them about now, [but I don't think] that I shall ever reach the kicking stage."[24]

Above all, he discovered opportunities to reaffirm his middle-class faith. This faith emphasized that the core of true masculinity was individual integrity and freedom from subjugation to the will of others. The most fundamental demonstration of masculinity was the capacity to establish, protect, provide for, and control a home.[25] He had established the foundations for his future domestic life before leaving Scotland, promising himself to Clementina Trail. He would spend the rest of his life working to build and provide for a home over the distance between Yunnan and Edinburgh. In India and Burma, he quickly learned to reframe his masculine sexual virtue along racial lines. At night, a new acquaintance took him to Bombay's brothel district. There, the filth, nakedness, and wretchedness of racially other bodies that had assaulted him in the daytime streets were concentrated in experiences of unregulated transgress of boundaries, moral and physical: "In the native quarter . . . the stench is indescribable in places . . . a mixture of sweaty bodies and all sorts of reeking abominations. . . . I never thought it possible for vice to be paraded so openly anywhere. We were continually being tackled by the women, some of whom even went the length of trying to get into the carriage beside us. . . . I pity the poor wretches." In Bhamo, the last steamer port on the Irrawaddy before the frontier with China, he was advised to buy a Burmese girl to take with him. "All the officers in the regiments stationed here . . . keep them. . . . I could get a dozen tomorrow if I wanted them. . . . but I wouldn't touch any of them with a tarry [that is, tar-covered] stick. There is only one woman in the world for me and that is Clem, and she is white all through. . . . I have kept straight all my life and I have every reason in the world to keep straighter than ever."[26] It was a full and satisfying conflation of racial and sexual purity.

He was reorienting his sense of self in relation to different others, making decisions, some fleeting, others enduring, about how to negotiate the lines of difference that fissured colonial society. As he approached the border with China, he discovered new uncertainties and new opportunities. European discourses about China and the Chinese had long been more multivocal than discourses about race in India or Africa. On the one hand, since

the eighteenth century an influential strand of writing about China had described it as an advanced civilization with a venerable history, literate populace, and wise government. On the other hand, during the nineteenth century China had been incorporated into European racial theories, giving rise
to a discourse about "racial character," centering on images of Chinese as
a degenerate race, enfeebled, corrupt, deceptive, filth-ridden, and prone to
disabling disease.[27] After the Boxer rebellion and the punitive military operations that followed (1898–1901), florid racial invective became a commonplace feature of traveler's accounts, ethnographic descriptions, and fic-
tional treatments of China. Yet older visions of China as an advanced and
enlightened civilization were not entirely displaced. George Steinmetz has
shown how some European colonists in China indulged in imaginary iden-
tifications with the Chinese as they negotiated these differences. "The pre-
vailing European fantasy involved projecting oneself into the role of a Chinese mandarin or philosopher-king ruling over a literate and civilized
people."[28] The most popular way to play with such fantasies was to make
photographic portraits of oneself in the dress of a Chinese official. As he
neared the border, Forrest tried on such imaginary cross-identification.
"Once I get right into China," he wrote his mother and sisters, "I shall put
on the regulation Chinese dress, big baggy trousers reaching to the calf and
a loose blouse with a big hat and Chinese shoes."[29]

He experimented with attitudes towards officials and soldiers he met
along the road. He regarded them all as "Chinese"; later he would learn to
make finer distinctions: on both sides of the border he passed through what
he would come to know as Shan states. Near the border, the party waited
several days while the commander of a Chinese battalion stationed in the
region requisitioned men from surrounding villages to build a bridge across
the Namsa (Namwan) river for the caravan.[30] He entertained the region's
elite, including the district military commander and a "rather curious old
cove" who was probably the hereditary ruler *(sawbwa)* of a Shan state. They
exchanged calling cards, eggs, whisky, biscuits, chickens, cigarettes, tinned
plums, and cherry brandy. "The soldier one . . . is a most kindly and polite
little fellow, and I liked him best of all. He at once started to try to learn me
Chinese."[31] It was among the most genuinely good natured exchanges with
local elites he would ever record.

Across the border in Tengyue, a tiny cadre of Europeans struggled to im-
pose a regime of difference modeled on India's. Tengyue was a thriving town
of about three thousand with a large merchant class involved in the cross-
border trade. Forrest settled in at the consulate and held dinner parties
"amongst ourselves of course" with the only other Europeans there: Act-

ing Consul G. Litton and the British customs commissioner and his two assistants. When he went out, they insisted he take a soldier:

> He goes in front and by continual shouting and pushing gets the people to clear out of the way. He makes no bones about shoving some of them almost on their faces. It seems nasty, but it is really the only way and by doing this the people seem to respect one more.... Mr. Litton says that if it wasn't for the punishment which they know would be meted out to them our lives wouldn't be worth a moment's purchase, and I believe it from the look of some of them. Of course we always go armed and as they are great cowards, this keeps them in check. In coming home from the dinners or going anywhere at night we always had an escort of 4 soldiers and all (or most) of our servants carrying huge paper Chinese lanterns, quite a procession.[32]

Only a few years before, combined European forces had put down the Boxer rebellion, occupied the capital, forced the empress dowager to flee, ransacked the Summer Palace, and forced huge indemnity payments on China. The uprising had not spread to the southwest, however, and remote Tengyue had seen no violence. Even so, anonymous notices sometimes appeared on the streets excoriating the weak Qing government, decrying foreign influence, and agitating against British plans to build a railway from Burma. Two months after Forrest's arrival, Litton telegraphed to warn the Foreign Office that the Chinese authorities had received a rhyming pamphlet stating that foreigners "practice the violation of Chinese women and children and destroy tombs." In fact the pamphlet (a copy of which Litton sent with his report) was an announcement in somber verse by the Tengyue Militia Office *(tuanlian ju)* of reforms that would eradicate abuses and introduce more discipline into training, with the aim of discouraging the foreign powers pressing at Yunnan's borders.[33] Forrest gradually came to understand that the townsfolk of Tengyue were more interested in peaceful trade than violent nationalism. In subsequent visits, he abandoned the shows of force to which the other Europeans in town seemed addicted.

Still, whenever pressed, he drew on the assessment of "Chinese character" he had begun to learn in Tengyue. After three weeks, Litton invited him on a journey east to Dali then north to Lijiang and to the Khampa town of Gyaltang (Zhongdian) in Yunnan's northwest. The route suited his purposes, traversing high ranges where he might find novel alpines. A caravan route took them to Dali, where Litton decided it was his business to investigate reports of tax corruption at a horse fair in Songgui, a day north. On the way, their servants, hired in Dali, balked at showing them a shortcut. "At last, we lost our tempers, and Litton started to bully them." That night, thieves took Forrest's pony and two mules from the temple where the party put up. The two British men gave chase, caught a man, and trussed him with rawhide ropes. In the morning they handed him over to the local magistrates, who ordered that he be given three thousand strokes with a bamboo rod while strung up by his hands, a fifteen-pound weight tied to his queue—a punishment he was unlikely to survive. Satisfied, they went to the fair: "I never saw such a beastly rabble in all my life. All the scum of China seem to have collected. . . . A crowd of ruffians . . . followed us about making jeering remarks. . . . Then one of them started to pick up stones with the intention of stoning us, but when we saw this we drew our revolvers, and then you never saw such a scatter in all your life. The sight of weapons was sufficient without going even further." The next day, "we stayed at an inn at Hoching (Heqing) and were troubled by the usual crowd of gapers who followed us about wherever we went."[34]

In November and December of the same year, Forrest made another journey, this time with no European companion. With Litton, he had made calculated shows of force based upon the idea that the Chinese were "great cowards." On his own, he tended to lose control of such demonstrations, falling into rages that he would recall with remorse. In a village near Lijiang, "we also had a row with the people . . . because they wouldn't unlock their temple door and let us spend the night there. Temper not improving you see! But I think there is some excuse as the average Chinaman would rile the heart of a wheelbarrow." He blew the lock off the door with his rifle, barricaded the door from the inside, and left early the next morning.[35] In March, before the passes had entirely cleared of snow, he attempted to retrace most of the route he had traveled with Litton. From Dali, he went north to Jianchuan, where he commandeered a temple to avoid the county government seat, or *yamen*, and its magistrate—"for I despise the officials as much as the people." A crowd collected; some people began to examine and handle his baggage, and he ordered them out. One, a student, wouldn't move: "taking him by the scruff of the neck and the seat of his baggy breeches, I heaved him out of doors." The crowd yelled and threw stones at the temple windows. "Tired, hungry, and enraged as I was, this was more than I could stand and I let myself go . . . and I must confess that I was most heartily ashamed of my conduct. I seized a stick, rushed out amongst them, and began laying out right and left."[36]

He recorded these confessions in a diary in the form of letters addressed to his mother, who passed them around to Clementina and his sisters. Letters were his sole means participating in domestic life, positioning himself as a proper husband for Clem, whose parents, from a higher class, disapproved of him. He was feeling his way. Here, the mapping of difference was far more fluid than in India, where it was supported by state institutions. Each outburst was touched off by the "insulting" behavior of people who appeared not to recognized distinctions he was anxious to assert. But there was more to it than his perception of insult or threat. In India, his interactions with different others had been accompanied by pity, loathing, and reaffirmations of racial and sexual purity. He had seized every opportunity to reassert moral and physical boundaries: he might gaze at vice, but he would never let it into his carriage. Here, confrontation with difference inspired rage and hatred instead: he had so quickly reached the "kicking stage," mixing it up with all the sweaty bodies.

How real was the threat he felt? Certainly antiforeign nationalism was rife in those years preceding the 1911 revolution. And just as surely his posture incited hostility. One can imagine the anger of townspeople thrown out of their temple by an arrogant, rifle-wielding foreigner. But for many in the crowds, courtesy rather than hostility may have been the prevalent mood. These were the same looks imperial officials expected to receive as they traveled through towns with retinues of servants and gaudy palanquins, and they took these looks to evince fear, respect, and obeisance to authority. For Forrest, however, the looks were the problem. Why would the "average Chinaman rile the heart of a wheelbarrow"? It was a matter of ocular posture—what he called "gaping." It was not that these open looks were always evidence of hostility, though they sometimes were. It was not that they always challenged his superiority, though they sometimes did. It was that in gaping crowds he had no means of dialog, no words to learn, no gifts to exchange, no calling cards to accept, no language at all, whether about difference or identity. Mute gaping stilled exchange, reducing relations to the purely visual, a terrible mirror. Kicking, laying about with a stick, grabbing people by the seats of their pants, were ways of breaking that mirror and establishing contact.

GAPING AND PHOTOGRAPHY

Conditions of visibility were a persistent concern for colonial scientists and travelers in the nineteenth and early twentieth centuries. The power to frame what was seen and to regulate affect attached to seeing were at the heart of many imperial projects.[37] Imperial men and women created cultures of display, in which exhibitions and photographs represented the colonized world as objectively visible.[38] Historians of colonial photography have explored many ways in which photography structured fields of vision to emphasize the subjective, empirical gaze of the imperial viewer and deflect the power of returned looks.[39] British colonial travelers reproduced the "world as exhibition" by following established itineraries, seeing famous sights in predictable ways, and structuring their visions of the landscape around specific aesthetic principles: the sublime, the picturesque, the romantic, the realistic.[40] Relations of visibility were always entangled with relations of difference. Deborah Poole has pointed out that the production and circulation of millions of photographs of colonized peoples "lent support to the emerging idea of race as a material, historical, and biological fact."[41] Theories of race created racial categories much the same way Linnaean taxonomy created species, selecting particular features to use as criteria. Racial theorists chose photographs to use like botanical type specimens as standards for comparison. And photographs of colonized peoples circulated through public and private archives, where they were categorized and organized through the principles of an archival regime. Exhibitions, photography, and science were machineries for framing and structuring fields of visibility—the gazes of "imperial eyes."[42] In the ambiguously colonized peripheries of empire, however, these were cumbersome machineries, laborious to master and difficult to apply to daily experience.

During his first expedition, Forrest struggled to reconcile the landscape of China with the world that exhibitions and photographs of the colonized world had conditioned him to expect. Eventually, he learned to coordinate different modes of visual exchange with a typology of race and map them onto the aesthetic, affective, and botanical qualities of the landscape. The possibility of such a cartography began to dawn on him during his first journey with Litton. North of Lijiang, the travelers walked for three days through the deep gorge of the Jinsha, or upper Yangtze, in a "magnificent heat." They climbed out of the gorge to a boggy pass, where they camped in heavy rain. Forrest happily collected some of his first alpine flowers: *Gentiana, Primula,* and *Saxifrage.*[43] The next day, the party descended to the Zhongdian plateau, a triangle of very high land about ninety miles long, surrounded by mountains and the Yangtze gorge to the south and bordering on the Tibetan region of Kham to the north.

This plateau, known in Tibetan as Gyaltang, was a strand in the network of roads linking Yunnan to Tibet and Southeast Asia, known as the Southern Silk Road; it was the main corridor for the trade in tea and horses between Yunnan and Tibet. It had been passed back and forth among polities for centuries.[44] From the mid-seventeenth century, the plateau had been ruled by a hereditary nobility of two lineages who traced their ancestry to the ancient Tibetan kings. In 1723, in the context of a widespread rebellion against Qing rule in eastern Tibet, the Yongzheng emperor ordered armies from Sichuan and Yunnan to occupy the plateau. It was formally absorbed into the province of Yunnan as Zhongdian; appointed officials were sent to govern it; a garrison of some eight hundred imperial troops was installed in its largest town, and the hereditary rulers were demoted to district commanders.[45] By the time of Forrest and Litton's visit, then, the plateau had been within the fold of imperial administration for a century and a quarter. Nevertheless, the hereditary officials had retained their armies and much of their influence, and imperial officials seemed to have great difficulty asserting administrative control. In 1905, the subprefect *(tongpan)* subordinate to the prefect of Lijiang controlled only forty soldiers, remnants of the garrison installed in 1723. This tiny army had recently been faced with rebellions in seven of the plateau's eight districts.[46] Three miles north of the plateau's center of government was an important Gelugpa monastery, housing 1,226 monks, another competing source of political authority.

Forrest and Litton walked over the plateau in the rain. It was, Litton wrote in his report to the Foreign Office, "most picturesque . . . the wide sweep of barley, wheat, and oat fields interspersed with plots of marsh, the dark pine woods on the lower slopes, the bare mountain tops above, and the sparkling mountain streams below reminded my companion, a Scotch botanist, of a cultivated highland valley in his native land." *Rhododendron* choked the foothills; blue *Gentiana* and red *Euphorbia* carpeted the pastures: to the "Scotch botanist" it was "one huge flower garden."[47] In the late afternoon, they came to the southernmost of the plateau's two towns, called, in Chinese, Xiao Zhongdian. The barley harvest had begun, and the harvesters were holding a festival, eating boiled beef and piles of buckwheat cakes. The whole place seemed prosperous: shining fields, large white houses, well-fed people. While the land appeared fertile, most of the plateau's wealth came from trade. Caravans from the plateau moved through the length of Yunnan to bring tea, steamed and packed into large bricks, from the Tai Lü (or Shan) states in the far south of the province to Kham and central Tibet. Most households bred horses, and traders from the plateau supplied animals to Tengyue and Burma for the cross-border trade.

As they entered the town, a large man greeted them, finely dressed, with a sword. He was, Litton later learned, the occupant of a post called, in Chinese, *huotou*. This was an arrangement for local administration found in scattered parts of Yunnan at the margins of the imperial administrative system. *Huotou* systems developed mostly in larger villages lying on routes frequently traveled by traders and officials. They were a communal defense against the expensive obligation to host official travelers, the responsibility of local elites. Xiao Zhongdian was two stages south of the plateau's administrative center, Zhongdian, and two stages north of the nearest administrative center to the south in Judian, so it was a natural stopping place for every traveler on this route. Both Tibetan noblemen and Chinese officials had the right to requisition hospitality and transport from its populace. The elite families of the town elected one household of their member to undertake a series of ritual and administrative duties. Among these was the obligation to feed and house important travelers and to provide men and animals to transport them and their baggage to the next stage. Since, as Litton noted, the post of *huotou* was "far from lucrative" it was usually held for only one year.[48]

The *huotou* escorted the travelers to his house. They lived there three days while Litton recovered from a fever. It was a "common meeting place of the people" in Forrest's words, "practically a club" in Litton's. To his enormous surprise, Forrest found that he liked the people who gathered there. What distinguished them from "the Chinese," whom he emphatically did not like, he decided, was the way they used their eyes: "The people came riding in from miles around to see us. However they are much pleasanter than the Chinese, they do not stand and gape as those do. They are very curious regarding things they don't understand, but once you explain the article to them, they are satisfied and go away. On the other hand the Chinese simply stand and stare with the most ignorant expression imaginable on their faces, and will not clear unless you really chase them."[49] The articles that most interested these traders and farmers were the travelers' guns. Almost as soon as they arrived, their host urged Forrest and Litton to go shooting with him. Wet and cold as they were, they put it off until morning. The next day, they walked out into the barley fields that surrounded the town and shot more than a hundred pheasants. After this, Forrest became attached to his host. He described him in unusual detail to his mother and sisters:

> A regular savage beauty, about 6 ft. 6 ins. tall and 3 ft. 6 ins. broad, I
> should say, clad in a coarse scarlet Tibetan cloak . . . open at the neck
> and strapped round the waist, he had scarlet putties on his legs and huge
> Tibetan top boots over these. Matted hair hanging down to below his
> shoulders (no pigtails), and dirt ad libitum, completed his costume. He
> looked like a pirate out of one of Gilbert and Sullivan's plays. However,
> he was a right cheery sort, and he and I got on like a house on fire.[50]

As the description makes clear, this was a wealthy man, dressed in finery. Forrest coveted his beautiful sword with a scabbard worked elaborately in silver and turquoise.

When Forrest returned to the plateau again in November and December of the same year, it was very cold. The oil in his shotgun froze, and he had to take it apart and clean it before he could shoot pheasants on his way up the plateau's gentle valley. He stayed again in the *huotou*'s house. "I really like the fellow in spite of all his dirt," he wrote. "He is such a big man, and yet as simple, jolly, and kindly as a child."[51] Again, their companionship centered on shooting. The *huotou* showed Forrest his Tibetan-made muzzle-loaded flintlock rifle. It was quality workmanship, the best rifle available in Tibet before the 1911 revolution.[52] He took it outside, aimed at a tree on a cliff some eighty to one hundred yards away, and missed. Determined to "give him an eye-opener," Forrest loaded his Winchester with its full eleven rounds and blasted away at a limestone patch on the cliff. "There was a crowd around us by this time, and I can tell you that the group of faces would have made the fortune of any photographer. You never saw such astonishment depicted in all your life." The two then competed with Forrest's rifle, emptying some fifty rounds into the cliff.[53]

It is not surprising that Forrest imagined this as a photographic moment. The association between guns and cameras was commonplace in colonial situations. As Paul Landau observes, the technologies of gun and camera evolved in parallel after the 1860s. Chemicals developed for ready-made cartridges were adopted for use in dry-plate cameras by Eastman and Kodak; elements of the designs of the best cameras were based on the mechanism of the Colt revolver; both technologies moved toward the capacity to make clean, rapid, repeating shots.[54] By the end of the century, game hunting had become a largely visual enterprise, incorporating cameras to create visual trophies. Later, Forrest would photograph long rows of Lady Amherst and white-eared pheasants hung up by their feet from the eves of his house in Nvlvk'ö—both hunting trophies and scientific specimens. In addition, travelers who ventured out of the reach of the armies and bureaucracies of colonial administrations had few resources with which to demonstrate their superiority. Forrest had his sense of personal hygiene: thus his repeated remarks about dirt even at a moment of profound camaraderie. But it was difficult to stay clean and even more difficult to show one's cleanliness to others in a convincing way, especially when they were far better dressed than he. So his gadgets—his Winchester, his camera, his telescope—had to bear the burden of demonstrating difference.

The posture of astonishment Forest elicited in this crowd of Tibetans was nearly identical to the "gaping" of which he accused "the Chinese." Yet while those looks made him miserable and violent, these delighted him—and made him think of photography. In other situations, he and his competitors would use their guns to similar effect. Clever Francis Kingdon Ward didn't even have to shoot his: in a village on the Mekong he showed a crowd "what wizards the English are" by making the cartridge pop out of his breech-loading shotgun in response to his whistle. "The men were so obviously taken aback that they merely stared incredulously." "My gun and my camera," he observed, "were always a source of great interest to the Chinese."[55] Later, in Tibet, Kingdon Ward, frustrated in his desire to photograph the shy women of Jana, set up his gramophone in the street, played an operatic prelude, and whipped out his camera to photograph the "enchanted crowd."[56] Chapter 5 shows how Joseph Rock resorted to the same cliché in moments of melancholy.

Michael Taussig argues that Europeans and Americans found in the astonishment they provoked with such antics a reflection of their own deep fascination with technologies of mechanical reproduction.[57] It is certainly true that these men were fascinated with the wonder they cultivated in others with guns, cameras, and gramophones. But they hardly seemed awed at the mysteries of these technologies. They were all deeply troubled by the sensation of plunging into a sea of looks, where the eyes of others framed their sociality in alien terms. Their attempts, in practice or imagination, to generate astonished looks and freeze them with a camera were efforts to turn the power of "gaping" back on itself. Martin Jay observes that the eye of the camera, unreadable by those at which it gazes, holds the viewer still, "static, unblinking, and fixated."[58] To generate and capture fascinated looks was to draw "gaping" into the familiar field of mechanical reproduction—the field of the "world as exhibition," through which the sights of colonized places were reproduced and exhibited in the metropole.

·   ·   ·

Not being gaped at unless at his command became for Forrest a precondition for social exchange without violence. After his adventurous excursion up the Salween gorge with Litton, he settled down to collect with increased determination. Though he had traveled with cooks, muleteers, and servants, he had until now done his collecting himself. In the summer of 1906, with the aid of missionaries of the China Inland Mission in Dali, he hired two men and three women to collect in the Cangshan range, just west of that city: this was the only time he would ever employ women as collectors. Leaving these five in the charge of one of his servants, he went north to spend the season on the great Yulong range. He took the southern route into the range, straight up the main street of Nvlvk'ö. His letters make no mention of the people he met there. But fourteen years later, he would recall that it was that summer when he first began to hire men from that village as collectors. Over the next thirty years, the men from Nvlvk'ö amassed one of the most extraordinary collections ever attributed to a single botanical explorer. They spent years at a time away from home, eating what they could, sleeping in the open, enduring all the physical and emotional hardships that travel over this immensely rugged landscape entailed. He tyrannized them; he fired them for insubordination; he fumed at them as "Bolsheviks" when they asked for raises; he scoffed at their ritual life when funerals drew them back to Nvlvk'ö. But for all this, it is clear that he developed profound camaraderie with many and sincere respect for a few.

They had learned as children many of the plants of the Yulong range. But there were many skilled herbalists among the "Chinese" population of the province who may have made better collectors, had he the imagination to look for them and the linguistic skills or social connections to find them. What made it possible for him to employ these farmers and live with them lay in how they used their eyes. He never once mentioned being stared at in Nvlvk'ö, and given his sensitivities, this is evidence enough. Like the Tibetans of the Gyaltang plateau, Naxi farmers and traders found staring profoundly impolite. Looking directly at a stranger, particularly one of high status, no matter how strangely attired, was deeply disrespectful. Meeting an important stranger on a path or village street, one looked away and went about one's own business, satisfying one's curiosity later from within a doorway or behind a fence. Habits of looking were distinctly gendered as well. For a young woman to look directly at a man not of her household out of doors was improper; older women could manage more direct looks, but again not at high-status strangers. These were widespread habits of ocular comportment, shared with most of the highland Tibeto-Burman–speaking peoples of the province.

As it happened, they were also familiar habits, close enough to those he had grown up with. Living amongst people who, however curious, did not stare, he immediately found ways to initiate social exchange even in the absence of a common language: trading cards, shooting his gun, producing specimens. When such deeply installed, tacit habits of ocular exchange were broken by "gaping," possibilities for social exchange were also broken, "blind rage" ensued, and physical violence became his only means of establishing social relations. In addition, "gaping" profoundly fractured his visual field. Under the unblinking eyes of different others, he found it difficult to look at anything but himself, and his nakedness often came to mind. In 1905, complaining to his mother of the "ignorant expressions" of "the Chinese," he commented, "I have become so callus that I can stand now and take a bath with a crowd around me." In company of people who turned their eyes away, he could refocus his own, away from himself, onto the landscape, to find in it the qualities that made seeing productive. In Nvlvk'ö, he learned to turn the eyes of others toward that common object as well, initiating the long and intensive process of ocular instruction through which others' eyes were trained to see in the ways demanded by Linnaean taxonomical science. He could only begin this process once he had broken that mute circuit of eye against eye that gaping entailed.

Scholarship on imperial travel has emphasized looking as a mode of knowing and possessing. "The idea of the 'the gaze,' especially as exercised by the white male explorer, missionary, administrator, or itinerant naturalist," writes David Arnold, "has rapidly come to be seen as one of the principal expressions of the wider colonizing process. . . . 'the gaze' is the prelude to possession in more material and institutional forms, just as travel is more about imposing upon, than learning from, the landscape subject to the itinerant gaze."[59] Travel made the landscape a spectacle, replicated after the mid-nineteenth century in exhibitions for other eyes, in both metropole and colony. Photography provided a convenient model for this mode of knowledge production: a clear-seeing eye, unimpeded by the vagaries of subjectivity, rendering fields of sight into real objects, which could be classified, archived, and exhibited.

The temptation for scholarship has been to write as though imperial gazes could, in reality as well as in ideology, be modeled on the kind of vision commonly attributed to the camera. As though vision could ever simply be that of a sovereign master of two eyes, surveying the world and rendering it as representation. As though acts of vision are not always embedded in social fields of vision, composed of intersections of multiple pairs of eyes, intimate and distant, immediate and deferred, real and imagined. The visibility to others of one's own gazing body is constitutive of one's gaze; one can only see as a social being if one sees oneself as visible and one's own vision as including, reflecting, embracing, or deflecting the gazes of others. In the case of imperial scientific travelers like Forrest, this is easier to notice in places off the edge of the map of empire where travelers had fewer resources, material and ideological, at their disposal, to insulate their eyes from those of others.

2. Farmers and Kings

From the cobbled streets of Lijiang town, or Dayanzheng, Forrest could see the magnificent Yulong range gleaming in the northern sky. He walked toward the peaks, passing through several hamlets. The last and most substantial was Baisha, where a periodic market was held. After this, the valley narrowed and grew barren. The road skirted the eastern flank of the range, climbed a pass, and then descended to the gorge of the Yangtze river. This was the road toward the Zhongdian (Gyaltang) plateau and beyond. But if he walked straight toward the peaks instead of around them, he took another path, toward a long spur that extended south from the range. He passed three small Karmapa Buddhist monasteries and walked through the hamlet of Dügkv (Yulongcun). Nvlvk'ö (Xuecongcun) was the last village on this path, some fifteen miles from Lijiang town. After passing between the village's rock and mud-brick houses, the trail ascended into the alpine meadows.

Nvlvk'ö was the highest village in the Lijiang valley—over nine thousand feet in altitude. It is difficult to know exactly how many people lived there in 1906 when Forrest first began to explore the range. Forrest estimated "about 90 houses and huts" in 1913.[1] Joseph Rock guessed "about 100 families" in the 1930s.[2] The houses were actually family compounds, comprising two or more buildings facing a central courtyard enclosed with walls or fences. The main residential building usually contained a larder, a cooking area, and sleeping quarters for a married couple who headed the household and their unmarried children. Across the courtyard was a stable, housing horses, sheep, goats, cattle, or *maoniu*—yak-cattle crosses. Here and in nearby villages daughters married out and elder sons, when they married, were provided with houses of their own as soon as possible. Youngest sons and their wives remained in the parental house and inherited it upon the parents' deaths. So most houses contained only one married couple; a few contained couples of two generations: parents and a youngest son with his wife.[3] Behind or to the side of each house was a vegetable garden, beyond that, fields. The cultivated lands were poor and stony; no rice could be grown, and farmers planted wheat, maize, oats, peas, beans, barley, and potatoes.[4]

The Yulong range presented a dramatic contrast in biological diversity to the valley below. The valley was dry, particularly near the range, and supported relatively few species. The mountains rose steeply from the valley floor to a height of over eighteen thousand feet and descended as steeply on the west side to the deep gorge of the Yangtze. These differences in altitude and the contrast in precipitation between the wet western slopes and the relatively dry eastern slopes created a great diversity of ecological environments and a corresponding diversity of flora and fauna. Nvlvk'ö was a natural gateway to the range. Every household pastured animals in its meadows, some fifty to sixty head. Bamboo was abundant, and most households crafted bamboo baskets and brooms to sell in the Baisha market.[5] Most also harvested edible greens, roots, and fungi from the mountain and many gathered medicinal plants to sell in Baisha or Lijiang. In short, most villagers, particularly when young, spent a great deal of time on the range. They knew its southern peaks intimately, and they had names for every crag, gully, and meadow and histories for many.

Culturally, Nvlvk'ö belonged more to the mountainous hinterland north of the Lijiang basin than to the basin itself. Within the Lijiang basin, particularly in its southern part, many Naxi had intermarried with Han settlers by the early twentieth century, and kinship practices and terminologies were similar to models practiced by Han further to the south. In these northern mountains, people organized themselves into exogamous, ideally locally resident patrilineages, called *coqo* in Naxi. There were four surnames in Nvlvk'ö, the He, with four *coqo,* the Zhao, with two *coqo,* and the Li and Lü with one *coqo* each.[6] The male members of each *coqo* assembled twice a year to remember their ancestors and mark their participation in the lineage in a rite called *muanbpò.*[7] *Coqo* were involved in the crucial transitions in the lives of their members' households. Major property such as land, houses, livestock, and monetary wealth passed from fathers to sons upon the latter's marriages; the senior male members of a *coqo* supervised the public division of property among sons. The village as a whole owned and managed extensive meadows and forests around Nvlvk'ö, and one obtained rights of access to these resources through membership in a *coqo.*

Every three years, during the "fire torch festival" of the sixth lunar month, the male members of all the village's *coqo* gathered to elect a governing council of six to ten elderly men. The council was charged with managing the village's collective resources, establishing and enforcing customary law, and mediating civil disputes. Households applied to the council for permission to cut timber to build houses and marriage beds; a "mountain manager" appointed by the council supervised the cutting. Each spring after snowmelt, when the alpine vegetation was in its most delicate state, the council closed the mountain to firewood and pine needle harvesting for a period. At the end of this time, only one or two members of each household were allowed to harvest pine needles for compost so those with large labor forces would not have unfair advantage.[8]

Just north of Nvlvk'ö was a small artificial lake with an islet in its center, supporting a solitary tree. Called Yühu, or "Jade Jug," this lake was central to the historical imagination of people in Nvlvk'ö. It had been built by the native hereditary rulers of Lijiang, the Mu chiefs, as the setting of a pleasure villa and deer park. In the 1990s, Nvlvk'ö residents told ethnographers that their village had been founded by laborers and craftsmen recruited by the Mu chiefs to build the lake and villa. The ancestors of the He were said to have been serfs of the Mu chiefs of Lisu ethnicity, brought from Yongsheng, to the east, to work on the lake. The ancestors of the Zhao, Li, and Lü were said to have been stoneworkers and potters, brought from Yingtianfu in Nanjing (named by many in Yunnan as their place of origin) to construct the villa.[9]

A little way from the lake, standing in a grove of spruces, was a small Buddhist temple, which had stood beside the Mu chiefs' villa. It was adorned with frescos and a marble slab inscribed with the date Ming Wanli 36 (1608). According to the genealogical chronicles of the Mu lineage, Mu Gong, one of the aggressively expansionist Mu chiefs of the Ming dynasty, had spent his old age in the villa, buried in his books, "seeking to understand thoroughly the ancient literary allusions."[10] His son, Mu Gao, a great military strategist and a poet, composed a paean to the lake, "Journey to Yühu" *(Yühu youji)*. And in the Qing, two well-known scholars, Ma Ziyun and Yang Chang, who may have enjoyed the villa as guests, also wrote poems about the lake.[11] Just behind the temple was a limestone cliff. After the Mu chiefs were deposed in 1723, the first regular magistrate of Lijiang, Yang Bi, inscribed his name there and, in gigantic letters, "jade pillar upholding heaven," a metaphor for his rule.[12]

KINGS

The history that left traces in the lake, temple, and cliff of Nvlvk'ö was a heroic history of divine kings, in which the ancestors of people in this village were nameless laborers and soldiers at best. This history, of the Mu dynasty and its militarist kingdom, is important to my story for two reasons. First, the Mu kingdom's conquest of northwest Yunnan reshaped the ethnic and linguistic geography of the region. In particular, this conquest produced the distribution of Naxi people, centering in the Lijiang basin but scattered in pockets throughout the rest of the region, and this distribution powerfully influenced the geography of the botanical discoveries attributed to Forrest. Second, the history of the Mu conquests shaped what the collectors from Nvlvk'ö knew and felt of this landscape in myriad ways. It reverberated through the ritual practices through which they learned of the real and mythic geographies of the regions they explored. It was the substrate for all their collaborations, and it ultimately guided Forrest's own long and intensive project of botanical mapping.

The central written sources for this history are the genealogical chronicles of the Mu chiefs. There are two versions. The highly literary Mu Gong, who spent his old age in his Nvlvk'ö villa, authored both in Chinese. He finished the first in 1516. Twenty-nine years later, he completed a revised version under the influence of a famed Confucian literary scholar named Yang Shen. This chronicle, to which Yang Shen wrote a preface, was the official one, which Mu Gong's successors kept up to date, and which they showed the Ming and Qing emperors who ratified their rule. Another important source is a gazetteer compiled in 1743 by Guan Xuexuan, one of the first magistrates of Lijiang after it was brought into the regular system of administration. Stone inscriptions, genealogies of the Mu chiefs from other sources, and a printed copy of letters of appointment from the Ming emperors supplement these sources. The "Veritable Records" of the Ming emperors, the *Ming Shi-lu*, contain many references to the Mu kings, compiled and analyzed by Xin Fajun.[13] And historians of Lijiang have also found references to the Mu kingdom in other county gazetteers, in the Yunnan provincial gazetteers, and in other sources on the Southwest.[14]

In Mu Gong's chronicles, the Mu lineage originates with the creation of the cosmos. The first chronicle begins with eleven lines of Chinese characters that make no sense in that language. Joseph Rock discovered them to be a phonetic transcription of a Naxi-language creation myth, told as a lengthy epic in dongba texts.[15] Mu Gong's version is succinct. Heaven laid an egg, white as a conch shell; the egg emitted a hot vapor; the vapor

Figure 9. Ajia Ade (Mu De), first Ming dynasty chief of Lijiang, 1382, from the genealogical chronicles of the Mu chiefs, reproduced in Joseph Rock, *Ancient Na-Khi Kingdom of Southwest China.*

emitted hot dew; a drop of dew fell into the sea; from the sea the first being emerged; another two generations produced Muanza ("Heavenly Man"), the Mu kings' first ancestor. Muanza married a celestial princess, begetting eleven generations of sons; the first seven married celestial women, and many lived for more than a thousand years. Ye Gunian is the first "historical" ancestor—his dates correspond to the 618 Tang invasion and occupation of much of Yunnan. Seventeen generations intercede between Ye Gunian and the next named ancestor, Qiu Yang, who became the ruler of Sandian (Satham, the Tibetan name for Lijiang) in 670, corresponding to the subjugation of what is now northwest Yunnan by the newly powerful Tibetan kingdom.

From around 750 until 900, Yunnan was dominated by the Nanzhao kingdom, which took northwest Yunnan from Tibet, expanded west into Burma, and presented a potent military threat to the Tang dynasty.[16] Mu Gong names ten generations of ancestors in this period; all are said to be rulers of the Lijiang area subject to, or allied with, the Nanzhao. By the end of the

Tang, sustained military conflict and internal rivalries had weakened the Nanzhao considerably. The next stable state to succeed it was the Dali kingdom, which lasted more than three hundred years, until the Mongol invasion of Yunnan in 1253. The Song emperors (960–1253) found Yunnan not worth the trouble and risk of military conquest, and relations between the Dali and the Song were relatively restricted and peaceful. For this reason, much less is known about the Dali kingdom than about the Nanzhao. Mu Gong gives only four generations of ancestors for this three-hundred-year period, rulers of a kingdom called Mosozhao.[17] Though pressed hard by the Dali kingdom, Mu Gong notes, these rulers never submitted. The last was Moubao Acong, a sage who could read Chinese characters at the age of seven, who invented the Naxi-language writing system, and who unified the warring clans of his family, laying the foundations for the royal Mu throne. In 1253, the Mongols invaded and conquered the Dali kingdom in their campaign to take Song China. Moubao Acong's son Acong Aliang welcomed the Mongol army as it entered his territory. As a reward, Kublai Khan confirmed him as the ruler of the Lijiang area.

Thus the early history of the Mu kings and the peoples they ruled is obscure. Even after the mythological first twelve generations, Mu Gong's chronology is strange. It has many gaps and impossibilities, and there are many inconsistencies between the two chronicles and between them and genealogies from other sources. As Christine Mathieu points out however, it is at least clear that Mu Gong used the chronicles to make some forceful assertions. In the 1516 chronicle, he implicitly claimed that his dynasty's mandate for rule did not ultimately lie with the emperor. Having descended from heaven, the Mu ruled Lijiang since long before Yunnan was absorbed into the Chinese empire. In his 1545 revision, Mu Gong founded his ancestry in a more acceptable historical truth, beginning the chronicle with Moubao Acong's son Acong Aliang, who was installed by the Mongol army as ruler of Lijiang after having welcomed it to Yunnan. Yet the preface by Yang Shen gives the earlier genealogy as well, repeating its implicit claims of autochthony and a divine mandate.[18]

In a careful and clever rereading of the chronicles, Mathieu reconstructs the early history of the Lijiang rulers as follows. During the Nanzhao kingdom, Lijiang was controlled by a clan named Ye from Nanzhao. The names of the rulers and their wives during this period give evidence that this clan used marriage alliances to extend a federation of tribes through northern Yunnan. The murder of the entire Meng royal family of the Nanzhao in 901 and their eventual replacement by the Duan lineage, which founded the Dali kingdom, coincided with a dynastic change in Lijiang. The region was conquered by the Mou (Moso) tribes, the ancestors of the Naxi who reside there today. The Mou chiefs abandoned their predecessors' tribal federation, married within their own clan, and eventually disunited into warring factions. In 1253, another dynastic change occurred. The Mongols swept through northern Yunnan, leaving in Lijiang an officer (or more probably several) with the clan name Ah. One officer (Acong Aliang) was appointed to rule the area; he married a local chief's daughter, unified the warring clans of the Mou, and founded the royal patriline of which Mu Gong was the direct descendent.[19]

Beginning with the Mongol conquest, the chronicles correspond more directly to the Chinese historical record. The chiefs of Lijiang in the Yuan dynasty (1279–1382) were all named Ah; royal power passed from father to eldest son. The Yuan imperial court instituted a system to control native hereditary chiefs. As in previous dynasties, through most of the empire the emperor appointed local rulers to posts outside of their region of birth, and their posts were not hereditary. In frontier regions where native hereditary chiefs had long ruled, the Yuan implemented new regulations governing their nomination, promotion, and transmission of power. At first, the Ah chiefs were granted territories within the Yangtze loop, most importantly the Lijiang basin and Heqing directly to the south. Soon, the Yuan emperor confirmed them as rulers of a much expanded territory, including the Yongning basin to the northeast, Yongbei and Yongsheng to the east, and Judian and Weixi to the west. The chiefs appear to have pursued marriage alliances with elite or chiefly families within their realm and on its peripheries.[20]

In 1382, the Ming Hongwu emperor mounted an armed expedition to bring Yunnan under his control. Hongwu issued a decree that conferred upon Ajia Ade the surname Mu. (Neither the chiefs nor the common people of Lijiang had surnames: their names were linked patronymics. Ah, the name of the Yuan dynasty chiefs, was an indication of inherited rank rather than a Chinese-style surname.) Ajia Ade, now called Mu De, was named magistrate of Lijiang. At the same time, the emperor shifted the territory of Yongning out of his control and confirmed its own local chief, also named Ah, as ruler. Yongning would maintain its relative independence under the Ah lineage for nearly six centuries until 1956.

Mu De immediately embarked upon military expeditions against neighboring native hereditary rulers, whom the chronicles describe as brigands or rebels. The Ming instituted stricter policies regarding native hereditary chiefs: they could be promoted to higher positions, but they could also be punished if they transgressed court regulations. In the sixteenth century, the Ming began a campaign to reform the system of native hereditary rule and to appoint regular magistrates wherever possible. Mu De and his successors mounted armed expeditions with Ming troops south to Dali, west into Burma, and north into Tibet. Many such forays were against native hereditary rulers who had been recognized by the Ming but who had been designated rebels after attacking a neighbor or disobeying an order. In effect the Mu armies became a vanguard of Ming campaigns to demote and replace native hereditary rulers. The Mu kings were savvy about the politics of this precarious enterprise, and they went to great lengths to display loyalty and obedience to the Ming court. They were good at producing heirs: most had several wives in addition to the senior wife named in their genealogies, and during the Ming each succession was to the eldest or only son of the previous chief. Shortly after his father's death, the designated heir sent envoys to the imperial court with gifts of horses and local products. The emperor returned letters of appointment, titles and honors, posthumous titles for previous chiefs, and instructions to keep the peace on the frontier.[21]

With the approval of the Ming court, the Mu kings carried out an aggressive and extended military expansion into territories occupied by "Xibo" (Tibetans) and "Yifan" (other Tibeto-Burman peoples) whose native rulers had little standing in the imperial system. Officially, they were suppressing "brigands and nomads." By the mid-fifteenth century, they were openly conquering village after village. They extended their territory east into the Yanyuan and Muli districts of Sichuan, or southern Kham, north into the Gyaltang (Zhongdian) plateau, and northwest into the vast regions between the Jinsha (Yangtze) and Nu (Salween) rivers. The Mu chronicles list villages conquered and numbers of brigands killed in battle, prisoners taken, and captives decapitated. In each key location they conquered, the Mu garrisoned soldiers, imposed corvée, impressed troops for the imperial armies, and created regional commands for members of their lineage, subordinate to Lijiang. In 1548, Mu Gong's successor Mu Gao ordered an inscription cut on a huge disk of stone, a "stone drum," erected where the Yangtze makes its first abrupt turn northward. The inscription, which celebrates victory in a "punitive expedition against the wild tribes," gives a sense of the aggressively military ethos of the Mu court at the time:

> Within two years we have carried our victory east and west of the
> river. The neighing of our steeds was heard thousands of li away. . . .
> Heads were heaped up like hills; blood flowed like rivers, and buff-coats
> and rattan shields filled hollows and ditches. . . . We have annihilated
> four hundred camps of dogs and sheep and swept away five thousand
> villages of barbarians. . . . Our military virtue was made to redound;

streams of blood eddied around our armor; the bodies of the slaughtered lay so thick that our horses could not move; our soldiers' prowess
was such that the western corner was pacified.[22]

By this time, the Mu kingdom sprawled over all of northwest Yunnan and
part of southeast Sichuan, and it controlled all China's southwestern routes
into Tibet. It held these territories for well over a hundred years.

The great traveler Xu Xiake visited Lijiang in 1639, when the Mu dynasty was at the height of its power. He stayed in Mu Zeng's villa near
Nvlvk'ö, where he was feasted at banquets featuring eighty dishes of exotic
delicacies. "The Mu family have lived here for two thousand years," he
wrote, "their mansions are as beautiful as the ruler's. Should the imperial
army come near [the family] meekly submits to being tied up. When the
army retreats, they reassert their power: consequently, the Mu family has
not suffered the ravages of the imperial army for generations. Moreover,
thanks to the uniquely prospering mining production, their region is the
wealthiest of all the non-Han regions."[23]

Mu military might disintegrated in the 1640s as the Ming lost control
of Yunnan to roving bandits. In 1647, rebellious peasants sacked the royal
compound in Lijiang, looted the imperial gifts to the Mu family, and burned
the chiefs' letters of appointment, forcing Mu Yi to withdraw his troops from
much of the northwest to fight them. The Qing regained Yunnan in 1659
by granting civil and military control of the province to Wu Sangui, the former Ming general who had invited the dynasty's founders into the capital.
Though the Qing ratified Mu Yi's appointment, Wu Sangui eviscerated the
Mu realm. The chronicles from this period are full of aggrieved complaint
against Wu Sangui, who imprisoned Mu Yi, replaced him with his heir, imposed unexpected demands for soldiers and back taxes, and ceded most of
the hard-won territories of the north and northwest to Tibet. When Wu Sangui rebelled against the Qing in 1673, Mu Yao raised troops and provisions
for the imperial armies. He was awarded with a ratification of his hereditary right to rule Lijiang. But the Mu family's realm had shrunk dramatically to the Lijiang basin and its environs on the two great bends of the
Yangtze—today's Lijiang County. The Mu retained their own troops, but
these had decreased dramatically as well, to only a thousand or so soldiers.

In 1720, Mu Xing and his heir both died of ailments contracted during a
large Qing punitive expedition in Tibet. The next succession did not go well.
When the energetic and expansion-minded Yongzheng emperor succeeded
to the imperial throne in 1722, he launched a reinvigorated campaign to incorporate the Southwest more fully into the empire by abolishing native

hereditary rule wherever possible. Mu Xing's successor Mu Zhong had still not received his final ratification in 1723 when five local leaders, four from Mu patrilineages, traveled to the provincial capital with a list of accusations against Mu Xing and a petition to make Lijiang part of the regular administrative system. The new governor-general of Yunnan and Guizhou seized this opportunity to further Yongzheng's policies by appointing a regular magistrate *(liu zhifu)* to Lijiang. The magistrate confiscated all of the Mu ancestral lands and residences to pay for back-taxes, burned the Mu family records and accounts, and demoted Mu Zhong to a powerless official position, ending the Mu family's 470 years of rule.[24]

·   ·   ·

The conquests of the Mu chiefs redistributed ethnic groups and languages throughout northwest Yunnan. The ancestors of the people who under Mu rule came to know themselves as Naxi probably moved to the Lijiang basin near the end of the Tang dynasty, brought together in a federation of tribes under rulers who married into the families of other local chiefs. During the Song and Yuan, these peoples probably occupied only the areas within and immediately around the Lijiang valley and the Yongning basin on the other side of the Yangtze. As the Mu kings expanded their domain, they garrisoned troops in key locations throughout northwest Yunnan and southwest Sichuan. In some locations, like the southern Gyaltang plateau, they also encouraged people from Lijiang to immigrate and open up uncultivated lands.

In others, like Eya in the Muli district of Sichuan, and Fungdizhuang near Yongsheng, Yunnan, they established personal feudal enclaves, where serfs from Lijiang worked royal lands. Many of the descendants of these soldiers, settlers, and serfs were likely absorbed into the populations of Tibetans, Lisu, Yi, and other Tibeto-Burman peoples who inhabited these areas. But many were not. They created pockets of people over the entire Mu kingdom, particularly along major routes of trade and conquest, who spoke the Naxi language, practiced the dongba cult, and maintained strong cultural affinities with people in the Lijiang valley. Linguists now identify six major dialects of the language they have agreed to call Naxi. The pockets of Naxi speakers along the routes to the northwest, descendents of the garrisons the Mu had established to watch over "brigands and wild tribes," all speak the dialect spoken in the Lijiang valley.

The region's demography underwent further dramatic changes in the eighteenth and nineteenth centuries. The Ming stationed many garrisons of troops from central China in the region, whose descendants established towns and farming communities. The Ming also encouraged settlers from other provinces to populate Yunnan, making Han Chinese the majority in the province by the end of the dynasty. Following the collapse of Wu San-gui's rebellion in 1681, the Qing launched an aggressive program to colonize and develop the southwest. Huge numbers of troops were stationed in Yunnan; a network of new roads was built; new walled cities were constructed; and large numbers of landless colonists were encouraged to immigrate with tax remissions and grants of land and seeds. Some three million people moved to Yunnan between 1700 and 1850.[25] Han Chinese settled in large numbers in the southern Lijiang basin and diffused through the rest of the former Mu domains.

Even in areas where they were vastly outnumbered, however, Naxi speakers were not often absorbed into other ethnic groups. Marriage practices are probably the central reason. In distinct contrast to Han, Chinese, Tibetans, and all other Tibeto-Burman peoples, the Naxi scattered through the mountainous parts of the Mu realm preferred, or prescribed, patrilateral cross-cousin marriage. That is, a young man was expected to marry his father's sister's daughter; or, as Naxi put it, a young woman's mother's brother had the right to bring her into his home.[26] Mathieu points out that from the beginning of the Ming onward, the lineage names of the Mu kings' first wives alternate by generation, a pattern characteristic of patrilateral cross-cousin marriage.[27] Ethnographies written in the twentieth century suggest that Naxi in the mountainous hinterlands of the Lijiang basin continued to take a man's rights over his sisters' daughters very seriously even after the socialist state proscribed any form of cousin marriage.[28] As an ideal type, patrilateral cross-cousin marriage creates alliances between at least three patrilineages which alternate each generation. I give my daughter in marriage to my wife's lineage; her brother gives a daughter in marriage to his wife's lineage; they give a daughter in marriage to my lineage; and in the next generation the circuit is reversed, as each lineage takes a wife from the lineage to which it gave a daughter the generation before. Patrilateral cross-cousin marriage is rare: anthropologists used to argue about whether it could exist at all. In contrast to matrilateral cross-cousin marriage, which draws a potentially infinite number of groups into exchange relationships, patrilateral cross-cousin marriage tends to link lineages in stable exchanging pairs.[29] Life could never correspond very well to this ideal type: siblings did not always get along; sons and daughters could not always be bent to their parents' will; and death, infertility, and migration often interfered. Still, even in real life, patrilateral cross-cousin marriage tended to create stable sets of allied patrilineages where status hierarchies between wife-giving groups and wife-taking groups could not develop. The other Tibeto-Burman peoples in this area preferred matrilateral cross-cousin marriage, establishing diffuse marriage alliances across the region and often across ethnic groups. But Naxi people, especially in mountainous areas outside the Lijiang basin, tended to trade wives only within stable groups of lineages, rooted in a single place, which passed property, language, and traditions along the extremely strong "bone" links this form of marriage encouraged: the male bloodlines of the patrilineage. This, very likely, is the chief reason the small Naxi settlements north and northwest of Lijiang did not melt into the overwhelming majority populations of Tibetans, Lisu, Yi, and Han.

Whenever Naxi from Lijiang traveled north or northwest, most people they met were Han, Tibetan, Lisu, Yi, or Nu. But along every important route they also found communities of people who spoke their own dialect of Naxi as their first language. In some of these places, the local rulers were also Naxi—native officials *(tuguan)* who had descended from hereditary rulers installed by the Mu chiefs. These people shared more than language with the Naxi of the Lijiang valley. They shared a powerful sense of the history of this landscape and their own part in it. They shared names for every mountain, river, and town; names for villages that had once existed in these valleys but had long since disappeared; stories of the exploits of their illustrious forbears, rooted in ridges and river crossings; and a sense of the routes their ancestors had once taken through these mountains and the paths their own souls would take after their deaths.

PRIMULA

By the time he finally rode out the west gate of Tengyue at the end of his first expedition, Forrest had shipped some 5,500 specimens and seeds of hundreds of species to Edinburgh. Bees Ltd. advertised *Primula forrestii* on the cover of its catalog and featured his plants in the annual Royal Horticultural Society show. His specimens went to the Royal Botanic Garden, where Isaac Bailey Balfour and his staff had begun to devote much of their time to naming and describing them. Arriving home in late 1907, Forrest arranged to work in the Garden's herbarium again, earning two pounds a week to sort and label his collections. He quarreled with Balfour over what his hours and pay should be and who owned his specimens.[30] The *Gardener's Chronicle* published full-page images of his primroses and orchids and a dramatic article on his escape from the Tibetan attack on the mission at Cigu. He visited Bees Ltd., where he was distressed to find his best plants dead, butchered, or eaten by slugs; he worried that he would be seen as "an incompetent ass." It simply didn't seem as though he'd done enough. He needed another chance in the same territory.[31]

He and Clementina married, and she was soon pregnant. He entertained an offer from Charles Sargent, director of Harvard's Arnold Arboretum to return to China for £400 in expenses and a salary of £300 per year. But Sargent wanted him to depart before Clementina was due, so Forrest decided, reluctantly, to turn him down. Clementina delivered George Jr. in March 1909 and nearly died of an infection.[32] Forrest accepted an offer from Arthur Bulley of Bees Ltd. to go to Yunnan for £650 in expenses and £200 per year in salary, a sum that shamed him.[33] He planned a three-year expedition; financial disagreements with Bulley would cut it short, to a year. In January 1910, a few days before he set off, his mother suddenly died.[34]

He spent most of the year on the Yulong range. The Abbé Jean Marie Delavay, stationed near Dali in the 1880s by the Missions Étrangères de Paris, had amassed an enormous collection of alpine species here, including many *Primula*. Forrest attempted to find, photograph, and collect seed of each. *Primula* seemed very local: each species seemed to inhabit a specific territory on a ridge, slope, or watershed. Each, he thought, might be traced back to a "source"—the actual place on the range where it had differentiated from a predecessor species and from whence it had distributed: "Pulchella Fr., of which I saw occasional straggling plants during my last visit I traced gradually over fully five miles of country to an open pasture where acres of the hillside was literally a mass of blue with it. The sight was worth all the weary toil, and now I can get heaps of seeds at my leisure. I'm sorry the photos I took of this sp. are all failures."[35] The idea that a species or genus might be traced back to its origin in space where it would exist in exuberant abundance would become central to his imagination of the wider geography of northwest Yunnan.

Balfour had given him a camera and developing equipment. In return, he agreed to deliver a print of each species he collected, growing *in situ*, with a descriptive note.[36] He had planned to buy paper, glass-plate film, and chemicals for developing and toning in Rangoon, but Bees Ltd. had not sent his funds. Now he was handicapped, especially in making decent prints. Still, he tried to photograph every new flower, and he enclosed prints in his letters to Balfour, along with entreaties to send plates, paper, and chemicals. Photography was "a heart-breaking business. . . . To secure P. Pinnatifida, I have returned to its habitat five times, each journey representing a climb of fully 4,000 feet—in all to expose a dozen plates, and so far I haven't a negative worth showing. All of them show movement . . . or are underexposed due to wind and fog. . . . I feel at each moment that I could hurl the camera into space and be done with it."[37]

To this visual record, he added a textual one. He wrote a rough description of each species as he collected it, with notes on the soil, altitude, surroundings, and plant associates. He numbered these notes and used them to compose a fuller description in his field book, marked with the same numbers. He bought reams of cheap dwarf-bamboo paper, folded each specimen into a large sheet, stacked and tied the sheets in bundles, and pressed them between boards weighted with large rocks. From every species he thought particularly desirable he laid aside samples to which he attached another label; he used these samples to write up full technical descriptions.[38] In theory, each species might be identified with a territory; the landscape above eleven thousand feet was an interlocking, overlapping mosaic of such territories. He was papering this landscape with photograph, specimen, and text—a dense, cross-referenced archive.

He wrote always as though he were alone. And indeed he was lonely, explicitly wishing himself at home with his wife and new child. But he was never alone. Tracing his footsteps on each toilsome search and climb were three or four unnamed men from Nvlvk'ö, carrying his camera, tripod, rifle, and specimen baskets. More joined him in his camp near the village where he taught them how to press and dry specimens. In the autumn, when the plants were seeding, he brought out the samples he had laid aside as representing fine species and discussed their characteristics and locations with the men. In parties of four or five they walked the range again, gathering seed. They brought it to camp in baskets, packed it into paper packets with numbers and labels and into crates for shipping.[39] They shipped some two thousand plant specimens and seeds of about a hundred species. They also collected about two thousand species of insects, especially butterflies, some bats, birds, frogs and leeches, and eighteen snakes, which they bottled with spirits.[40] They had all grazed goats and cattle in the range's meadows, hunted pheasants and musk deer in its forests, and gathered wood, bamboo, pine needles, medicinal herbs and plants for ritual use on its slopes. They knew Naxi names for many of its plants and Chinese names for those that could be sold as medicines. Forrest was following the flowers as he traced its *Primula* to its "source," but he was also following these men, picking up their hints about where a flower of this color grew and where meadows dense with that species might be found. The process of archiving the plant geography of the range was intensely collaborative. Still, Forrest controlled the

Figure 10.   On the Yulong range with Forrest's rifle and camera. Courtesy of the Royal Botanic Garden, Edinburgh.

product, translating their knowledge into his voice, erasing all their names for plants, and retaining only a few general Chinese names for places.

.   .   .

In October 1911, as Qing power disintegrated in the face of nationalist uprisings throughout China, the Yunnan Revolutionary Party, having seized control of military forces in the province, created a military government, with Cai E as governor-general of Yunnan and Guizhou. Revolutionists deposed appointed officials in west Yunnan's cities and towns and competed for local power, sometimes violently. British and Chinese negotiators had still not settled on a decisive frontier line along portions of the border with Burma. Mindful of the importance of the frontier and determined to prevent British incursions, Cai E assigned a prominent revolutionary, Li Genyuan, to take control of west Yunnan with a large military force. Li Genyuan was a native of Nandian, one of the Shan polities on the border near Tengyue, and his father was prominent in the affairs of that town. Li arrived in Tengyue in March 1912 with a force of around three thousand infantry and cavalry, armed with machine guns and artillery. He installed a new border commissioner *(daotai)* to govern the city and manage diplomatic affairs with the British consulate.[41]

Peaceful in the years preceding the revolution, Tengyue had been chaotic in the months since. The native hereditary ruler, or *sawbwa,* of Kanai, a Shan polity on the road from Tengyue to the border, was a man of revolutionary ambition. He had traveled to Japan and returned with a group of Japanese advisers charged with modernizing his state's administration. Now, with the help of an armory of Japanese Murata guns, he made a brief bid for power in Tengyue. Also vying for power was a local merchant named Zhang Wenguang, now raised to the position of military governor *(jujnwu shisi).* The Kanai prince eventually retired from the scene; "the respectable people of Tengyue and the neighborhood not prepared to submit to the rule of a barbarian Sawbwa," as the British consul put it. But Zhang Wenguang made plans to take power over the entire Western Circuit of Yunnan. He had marched troops from Tengyue to Dali, where the Republican general in charge of that city roundly defeated them, inflicting some two hundred casualties.[42] Against Li Genyuan's three thousand men, however, Zhang had no recourse, and he did not resist the revolutionaries' entry into the city.

Li Genyuan imposed stability on his native place with extreme violence. Four of the city's gentry composed a petition to Sun Yat-Sen charging Li with "fratricidal tyranny and inhumanity." They accused him of arbitrarily executing 163 people, including eleven women, allowing his soldiers to loot and kill at random, robbing and intimidating the gentry and merchants, seizing land and property for himself and his father, killing or dismissing every capable and popular official in western Yunnan, and erecting monuments to his ego. "It is as though Li Zicheng [the bandit leader from Yunnan who occupied Beijing at the fall of the Ming] has come to life again, so that to the women and children of Tengyue and Yongchang the names of General Li and his father are like a frightful nightmare. . . . What have the people of Tengyue done so that they should suffer so bitter an affliction?"[43] Though Li departed the city after six months, the petitioners claimed that the people awaited his possible return with great fear. Li's Border Colonization Army *(zhibian dui)* marched through west and northwest Yunnan, creating new nationalist governments in each major town. In a campaign sustained over many months, the army entered the gorge of the Nu (Salween), deposed the six Lisu native hereditary chiefs who had held power there for centuries in near complete independence from the Qing, and installed new local governments to administer the area directly, subordinate to Lijiang Prefecture.[44]

. . .

Forrest had arrived home again in January 1911. He had felt badly treated by Bulley and Bees Ltd., and he declared that he would never work for the firm again. Bulley hired the young son of a botany professor, Frank Kingdon Ward, as a replacement. Balfour introduced Forrest to J. C. Williams, a wealthy *Rhododendron* lover with an estate and castle in Cornwall. Williams had been growing Chinese rhododendrons from seeds Ernest Henry Wilson had collected for Veitch and Sons. He had paid Bulley three hundred pounds for all the *Rhododendron* and conifer seed Forrest had collected on his last expedition. Soon, he and Forrest were planning another expedition, to last three years. Williams would pay for all equipment and expenses and more than double Forrest's salary, to five hundred pounds. In return he would receive a portion of all the seed; the specimens and the rest of the seed would be shared among the Royal Botanic Gardens at Kew, Edinburgh, and Dublin. Forrest would plan his own routes, but he would concentrate on *Rhododendron.* Forrest prepared assiduously: he studied seven hundred photographs of rhododendrons taken by Wilson in China; he traveled to Kew to review its collection and to France to study the rhododendrons at the Paris Herbarium. Williams was far freer with funds than Bulley had been: Forrest bought two shotguns for birds, a Winchester repeating carbine, a Colt 45 pistol, and a Ruby Reflex camera with a wide-angle lens. John Eric Forrest was born January 7, 1912. A month later, his father sailed for Rangoon.[45]

The "real home" of the genus *Rhododendron,* Forrest had written in an article for the *Gardener's Chronicle,* "is, unquestionably, those high alpine regions on the Chino-alpine frontier, which form the basins and watersheds of the Salwin, Mekong, and Yangtze.... There, somewhere about 98° to 101° long. and 25° to 31° N. lat., the genus reaches its optimum."[46] He arrived in Tengyue with plans to travel north towards that gigantic shining range he and Litton had seen on their journey up the Salween in 1905. But in Kham, north of Yunnan, Tibetans in the towns of Batang and Litang had seized the opportunity the revolution offered to revolt against Chinese rule. The revolt spread into Yunnan, and Tibetans in the Lancang (Mekong) and Nu (Salween) valleys joined with Lisu to fight Li Genyuan's Border Colonization Army. A military expedition was launched from Sichuan to regain control of Kham, and there was fighting throughout the region. Forrest decided he could not explore the north: instead, he would spend the year in Tengyue. The city was still in a funk from Li Genyuan's visit. Li had "turned the place into a perfect shambles," he wrote. "Affairs have reached such a pass that now the bulk of the population is realizing that they are now ten times worse off than before the revolution, and now there is quite a revulsion of feeling in favor of restoration of the Manchus."[47] He was inspired to flights of imperialist fancy: "I wish, all Europeans here wish, even the people themselves I should say, are desirous that Europe should intervene and partition the country."[48]

Summoned by letter, Zhao Chengzhang and several others from Nvlvk'ö had met Forrest in Tengyue. They set to work just east of that city, in the southern portion of the Gaoligong range, which divides the Nu (Salween) from the Long (Shweli). They worked in groups of three or four; they were out weeks or months at a time, often several parties at once. They carried their bedding and slept on the ground. The Mu kings had campaigned in the Tengyue region, but they never garrisoned it, and there were no communities of Naxi speakers in the area. Zhao Chengzhang and his companions used Yunnan Chinese to communicate with the people they met—Chinese and Tai (Shan) speakers in the valleys, mostly speakers of Jingpo (Kachin) languages in the mountains.

With Forrest, they forged a style of working which they would continue for the next two decades. They already had an excellent idea of the genuses Forrest wanted and the altitudes at which they might be found. Forrest would instruct them to travel to a particular range or pass, and they would collect along the way. On their return, he would go over their collections with the head of each party, discussing each specimen's location and surroundings. He would pick out the finest species, and they would discuss where they might search for large quantities during the seeding season. Often he would launch searches for particular plant: a blue-flowered *Primula* (*P. winteri*) he had seen on his journey up the Salween in 1905, a variety of the spectacular *Rhododendron sinogrande* with flowers of deep yellow blotched with maroon instead of the usual white. I say "discussed," but in fact they did not share a fluent spoken language. Forrest's Chinese was rudimentary; the Nvlvk'ö men spoke Chinese as a second language, some better than others; and neither party knew anything of the others' native language. Their specialized language of communication was highly indexical, founded in specimens and maps.

By September, Tengyue was in a panic. Rumors of impending doom were everywhere: disbanded soldiers were gathering with Shan villagers under the Kanai prince, readying an attack on the city; Li Genyuan was marching back to burn and loot the town; Qing loyalist militia were to begin an uprising. Many citizens moved outside the walls; the new border commissioner prepared to flee. The British consul ordered Forrest to leave. The Nvlvk'ö men packed up 1,758 specimens and 130 pounds of seed on twelve mules and thirty porters, and the party set off together for Bhamo. It was a hard trip: execrable weather, undersized mules, opium-smoking porters, and frightened charges (a British customs officer with his wife and three children). Forrest waited in Bhamo, listening to rumors and watching British troops move along the border.[49] Zhao Chengzhang and his colleagues returned to Tengyue to continue working without him. Forrest had begun to develop deep respect for Zhao: "a man I put more faith in than myself. He has been tried many times now, and never failed me yet."[50]

Nothing at all happened in Tengyue. In less than a month, Forrest was back. He worked out a rough division of labor with the Nvlvk'ö men. They ranged northwards over the Gaoligong range. Upon each return they brought "heaps of good things": *Rosa clematis, Gentiana, Jasminum, Primula, Magnolia, Crawfurdia,* and, most of all, *Rhododendron.* Forrest spent his time in the city, drying, cleaning, and packing seeds, and arranging and writing up specimens. When he did get out, it was to make photographs. In the winter, he returned to Bhamo for supplies, where he watched four divisions of the Indian army move towards the frontier. British troops patrolled all the roads across the border; 140 military police with two machine guns occupied the disputed village of Pianma. "If they mean the occupation of Yunnan," he fretted, "why don't they do so at once." Tibet had declared its independence from China. "Poor China; I think the end is very near. I trust, however, that this will not affect my plans."[51] In April, four of the Nvlvk'ö men met him in Bhamo, bringing with them specimens of 190 species and seed of seventeen. They returned with him to Tengyue, where another 120 species awaited him. Zhao Chengzhang had returned to Nvlvk'ö and hired three more men; he anticipated having twelve in the coming season. "He is a jewel!" Forrest wrote. "I would willingly give him treble the salary he has, but he is quite contented and I fear to spoil him."[52] The consulate gave him permission to travel to Lijiang on condition that he stay away from the fighting between Li Genyuan's troops and Tibetan forces in the Mekong and Salween valleys. Two men from Nvlvk'ö stayed to continue exploring the Tengyue area, making the customs house their base.

· · ·

He rented a house in Nvlvk'ö. He lived on the second floor, one long room he called his "garret."[53] For Forrest, it had the advantages of a wooden floor, in contrast to the mud floor below, windows looking out over the courtyard, and a degree of privacy. But for the men who worked in the courtyard on specimens and seeds, it must have seemed a strange choice. Most homes in

Figure 11.   Forrest's "garret" in Nvlvk'ö. Courtesy of the Royal Botanic Garden, Edinburgh.

Nvlvk'ö had only one story, with roof beams directly above the ground floor. In the few two-story homes, the second floor—really more of an attic—was used to store grain and meat and sometimes to stash an unmarried son or two. The household's master and mistress always slept on the ground floor: a household head would never have chosen to sleep in the storage space of the attic.

Houses centered on a raised hearth on the ground floor. Sleeping and sitting areas occupied three sides of the hearth, where a family ate and entertained. A male household head slept on the innermost side, away from the door, usually along the south wall. If he was married, his wife slept to his right, along the east wall. Between them was a small shrine. The arrangement was iconic of the most significant division in Naxi kinship relations, between patrilateral "bone" relations and matrilateral "flesh" relations. Male visitors from the household's patrilineage—his "bone"—were expected to sit along his side of the hearth. Women and male visitors from his wife's patrilineage—his "flesh"—sat on his wife's side. Non-kin guests sat opposite the "flesh" side. On formal occasions, people sat according to their seniority in their agnatic groups, the most senior closest to the shrine. The open side of the hearth, closest the door, was where the work of cooking and serving took place. Most of this work was performed by women—typically by unmarried daughters, who had an ambiguous status, never quite full members of the "bone" they were to leave nor that they were to join.[54]

Household space was a resource for imagining social relations. The position one occupied by the hearth offered one a perspective on an ideal diagram of an extended network of kinship relations. But it was a flexible resource. In the informal activity of everyday life, even a household head often found himself in the position of another, gazing at the side of his own "bone" from the "flesh" side of the hearth, or surveying the entire diagram from the "tail" where his daughters cooked and served. The differentiated spaces of a house offered every member and visitor a variety of perspectives on the array of social relations that entwined him.

In Forrest's house, his cook and servants occupied the hearth, climbing the ladder to the attic to bring him food and tea. His collectors filled the rooms and courtyard with the activities of drying seeds and pressing specimens. He had neither bone nor flesh relations, and his place in the attic offered him no opportunities to create fictive or ad-hoc substitutes. The work of producing and ordering social relations went on below, outside of his awareness. Most important, his place in the attic offered his collaborators no comprehensible perspective on his relations with them: it gave them no opportunities to see themselves from his place. Perhaps he sensed and cultivated this. In Xiao Zhongdian, he had slept in a similar house, where the diagram-like character of the central hearth area had inspired him to draw it as a diagram. Yet even as his feelings of companionship with his hosts were flourishing, he chose to make a place for himself outside the diagram, in the bitter cold: "They all sleep higgledy-piggledy over the floor, each one choosing his own particular spot. I slept in a [fireless] room by myself."[55] In Nvlvk'ö, he arranged his attic space to suit him: a table before the open window for reading and writing, his camp bed with its head to the opposite wall, another table laden with bottles of alcohol and medicine against a side wall, a line for hanging photographic negatives to dry. This was a space for nurturing social relations of another order—with the family, scientific patrons, and financial sponsors who received his letters, photographs, and specimens.

This house as his base, he conducted two explorations: an intensive search of the Yulong and Cangshan ranges and an extensive search of territories to the northeast and northwest. He walked the Yulong range himself, with two men from Nvlvk'ö, thinking of the origins of *Primula*. In one meadow, he found flowers with the crimson-chocolate eye of *Primula pulverenta* but in other respects the coloring and habit of *P. beesiana*. In another, he saw "numberless hybrids" of *P. bulleyana* and *P. beesiana*.[56] He began to expect that all the *Primula* growing in such abundance on this range were hybrids. He had been trained in a botany that depended upon creating strict and exclusive correlations between species and referential language. His mentors at Edinburgh effusively praised his skill at this—the detail and precision with which he described his specimens. Because so many of his finds were new, many of his specimens, along with the words he attached to them, became type specimens, the standard for the species. As he wandered the *Primula*-studded meadows of the Yulong range, however, such firm correlations seemed to melt away. Each individual plant seemed to be a point in a continuum rather than a token of a type: species and varieties became, in strict terms, indescribable. His descriptive language detached from individual specimens and spread out over the meadows around him to express an impression of infinite modulations of color: "From the richest and deepest crimson scarlet, through flame reds, and the richest of oranges and buffs, to every conceivable shade of rose and pink-salmon, geranium and the bright pinks one finds in aniline dyes, back to the soft crushed strawberry and buff-rose, if you can imagine such a shade!"[57] Slipping through continua rather than arranging itself into orderly taxonomies, his language pushed him toward intense feeling, which seemed to him to exceed anything language could express. It was as though language had only two poles: strict reference, in which words correlated precisely with the world's material attributes, and pure affect, in which the loss of this precision of reference to the outer world generated ecstatic inner states. As it moved toward the latter pole, language was pressed to its limits, or beyond, to disintegration. "By now," he wrote Williams, "you must be aware of the fact that I am no writer. . . . It is not that I am unobservant, or that I do not appreciate the beauties of nature, quite the reverse. I think most often I feel too intensely for words, on such occasions words fail me."[58]

He came to a conviction that one more *Primula* must exist—a key to that amazing diversity of color. He wrote J.C. Williams, his sponsor, and William Wright Smith, who was identifying his *Primula* at Edinburgh: "Now I feel equally certain that there must be a yellow Primula somewhere. So sure am I of this that I and all my men are on the lookout for it and there is a reward proclaimed for its capture! The presence of such a species is the only way to account for the buffs and oranges of *Primula Bulleyana* and the rose shades of *Beesiana* and *pulverulenta*. Of course you may consider all this reasoning very silly, but do not laugh too much."[59]

Two months later, he announced that the species had been found: a robust *Primula* with foliage closely resembling *bulleyana* and flowers of "a real good yellow." After this find, he abandoned his ecstatic investigations of color modulations and returned to describing individual species and varieties. And the hybrids themselves seemed to disappear. Most, especially those tending toward yellow, failed to seed that autumn; Forrest shipped seeds of only a few to Edinburgh; fewer still were successfully grown there; and none appeared in the taxonomies the Garden eventually produced.[60] Other searches, similarly inspired, would be more open-ended.

In that season's extensive explorations, villagers organized parties of two or three collectors, a mule, and a muleteer to make expeditions of about three weeks. Up to three parties were out at one time. On their return, the men in a party would stay in Nvlvk'ö for a few days, reviewing their collections with Forrest and helping with the work of drying and labeling. They would then set out again, often traveling further from Nvlvk'ö in search of specific species that, based on his review of previous collections, Forrest hoped to find. In September and October, all the parties turned to gathering seed.

These journeys covered a large part of northwest Yunnan. The Nvlvk'ö men searched almost exclusively in areas that the Mu chiefs had conquered, where they encountered other Naxi, speaking their own dialect, with familiar understandings of the region's history and geography. To the west, they walked around the first great bend of the Yangtze River at Shigu, where Mu Gao had inscribed his giant stone drum. They continued along the Yangtze's hot gorge, with its poor, scattered Naxi villages, and over a high pass to Weixi on the east bank of the Lancang (Mekong). Mu Chu had annexed Weixi in 1406 on orders from a Ming general to establish officers there to urge the wild tribes to pay tribute to the imperial court. He created a garrison of Naxi troops there and appointed a Naxi official to govern the district. In the early twentieth century, the district's inhabitants were mostly Tibetan and the townspeople mostly Han. But Naxi occupied the town's outskirts and the surrounding villages, and because of their origins as a garrison, they spoke the Lijiang dialect.

To the north, parties of collectors crossed the Yangtze into Baidi (Bberdder in Naxi), where the most prestigious Naxi ritual specialists were trained, then walked eight days north to Zhongdian. The Mu had begun pushing into the Zhongdian plateau at the beginning of the Ming dynasty. They garrisoned Naxi troops in large numbers in both the plateau's towns to guard the strategic route to Tibet, and they encouraged peasants from Lijiang to settle in the south. Attracted by the pleasant surroundings of the plateau's southern town, Mu De built a summer retreat there in 1529. The peasant rebellion in Lijiang in 1647 forced the Mu chiefs to withdraw their forces from the plateau. In the early twentieth century, the plateau's population was largely Tibetan, but descendants of Naxi settlers lived in villages in the south, and both towns had small Naxi populations descended from garrisoned troops. These people too spoke the Lijiang dialect, the native tongue of Nvlvk'ö.

To the east, parties explored the Dongshan range, which rimmed the Lijiang basin on that side, then crossed the Yangtze into the mountains of Yongbei (now Yongsheng). East of the mountains was the populous Three River Basin, granted to the Mu chiefs in the thirteenth century by the conquering armies of Kublai Khan. Later, the Ming carved this temperate and fertile basin out of the Mu realm as a base for garrisons of troops from Changsha, in central China. Its people were largely Han, descended from these troops. Yet it also had small settlements of Naxi families from the Baisha area, who had moved there to set up leather shops.[61] Zhao Chengzhang and one companion pushed further north and east to Yongning. This was a small basin around a lovely lake, which the Mongols had granted the Mu chiefs. It was populated by people closely related to the Naxi of Lijiang who spoke a dialect incomprehensible to the Nvlvk'ö collectors. Yet at the beginning of the twentieth century some Naxi families from the Baisha district had moved there to set up leather shops. Their street, in the basin's main village, had become the nucleus of a small business district and the region's center.[62]

Forrest seemed neither to know nor to care how the men from Nvlvk'ö slept and ate in the field. They carried little or nothing on their backs, and each party led only a single mule for paper, specimens, and presses. In the best weather they might have slept in the open on their goatskin capes, wrapped in felt blankets. Most often, however, they were guests in the houses of local people. When these were Naxi houses, the men entered familiar spaces, where relations among household members were laid out comprehensively around the hearth and where, as they sat opposite the flesh side of the hearth representing the bones of potential marriage partners, their own status was also clear. In the houses of Han, Tibetans, and other Tibeto-Burman-speaking peoples, relations must have been more occluded and ambiguous. The patrilateral cross-cousin marriages to which mountain-dwelling Naxi were devoted were anathema to all these peoples, and for the Nvlvk'ö botanists, this must have made the easiest subject for casual talk with strangers—marriage and kinship—more difficult. Moreover, conversation had to take place in the lingua franca, Yunnan Chinese, which many people, particularly Tibetans, spoke imperfectly or not at all. And in Han areas it was difficult and embarrassing to ask for hospitality, since most local Han looked down upon Tibeto-Burman-speaking peoples as inferior and uncivilized.

Figure 12.   After a journey, with specimens (Zhao Chengzhang is sixth from left). Courtesy of the Royal Botanic Garden, Edinburgh.

In all this, the Nvlvk'ö collectors were working through strata of the landscape invisible to Forrest. They followed the ancient routes of conquest of the Mu chiefs, where every mountain, stream, village, and former village site had a Naxi name. Their searches were based in the small communities of descendants of Naxi garrisons, stretching far to the north and northwest, where hearths, tongues, and social relations were familiar and comprehensible. All Forrest's letters, diaries, and field notes show him to be undertaking a systematic search of the province, concentrating on areas where the most species were to be found, gradually developing an increasingly comprehensive sense of the region's plant geography. But other evidence contradicts this. In later expeditions, Forrest would make many attempts to extend his operations to the east and northeast, beyond the former domains of the Mu chiefs. He wanted to explore the Liangshan, the mountains of southern Sichuan inhabited by Nuosu (Yi); he sent men to the mountains surrounding Yongbei (Yongsheng), inhabited mainly by Yi and Lisu. But all these attempts failed. In some cases, the Nvlvk'ö collectors returned from brief forays to declare the country barren and unproductive. In others they simply did not go where Forrest asked them to, wandering regions within the former territories of the Mu chiefs instead. The only exceptions to this pattern were along the main route from Burma—the hinterlands of Tengyue and Dali—which Forrest had begun to explore on his first expedition, before he began to employ men from Nvlvk'ö.

· · ·

Two archival regimes, then, were anchored in that house in Nvlvk'ö. From the garret, a new botanical geography of Yunnan was being added to the imperial archive, with the mass shipment of seeds, specimens, field notes, and descriptions to Edinburgh. From the courtyard, ancient connections between Lijiang and its vast hinterlands to the north and northwest were being renewed, as the botanists from Nvlvk'ö fanned out through the region, walking old routes of conquest, building relations of friendship and potential kinship with descendants of ancient garrisons, and working through lists of Naxi names for routes, villages, mountains, and streams. Up and down the ladder between garret and courtyard moved a stream of specimens, a steady current of cash, and a trickle of indexical communication. Each regime had, at best, a dim awareness of the contours of the other. Thin as they were, however, these flows of flowers, coins, words, and gestures energized each regime, setting it in motion. And, as it happened, over the next few years, still largely in mutual ignorance, these two regimes exerted force on each other, bending each other in new directions. Mapped onto the landscape, as was inevitable, these forces would pull ever more strongly towards the northwest.

3.   The Paper Road

They must have found his incessant scribbling strange. They did not write as they walked, and they wrote little if anything about what they collected. They knew that this landscape had already been written long ago and many times. Perhaps they sensed that he was getting it backwards. They could no more imagine the land without writing than he could: for them too texts were part of the earth. But in a different way.

For him, walking generated writing. This region of the earth, especially the parts that mattered to him most, was largely blank. Writing was the absolutely essential effort to bring the earth into social being. There was no social relationship with the earth not mediated by what he could put on, or between, sheets of paper. Without writing—and without collecting, which was merely another form of writing—all those days of walking, looking, and searching would have had no social meaning at all. Writing drew these experiences into the circles of social exchange that defined him: most centrally the circles of scientific and aesthetic patronage that funded his expeditions and the domestic circles that gave it all a point.

For them, there was no blank spot anywhere in this vast region: the spaces without names were off to the east and south, beyond borders they avoided crossing. These mountains were laced with myriad journeys which they had heard recited many times. In the eyes of the world they were merely farmers, and writing about their journeys would have done nothing at all for them—no more than creating their own herbaria of the plants they collected. Several knew how to write, but only in the practical, workmanlike Chinese script in which accounts were kept, specimens labeled, and letters composed. The writing that mattered to the earth and its hidden inhabitants they left to experts. They did know, however, that this writing was not generated as feet and eyes passed over the earth. Texts were not created; they were copied, transposed, recombined. It was as though the earth had always held these narratives and journeys, since their ancestors had walked down from the north, and the task of writers was to copy and elaborate them. In any case, the relationship between experience and archive—among walking, seeing, and writing—was almost exactly the reverse of that they saw him cultivating. If he wrote as he walked, they read as they walked—or rather they remembered texts they had heard recited or sung.

What did they know? What aspirations and assumptions structured their experiences of walking, looking, and gathering? What did they see when they looked out over this fractured landscape; how did they assemble what they saw into larger or more abstract visions? Since, for the most part, they wrote nothing, the only way to approach these questions is by examining the knowledge that was available to them. As it happens, there were piles of paper all about them, in school rooms and in their neighbors' attic libraries. This chapter picks up some of that paper, asking what knowledge represented there they might have had access to, and how it might have shaped their experience. Most were probably partly literate in Chinese and illiterate in the complex script used for ritual life. Yet reading, for them, did not require what we might narrowly think of as literacy. Reading was not silent and solitary: it was collective, vocal and aural. It is clear that they frequently took part in gatherings where texts were recited, and it is not difficult to determine which texts they might most often have heard read, or even recited themselves. These texts were about walking, looking, plants, mountains, journeys, obstacles, and returns—the stuff of their lives. They were voluminous, and it is impossible to examine all of them. Here I trace a route through a few, looking for structuring assumptions that might have made a difference for them.

TEXTS

Their people had long placed great value on formal education. The Mu chiefs had taken pains to demonstrate their literary prowess in order to cultivate the emperors who granted their titles and distance themselves from the *fan* barbarians they conquered. "Of all the native officials in Yunnan," the "Veritable Records" of the Ming states, "the Lijiang Mu family is the first in literary accomplishment, knowledge of the rites, and preservation of dignity."[1] Several of the Ming dynasty chiefs, particularly Mu Gong (1494–1533) and Mu Zeng (1589–1646) were accomplished poets, who left behind substantial bodies of literary work.[2] The Mu chiefs hired tutors and built schools

to educate the children of their lineage to be upright Confucian scholar-offi-
cials. After chaos and rebellion at the end of the Ming severely restricted
Mu military might, formal education began to become available to the common people. The last of the Mu chiefs, Mu Xing (1667–1720), a poet and
calligrapher, created a public secondary school *(xuegong)* and five free public elementary schools *(yixue)*, including one in Baisha, the market town
near Nvlvk'ö.[3]

Public education in Lijiang expanded dramatically during the Qing.
William Rowe shows how the Qing educational reformer Chen Hongmou
undertook to create public elementary schools in every subdistrict of every
county in Yunnan, building nearly seven hundred in total.[4] During this campaign, the Yongzheng emperor replaced the Mu chiefs with regular magistrates, tasked with turning "a frontier, barren and desolate, into a civilized
country," as magistrate Guan Xuexuan put it in 1743.[5] The method would
be to "educate the people and correct their customs" *(hua min cheng su)*.[6]
By the end of the Qianlong period (1796), some thirty-one free public elementary schools had been created in the Lijiang District *(fu)*, funded by rents
from school land *(xuetian)* granted at their founding. The goal of "mass literacy training" *(guangxing jiaodu)* in Yunnan was not to turn peasants into
imperial degree holders.[7] It was cultural change: "the understanding of moral
principles through familiarization with Chinese characters" *(shizi mingli)*,
and the reform of language, dress, kinship, and ritual practice.[8] One way to
make use of literacy gained in public school was to open private academies
(shishu) to teach students the *yixue* did not reach. Scholars opened *shishu*
in private homes and village temples, teaching the curriculum used in *yixue*.
By 1911, Lijiang county had thirty-seven public elementary schools, teaching more than 1,207 students, and many private academies. Both *yixue* and
shishu continued into the Republic, when many more schools at all levels
were built, and by the time the war with Japan began in 1937, Lijiang County
had 243 elementary schools.[9]

In the first decades of the twentieth century, people in Nvlvk'ö could send
their children to school in Baisha or in neighboring villages at low cost in
tuition. Zhao Chengzhang and several of those he hired were probably educated in the *yixue* and *shishu* of the late Qing, learning to recite and copy
the classical primers: the *Three Character Classic (San zi jing)*, the *Thousand Character Classic (Qian zi wen)*, and the *Hundred Surnames (Bai jia
xing)*.[10] In the life of farming, travel, and trade pursued by Nvlvk'ö villagers,
basic literacy had some limited utility. It allowed one to keep accounts, write
genealogies, inscribe gravestones, and deal more confidently with bureaucrats and tax collectors. Spoken Chinese (or Yunnan dialect), learned in school,

was the lingua franca of the marketplace throughout Yunnan, used in journeys south to Dali or east to Yongsheng. But most of the trading networks in which people in the Baisha area were involved were flung out to the north and northwest, in the former territories of the Mu kings and beyond, where the Khampa dialects of Tibetan were more useful languages of the marketplace.

Written Chinese had long been the language of political and legal authority. In the sixteenth and seventeenth centuries, the Mu kings had deftly wielded literary Chinese to legitimate their military conquests. In the eighteenth and nineteenth centuries, Qing imperial magistrates had used Chinese literacy to transform this periphery into a more acceptable extension of the empire. In the early twentieth century, people in Nvlvk'ö continued to experience written Chinese as an instrument of political authority: as schoolchildren tracing characters under their teachers' watchful eyes; as land owners and taxpayers watching their property and contributions being recorded in land deeds and tax rolls; as citizens reading proclamations pasted up on village walls; even as herders passing on their way up the mountain the upright script of Yang Bi, first imperial magistrate of Lijiang, inscribed in gigantic characters on the cliff behind their village.

· · ·

Chinese, however, was not the only written language in their lives. Baisha, ancient ritual and political capital of the Mu lineage, was a preeminent center for the spectacular writing used in Naxi ritual. This writing took two forms. The most common and best developed is conventionally called dongba, a Chinese transliteration of *dtomba,* one of the Naxi names for the ritualists who read and wrote it. Dongba script is usually described as pictographic, though it is a complex mix of iconic and symbolic elements. Dongba books—for nearly all dongba script is in books—were modeled on Tibetan books: elongated pages bound between covers, stitched on the left side. In most books, each page was ruled lengthwise into three lines; each line was divided into rectangular boxes; within each box was one or more verses of a text. Within the boxes, or panels, were figures of gods, demons, humans, animals, heads, limbs, hands, insects, plants, rivers, mountains, tools, weapons, foods, clothing, and ritual objects. The panels were read left to right and top to bottom. A minority of books, used for divination, were shaped differently: nearly square rather than elongated, bound at the top, with four to five lines per page.

The second kind of script is called geba (*ggobaw* in Naxi). It was written in books ruled into lines and panels, like dongba script, but it was purely phonetic: each syllable was represented by a conventional graph with no iconic qualities. Some graphs appear to be derived from Chinese characters; some resemble graphs of other scripts used in the region; some are idiosyncratic. Dongba script has no counterpart in the Sino-Tibetan region. None of the myriad other forms of writing of the trans-Himalayan area resemble it in the least, and, while a few other peoples of the former Mu domains adopted it for their own use, it never spread beyond northwest Yunnan. Geba script, probably older, is similar to scripts used by the peoples now called Yi.

There is no consensus as to the age or origins of either script. Mu Gong's chronicles state that his ancestor, Moubao Acong, last chief of Lijiang before the Mongol conquest, invented a script for the language of his place *(zhibenfang wenzi)*.[11] Dongba taught that Ddibba Shílo, founder of the dongba cult, brought ninety-nine sets of books on long-horned yaks when he descended from the skies to suppress demons. His reincarnation Ami Shílo taught the ancestors dongba script.[12] Only a few books can be firmly dated with textual evidence. These dates are mostly in the nineteenth century, when there was a surge in manuscript production. The earliest date on a manuscript about which scholars can agree is 1703.[13] One book, attributed to a group of closely associated dongba who lived in and around Baisha, is dated with the seventh cycle of the water-rooster year.[14] This date could be either from the Tibetan calendar (1396) or, more likely, the Mongol calendar (1630).[15] (The manuscript itself is a copy, produced in the nineteenth century.[16]) Matheiu cites some evidence that Naxi may have been writing with pictographic script during or shortly after the Mongol period (1253–1381). Some graphs show unmistakable Mongol influence. The graphs for *ka*, emperor, and *kadiu*, the imperial capital, both clearly derived from the Mongol word *khan*, incorporate the graph for Mongol, *gelo*. Matheiu argues that it is unlikely that these graphs would have been invented very long after the Ming conquest swept the Mongol khans out of power.[17] In addition, some dongba books refer to a hierarchical prestige order of four tribes (Per, Na, Boa, and Wu) that disappeared during the Ming when the Mu chiefs reorganized society around their own lineage.[18] On the other hand, there is no evidence at all that the script antedates the fourteenth century.[19] While none of the evidence is robust, it appears most likely that dongba script came into being sometime during the Ming, when the Mu chiefs were conquering northwest Yunnan. Geba script may have developed much earlier: it could be as ancient as the Yi scripts to which it is allied, which probably date from the Nanzhao period (750–900).[20]

Dongba had no temples or other centers of learning; they wrote and copied texts in their homes; and they read them at ceremonies in and around their villages, which they were paid small fees to perform. They learned to write and read in lengthy apprenticeships; they transmitted texts and knowledge through patrilineages, usually from father to son or paternal uncle to nephew. Some made pilgrimages to a district of some eight villages in the southern part of the Zhongdian plateau called Bberdder (Baidi), where they apprenticed with local dongba. Bberdder was where Ddibba Shílo was thought to have lived and was considered the sacred center of the dongba cult. An apprenticeship there conferred the title great dongba and the right to wear a white cape. By the nineteenth century, four distinct schools of dongba writing had emerged.[21] The most prominent centered in the Baisha region, which included all the villages around Nvlvk'ö. Dongba of the Baisha school were skilled and prolific in the script, and they could read and write geba script as well. In an analysis of the title pages of seven thousand dongba books in Western libraries, Pan and Jackson found 1,233 written by some forty-five authors from the Baisha school.[22] The other schools were centered in Bberdder; in Baoshan in the mountains in the northern part of the Yangtze loop; and in Tai'an, southwest of Lijiang town and Ludian, and west of the Yangtze.[23] The schools were distinguished by style of writing, by whether their dongba could read and write geba script, and by particular ceremonies. Since extant manuscripts can be identified with schools and their regions with some clarity, it appears that books were passed down along lineages of dongba; texts circulated by being borrowed and copied rather than being bought and sold.

Dongba were incredibly prolific writers and copyists. The approximately seven thousand manuscripts in Western libraries and twenty thousand in libraries in Beijing, Nanjing, Kunming, and Lijiang are but a small fraction of those extant in the Lijiang region before the Cultural Revolution, when red guards collected and burned tens of thousands. A few books were for divination and astrology, but most were written to be recited aloud at rituals to maximize the fertility of households and villages, to escort the dead to the lands of the ancestors, to suppress ghosts and demons, and to heal bodies, fields, and communities. Most rituals required several books; some used twenty or thirty; index books kept track of texts to be read and objects to be used. The Dongba Culture Research Institute in Lijiang counts some two thousand separate ceremonies in which texts were recited. Some 90 percent of books, however, were copies. Dongba copied texts "with the zeal of Buddhists in pursuit of merit."[24] Though most texts were written and copied by hand, a few, beginning in the late nineteenth century, were printed with wood blocks.

Dongba textual practice was, in a very concrete way, *archival* practice. According to contemporary observers, dongba ritual was in decline in the Lijiang basin in the early twentieth century. Even so, dongba comprised roughly 8 percent of the Naxi population, and nearly every village had at least one dongba resident.[25] Every established dongba household had a library of hundreds or even thousands of manuscripts. In the Baisha region, every *coqo* still required dongba for at least three important ritual forms: yearly "sacrifices to heaven," in which a patrilineage collectively renewed the health of a village; funerals to send the deceased to the land of the ancestors; and funerals to send the souls of victims of love suicide to a paradisiacal land on the Yulong range. Hundreds of ceremonies, however, were no longer staged. It is clear that some dongba had little use for the texts they did not perform. It is just as clear, however, that many continued to cultivate their libraries, to copy texts no longer in use, and to learn to read books for scores of defunct rites. Some could read and perform ceremonies that had not been enacted for generations.

By nature, dongba writing lent itself to textual connoisseurship. It drew far more attention to its material nature than did most writing systems. The first pages of many texts were illuminated with beautifully colored figures of gods, dongba, and demons; some divination manuals consisted of nothing but colored figures of dongba and spirits. Illumination aside, writing and copying required great skill and a highly developed sense of style: most dongba were masters of calligraphy. The figures in the texts were not matched, as in most writing, graph by graph to verbal sounds. The symbols enclosed in a panel did represent one (usually) or more verses of five, seven (usually), or more syllables. But this representation took many forms. Most often, the figures were arranged in rebus-like puzzles, combining graphs that were direct icons of the words or syllables they represented, graphs that were iconic of words or syllables phonetically similar to those they represented, and graphs that were more conventional symbols with little iconic value. Sometimes an iconic graph represented the name of a god or person that would not be voiced when the text was read; more often a single graph represented an entire verse, which had to be memorized rather than read directly from the graph. Some texts, while using iconic graphs to represent sounds, were entirely phonetic; some were phonetic representations of verses that were merely combinations of rhythmic and rhyming sounds with no legible referential content at all (though this was more common in geba texts).

A fairly typical panel, the first verse of a funeral song called "Song for the Dead, the Origin of Sorrow" *(Mún ndzer ấ là dzhu)*, to which we will return, illustrates a few of these possibilities (figure 13). The text reads "*O!*

Figure 13.   The first verse of "Song for the Dead, the Origin of Sorrow" *(Mún ndzer ắ là dzhu)* from Joseph Rock, *The Zhimä Funeral Ceremony of the Na-Khi of Southwest China.*

à̀ssì dtá là dzhu." The first graph is silent. It is a conventional symbol with no verbal equivalent placed at the beginning of many texts. The second graph, also silent, depicts the dongba, a human figure wearing a hat and carrying a bamboo staff used at funerals. Were it not silent, it would read *Lo ch'ùng ndaw khù*, a title given a dongba who officiates at funerals. Below the dongba's staff is a round heap, the symbol for grain, *ó*; here it is read as an exclamation, *O!*, indicating that the song is beginning. The next graph is a stylized mouth with a wavy line representing a voice; alone it is read *à̀*. It means "at first," or "in the beginning." Here it is the first syllable of *à̀ ssì*, meaning "father." Below the mouth is a figure of a man with a tree on his head. This is the second syllable of *à̀ ssì*, "father"; the tree is the graph for *ssì*, "wood" and serves as a phonetic element in this graph. Ordinarily, the word *à̀ ssì* is written with this second graph alone; the *à̀* is not written. The author may be playing on the meaning of the graph used to represent *à̀*, "in the beginning"; the graph is used this way a few verses on. The next graph, a square within a square, depicts a box, *dta*; here it is used phonetically for *dtá* (in a different tone), "to relate" or "to sing." The next graph depicts a musk deer, *lä*; it is used as an adverbial particle, here pronounced *là*. The final graph depicts a muzzle-loading flintlock musket with three bullets issuing from it; it is the symbol for *dzhu*, "to empty," since the bullets are being emptied from the gun; here, however, it is used phonetically for *dzhu*, sorrow. A translation of the entire verse might read "Oh, I sing of father's sorrow!" Yet there is much in the written text that is not expressed in the spoken verse: the initial graph indicating the speaking dongba, the addition of the graph for *à̀* representing the mouth with words issuing from it to begin the song, the choice of an emptying gun for *dzhu* to give the father's sorrow a sharp, desolate quality appropriate to the knowledge of death, the subject of this song.[26]

Dongba script lent itself to an intense focus on writing as material practice. Texts were not *composed* in the sense we usually think of; they were transcribed, transformed, and elaborated. Most probably existed in some form as spoken verses before being written for the first time. Any verse might be transcribed in many ways—a nearly endless series of combinations of graphs used as icons or for phonetic value—and each dongba borrowed some combinations and created others.[27] To transcribe a spoken text was to engage in an intensive process of recombining and reinventing the written language. Nor did copyists concentrate on merely reproducing the originals. Apprentices, learning how to read, often created precise imitations. But mature dongba, borrowing texts from other dongba, often of other schools, copied them in their own styles, substituting graphs, recombining verbal puzzles, and adding verses to elaborate or lengthen the recitation. Even the substrate of writing was not taken for granted. Many dongba households made their own paper in a lengthy process of beating, soaking, screening, and drying the bark of a particular tree. In most writerly traditions, placing characters on a page may be a means of composing thought, but once written the characters tend to be seen as transparent media with which to access the text's semantic or poetic qualities. Where calligraphic writing is a form of art, it is most often appreciated in the abstract, largely separate from the text's meaning: rarely are calligraphic art and referential content tightly linked. In fundamental contrast, dongba writing was invested with great significance in itself, a significance only partially captured in the texts' spoken forms.

Dongba kept huge numbers of texts alive with their writerly archival practice, even after they were rarely required for rituals. Among these were a large category of books read at sacrifices to spirits of the waters, rocks, forests, mountains, and wild animals, the original inhabitants of the land. These spirits were called *ssù* or *llü* (dragons); they were represented in texts as snakes, usually with featureless human heads, sometimes with the heads of other creatures, in combination with other graphs to represent their full names. The dongba corpus names about 530 *ssù* of different ranks: regional kings and queens, chiefs of mountain ranges, rulers of single valleys, and plebeians.[28] Sacrifices to *ssù* were performed to heal bodies, houses, lands, and waters. Dongba guided the *ssù*, as malign influences, back to their lands and homes, and they implored the *ssù*, as fonts of fertility, to bestow *nnù̀* and *ò*, male and female powers of fertility, on lands, waters, people, livestock, and households. Scholars of dongba religion, who agree on very little, tend to agree that *ssù* texts show much evidence of influence from Tibetan Reformed Bon. Some texts closely resemble Bon texts, and many names of *ssù* and other gods, demons, and apotheosized dongba seem to be cognates of Bon names. In the eighth and ninth centuries, Bon practitioners, exiled from central Tibet as a result of fierce struggles with Buddhist sects, established communities in eastern Tibet. One Bon mission reached Jang (a Tibetan name for Lijiang) in the ninth century, and Bon monasteries survived in neighboring Zuosuo into the twentieth century. Whatever their Bon influences, however, the *ssù* texts and rites are unmistakably an indigenous product; they embedded the rivers, mountains, and villages of the former Mu domains in a vivid, kinetic geocosmography.

The texts performed for *ssù* were concerned in a fundamental way with establishing the place of human beings in the world. In a series of myths they told of the primordial act of distinction that founded the community of humans. At the beginning of time, one text relates, humans and *ssù* lived together. The ancestor of the *ssù* and the ancestor of humans were born of one father, but they had two mothers, who "glowered at each other, and this is the origin of frowning." The humans and *ssù* fought, for "two wives in one house is not the custom," and like brothers dividing a household they divided up the earth. The *ssù* took the rocks, cliffs, trees, waters, and wild animals; the humans took the fields and livestock. Each *ssù* ritual uses texts that describe transgressions of this founding contract. A human, often a warrior, sometimes heading an army, invades *ssù* lands, killing animals, cutting trees, blocking and diverting streams, washing clothing in rivers, performing funerals at springs. The *ssù* retaliate; their armies attack the humans, imprison their souls and make them ill. *Ssù* rituals trace illness, infertility, and bad fortune to these ancient quarrels, mirrored in the present world. The rituals mediate these conflicts to restore balance between the households of these brothers, *ssù* and humans. Dongba meticulously repay the *ssù* with sacrifices of meat, grain, and medicine. They send each *ssù* off to its home, imploring it, as it departs, to grant the humans *nnǜ* and *ò*, male and female fertility.

The positions of humans and *ssù* in these texts are structurally similar to those of the soldiers and settlers of the Mu kingdom (humans) and the prior inhabitants of the lands they conquered *(ssù)*. The Naxi settlers who followed the Mu armies to places like Baidi, Weixi, and Ludian settled the most fertile land, where they could engage in sedentary agriculture. The more mountainous "wastelands" were left to former inhabitants, who often became swidden agriculturalists, nomadic pastoralists, or hunter gatherers. Matheiu suggests that the Mu kings conquered with arms and ruled with ritual. The *ssù* rites, she argues, may have been used to legitimize and balance relationships between Naxi soldiers and settlers and the peoples they defeated and displaced. Little in the dongba corpus, however, indicates that the Mu kings established an official cult on the model of the Ming or Qing imperial rites. Cosmic and earthly authority in the texts is too decentralized, heterogeneous, and chaotic for that. It is more likely that garrisons and communities of Naxi turned to the *ssù* rites to carve out, from their point of view, a habitable place in the political, geographic, and cosmological orders they found around them. The rituals do not commemorate or celebrate conquest. They work instead to strike a delicate balance between related, antagonistic communities. They speak of mutual interdependence, restitution for wrongs done, and the exchange of agricultural products for trees cut, springs disturbed, and wild animals hunted in the realm of the other.

The sense of attention to delicately tuned exchanges that the *ssù* rituals promoted with the world beyond the pale of house, garden, and field also permeated other rituals to which Naxi continued to attach great importance in the twentieth century. Nearly every ritual act included a meticulous accounting of and restitution for incursions upon the nonhuman world. The following example is from a text recited for the demons of love suicide. With the exception of ordinary funerals, this was the most widely and frequently performed major dongba ritual in and around the Lijiang basin in the 1920s and 1930s.

> I fell trees as ritual trees. There remains a grave burden of guilt. I hasten the souls of the trees with this bamboo rod. I send the trees back to their place; I send them back to the high mountains. If the trees do not take the wrong road, the burden of guilt will be lifted.
> I prize out rocks as ritual rocks. There remains a grave burden of guilt. I hasten the souls of the rocks with this bamboo rod. On the ritual rocks, I scatter offerings of popped grain. I send the rocks back to their place; I send them back to the mountain crags. If the rocks do not take the wrong road, the burden of guilt will be lifted.[29]

Dongba rituals strove to maintain the balance between human and nonhuman by paying for the countless, inevitable, invisible infractions upon nonhuman beings made in everyday life. And, as the above passage implies, this exchange required movement—here to guide the souls of the trees and rocks back to their places of origin.

It is a striking characteristic of dongba rituals that their enormous repertoire of linguistic and gestural tools was focused, in the main, on moving beings through geographical and cosmological space. In the *ssù* rituals, for instance, dongba invited *ssù* and *llü* to travel from their rocks, crags, and springs to the ritual ground, paid them restitution there, and then guided them back to their homes in vast, sweeping journeys. *Ssù* texts laid out these travels in detail, with names for springs, lakes, rivers, mountains, living villages, and villages long since disappeared. These texts were, more than anything else, incantations of lists of place names, often running into the hundreds. The conquests of the Mu kings had embedded an expansive geography of northwest Yunnan into the corpus of dongba texts. This geography extended far into regions that lay beyond the concern or control of nineteenth- and early twentieth-century authorities in Lijiang. And it far exceeded in density of detail any repository of official geographic knowledge in maps and gazetteers until the 1930s, when accurate military maps of northwest Yunnan were created. Dongba kept this other geography alive through their archival practices of reading, copying, and conserving their thousands of texts.

·   ·   ·

If dongba textual practice was archival, it was also *recitational*. Dongba writing could rarely maintain syntactic or semantic coherence separate from its vocalization. As noted above, a verse might be transcribed in many different ways; each dongba created his own combinations of graphs; many single graphs represented long names or even verb phrases. For these reasons, dongba writing was more like a complex of mnemonic tools than a full writing system: texts had to be memorized to some extent before they could be read. Texts were never read silently. They were recited, and this recitation brought out many features not visible in the script: meter, rhythm, rhyme schemes, word plays. Parallelism in particular was omnipresent. At the simplest level, adjacent verses were often parallel; alternate verses might be parallel; groups of four verses might form parallel relationships with other groups; large sections of a text might, through repetition and variation, form parallel relationships with other sections. In these ways, dongba texts were similar to the oral and written texts of many peoples in the trans-Himalayan region.[30] And, as in some of those traditions, complex architectures of parallelism often took precedence over semantic coherence.[31]

Taken to an extreme, these tendencies could empty texts of all semantic value while creating unusually tight edifices of meter and rhyme. One category of texts, called *hóalü*, used simple pictographs or geba script to represent long passages with no semantic meaning at all. Joseph Rock called these *dharani*, as he believed them to be direct transcriptions of Tibetan texts that were in turn direct transcriptions of Sanskrit tantric formulae.[32] One of the accomplished dongba of the nineteenth century invented a system to transcribe Tibetan letters with dongba graphs, and it is possible that some *hóalü* were transcribed from Tibetan. But most were probably not direct transcriptions of any kind, since their phonetics and syntax were constructed to sound exactly like other dongba texts, with some comprehensible words, even a few comprehensible verses, inserted into the flow of recitation. In his analysis of Victorian nonsense literature, Jean-Jaques Lecercle calls this *charabia*, "the imitation of one's own language."[33] The first few verses of a text from the *hǎzhip'ì*, the "gods' road," which I will discuss further below, can serve as an example.

> *ndò bo' dtü là ssú,*
> > *lo gv bpä llü shwuà*
> *ssì bbu dtü là ssú,*
> > *lo gv bpä llü shwuà*
> *muan bbu dtü là ssú,*
> > *lo gv bpä llü shwuà*
> *à bo' dtü là ssú,*
> > *lo gv bpä llü shwuà*
>
> *hä ddù mba mi ndshi*
>
> *là bbu dtü là ssú,*
> > *lo gv bpä (llü) shwuà*[34]

The single comprehensible verse is the one that breaks the pattern: *hä ddù mba mi ndshi*, "light the gods' lamps," an instruction to the dongba. The rest display monotonous repetition and obvious simple parallelism. In general, the meter, prosody, and parallel structures of *hóalü* were repetitive and homogenous, containing little of the variability, complexity, and inconsistency that the need for semantic coherence forced on most dongba texts. In common with Victorian nonsense literature, they displayed an exaggerated respect for the linguistic code, in which "a formal excess of syntax compensates for a semantic (material) lack or incoherence."[35] *Hóalü* are one extreme in a spectrum: my point is that the play of sound within and across verses was such an important component of dongba texts that it could even be made to stand on its own—and that this play depended upon the texts being recited, in addition to being copied and preserved.

JOURNEYS

In the early twentieth century, the most important contexts for the recitation of dongba texts were rituals of mourning. Naxi performed different mourning rituals for the souls of dongba, children, love suicides, people who died of violence, women who died during parturition or shortly thereafter, and ordinary people who died peacefully. Though they spent years away from home, the botanists from Nvlvk'ö participated in many such rituals. The deaths of bone or flesh kin were among the most important occasions in one's life—tectonic shifts in the world of relations with kin within and outside of one's village. These shifts required work: social labor to reorder relations along new axes and affective labor to learn to live in a world suddenly bereft of foundational elements. Their kin often went to great lengths to send messengers to find the sojourning botanists and inform them of deaths. On some occasions, such as the death of a brother, a ritual could not proceed until a traveler returned home or sent word that he was unable to return. The explorers probably participated in rituals of mourning more often, and with more investment, than in any other ritual acts.

The most common such ritual was performed for people who died peacefully—not by violence or suicide and not in childbirth. It was called *zhimǎ,* "teaching the road." Shortly after a death, on an astrologically propitious day, kin and friends of the deceased, often numbering in the hundreds, gathered in his or her house and courtyard. They brought gifts ranging from a small bag of grain to an entire sheep or horse, depending upon their relationship with the deceased's household and obligations incurred in previous funerals. They held a night-long vigil over the coffin, drinking, singing, eating, dancing, and weeping. In the morning, they brought the coffin in a procession from the house to the graveyard and buried it. *Zhimǎ* did its work by moving the deceased's soul through the geocosmological territory archived in dongba textual practice. The ritual released the soul from the purgatorial realm of demons, into which it had fallen at death, and moved it back along the road the ancestors had taken when they immigrated to Lijiang. To this end, most significant acts in the ritual took the form of speaking to the dead. Close affinal and paternal kin keened verses recounting warm memories of the deceased. Friends and neighbors sang biographical songs describing her life and death, the gathering of mourners, and the road along which she would walk on her way to the lands of the ancestors.[36] And dongba, in return for a small fee of meat or alcohol, recited texts.

In a *zhimǎ* that Joseph Rock attended in the early 1930s, dongba read around forty texts, all intended to help move the soul along the road toward the ancestors' home.[37] Other participants overheard these recitations with varying degrees of comprehension. There were many competing foci of attention: lamenting mourners, the work of butchering and cooking the sacrificial meat, one's own cup of alcohol. Some mourners listened intently; some heard verses go by with half an ear; some paid no attention at all. But ordinary mourners also recited some written texts themselves, in addition to many unwritten songs and laments. The most striking were a group of songs that recounted the origins of heaven and earth, of winged, hoofed, and clawed beasts, of winged demons, and of death itself. These songs were central to the experience of the ritual for the village's senior men. As persons of leading social and economic status, the botanical explorers were very likely to have performed them at many funerals.

Joseph Rock took careful notes on one such song. On the evening of the night vigil, in the central room of the house, the deceased lay in his coffin, head to the door and feet to the wall. A dongba sat at the coffin's head. On the left side sat men representing friends and neighbors of the deceased; on the right sat senior men of the deceased's bone and flesh kin. Before them were tables laden with food and wine. The dongba read the song's verses; the men repeated each verse. Some blew on reed pipes or folded leaves to accompany the singers. After singing twelve pages of text, the men filed outside to join the hundreds of mourners assembled around a big fire in the courtyard. They joined hands in an open circle, interlacing their fingers, and danced around the fire, one step forward, one back, at the pace of a slow walk. The dongba led them through the five remaining pages of text, which they sang while dancing. They repeated the song for the rest of the night, taking turns dancing and resting.

On this occasion the text was "Song for the Dead, the Origin of Sorrow" *(Mún ndzer ấ là dzhu)*. It told the story of Ssússä of Ssùlò, who, panning for gold in the river, saw his reflection and understood that he was old. He was a rich man; he had chests of grain and trunks of silver and gold. But he had no years. So he set off to the south to buy years:

> from the Ssùlò river's head
>   to the Ssùlò river's tail
> from the Ssùlò river's tail
>   to the fields of Bberdder
> from the fields of Bberdder
>   to the river's winter ford
> from the river's winter ford
>   to the river's summer ford
> from the river's summer ford
>   to Ndawgv village
> from Ndawgv village
>   to Gvssugkò meadow . . .

> *ssu gyì gkv nnü dtǜ*
>   *ssu gyì man lä t'u*
> *ssu gyì man nnü dtǜ*
>   *bber dder llü lä t'u*

bber dder llu' nnü dtü
  ts'u gu k'u lä t'u
ts'u gu k'u nnü dtü
  zhù gu k'u lä t'u
zhù gu k'u nnü dtü
  ndaw gv dù lä t'u
ndaw gv dù nnü dtü
  gv ssu gkò dù lä t'u . . . [38]

And so on, the verses reeling off in linked pairs, imitating the walking cadence of the dance. Ssússä's route was given in sixteen place names. He walked from the source of the Ssùlò (Wuliang) river in Muli County, Sichuan, to where it emptied into the Yangtze at the northern apex of that river's second great bend. Then he continued south to Bberdder (Baidi), crossed the Yangtze at the Naxi village of Ndawgv, and climbed through a string of meadows and passes that led over the Yulong range. In the town of Baisha, he circled the market three times, finding wood, grass, wine, and food for sale, but no years. In Lijiang town he circled three times and found wine, food, silver, and gold for sale, but no years. In Kunming, the provincial capital, he found satins and brocades for sale, but no years. He turned around. The market had dispersed; the great Dianchi lake had dried up; the black rocks had split; the green bamboo had turned yellow. This is the origin of sorrow, the understanding that, riches or no, death will come, and sorrow will follow. This was a journey in the realm of ordinary experience. The botanists from Nvlvk'ö traveled these routes many times.

This much the mourners sang around the coffin inside the house. Dancing in the courtyard, they sang the song's second part, about the dance for the dead. To dance was to participate in the soul's journey. As they guided the soul towards the lands of the ancestors, the dancing mourners trod on the demons blocking the path.

in sorrow we guide the dead
  dancing we tread on demons . . .

dzhu lä zhì bpú bbue
  ts'o lä ts'ù szèr bbue . . .

this crowd of sons of sorrow
  slim-waisted they sway
    by custom they sway and dance

dzhu zo ch'i ddù hóa
  t'ü ts'ù t'ü nyu nyu
    nyú lä ddü ts'o ndu

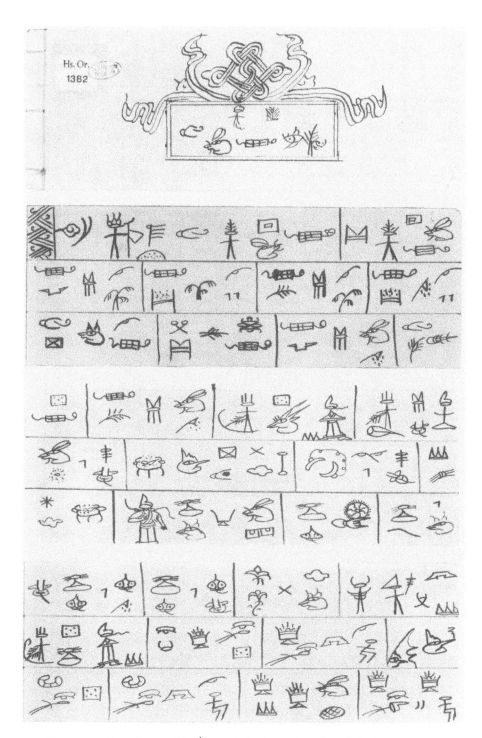

Figure 14. Pages from a *zhimǎ* funeral ritual text, from Joseph Rock, comp., *Na-Khi Manuscripts*.

we guide the crane to the clouds
    the tiger to the mountain peaks
        this ancestor up above
  to the dazzling white cloud gate
    flapping, the crane will return
  all the winged creatures
    guide the crane back to the clouds
        his power must not pass . . .

gko bpú gkyì gkyi bbue
    la bpú sso gkyi bbue
        yǜ bpú ggò gkyi bbue
gkyì k'u p'èr lǒ la
    gko ndzì lä bbue mä
ndu mun ndu ch'i dzù
    gko bpú gkyì gkyi ssä̀
        non ò khü muan chèr . . .

to the glittering golden mountain gate
    dancing, the tiger will return
  all the clawed beasts
    guide the tiger back to the mountain
        his power must not pass

sso k'u shì ghǚgh ghügh
    la ts'o lä bbue mä
dshì mun ch'i dzù la
    bpú sso lä gkyi
        non ò khü muan chèr

we guide this ancestor to the gods' land
    he will cross nine ridges
        his power must not cross
    he will ford seven rivers
        his power must not ford

yǜ bpú hä̀ dǜ t'u
    ssu bbu ngv mbù ló
        non ò mbù ló k'ü muan chèr
    p'ä̀ gyì sher hò nder
        non ò gyì nder khü muan[39]

Obviously, this part of the journey took the listener beyond ordinary experience to the mythical thresholds between earth and sky, associated with high mountain peaks. To dance was to winnow the dead soul, separating it from its "powers" *(non ò)*. These powers were the skills and capacities the deceased displayed in life. The song gives examples: the ability to sing and follow along when others are singing, to count numbers, to perform wed-

ding songs, to sing while walking in the mountains. These powers were to remain with the deceased's sons and daughters. They were to the soul as the clouds are to the sky, the grass to the earth, gold to the Yangtze, snow to fir branches, dew to bamboo leaves, a saddle to a horse, a yoke to an ox. They were a load to be borne (*gù*, written as a man with a sack on his back), and the soul could travel faster without them.

This song recapitulated a microcosmic version of the double movement through the archived landscape that the rite of "teaching the road" as a whole was intended to recall and effect. In dongba lore, the Naxi ancestors had immigrated from homes far to the north, perhaps in the grasslands of eastern Tibet. These texts make many references to this migration. Before the eighteenth century, mourning rituals for warriors included texts on the origins of weapons. One such text lingered over the lifestyle of the ancestors in their home in the north.

> they were rich and lacked nothing
> they herded white-footed yak on alpine meadows
> dwelt in white felt yurts in mountain flowers
> milked yak and gorged on butter and cheese
> led dogs to hunt over seven snow mountains . . .

They wore silk and satin brocade, the text continued. They wore silver and gold, turquoise and cornelian, tiger pelts and leopard skins. They doffed helmets with three flags and armor in two layers; they bore shields and lances, sharp swords and golden saddles. Returning to his homeland after death, a warrior wears garments of silver and gold and rides a fleet horse; a yak carries his many possessions.[40]

"Teaching the road" emphasized the power of the voice that compelled the soul to move. "Clean the wax from your ears with your fingers; strike your ears with your white palms to clear out what remains. Your ears are now clear; you can hear my good voice."[41] The journey completed a circuit of exchange, repaying the world the debt the ancestors had incurred as they walked south:

> you are not the tiger of the mountains
>   but you must repay the mountains
> you are not the stag of the ridges
>   but you must repay the ridges
> you are not a nomad of the road
>   but you must repay the road . . .
> go see your grandparents and parents
>   go see the lands of your lineage
> the heaven above, the houses and fields below

in places unknown to you
  feel the heaven with your hands
    feel the land with your feet.[42]

Other texts contained lists of place names mapping out the journey. A typical road for the dead might begin in the Lijiang basin, cross the Yulong range, ford the Yangtze river, and proceed north up the Zhongdian plateau. Interspersed with the place names were descriptions of obstacles or temptations and instructions on how to overcome them:

At Òyú Hägyìman, you will meet a green frog with claws
  you will be afraid,
you ride a horse, let your horse tread on it
  go on unafraid
At Pèrnà Nddügkanchúng, tie up your horse and shoe it
At Ts'ǎnyi Dtót'udzhu your horse will balk
  tie up its tail, whip it three times
  and leap across like a tiger[43]

To people in the early twentieth century, these texts grew gradually incomprehensible as they proceeded north, naming vanished villages or places whose Naxi names had been lost from memory. The final clusters of names were clearly mythological, as the soul wandered into the sky on the golden bridges and silver chains along which the earliest ancestor and his celestial bride had descended from heaven.

The structure of writing in these texts imitated the structure of walking. Each place name occupied a panel, followed by a name on the next panel. When the names were repeated, as in the "Origin of Sorrow," the repetition occurred within the frame of a single panel. This footstep-like pattern was emphasized in recitation. Voiced at a rapid clip, each verse went by at about the same pace as a footstep in a deliberate walk, and the simple syntactical and semantic parallelism of the verses lined them up in alternating pairs like walking feet. The written text made a list of place names into a visual map depicting the stages of the journey in successive panels.

Text-like maps accompanied these map-like texts. Dongba created, inherited, or copied elaborately painted strips of hempen cloth, thirty to forty feet long and about a foot wide called "gods' roads" *(hǎzhip'ì)*.[44] In the morning following a night of dancing, singing, and reciting, the coffin was placed in the middle of the courtyard, head to the north. A dongba unrolled a "gods' road" scroll, attaching one end to the coffin's head and extending the other to the north. Like dongba texts, the scroll was divided into panels. Within each panel was a painting of a place through which the soul would pass. At the coffin's head, the scroll began with the realm of demons, showing the nine black spurs of hell and the demons guarding them, the tree of spines on which souls were impaled, and the many other tortures to which demons subjected them. Souls were boiled in oil and stabbed with spears; their heads were pierced with spikes; their tongues were pulled out with pincers and plowed with oxen—all detailed in some forty panels. After the realm of demons came that of humans: thirty-three cities and villages. One panel mapped out the four directions and five elements of this realm; others showed temples, gatherings of dancing dongbas, and Ddibba Shílo, founder of the dongba cult, riding triumphantly on his white steed. Following the human realm was the realm of mountains and forests: seven golden mountains guarded by tigers and eagles. The final sections showed the thirty-three lands of the gods, arranged in hierarchies, ending with the highest gods.

Dongba made small dough figurines of demons and gods and placed them on panels to which they corresponded. Standing before the coffin, they chanted eight texts, moving a small effigy of the soul along the gods' road, and overturning the figures of demons as they were vanquished by the recited words. Sometimes dongba supplemented the gods' road with models of the geocosmological realms, made of earth, wood, and papier mache. It was as though a text, though bearing many of the iconic elements of a map, wasn't iconic enough. For this final enactment of the soul's journey, the text was stretched out in a line, a road in miniature, with miniature beings moving along it. It wasn't quite writing, for it could not be read directly. But it wasn't the actual road either. It was something in between—a mediator for these parallel acts: eyes and mouth moving along a written page, feet moving along a road. In all these representations of the road—texts, scrolls, and models—the threshold between the world of mountains and rivers and the world of the gods was Ngyùná Shílo Ngyù, the "vast mountain of Shílo," at which Ddibba Shílo had descended to earth, leading 360 disciples and yaks loaded with texts.

The landscape archived in dongba texts was central to Naxi rituals of mourning. To reconstitute the social world rent by grief was to plunge into this archive, to move through this textual landscape with eyes, voice, ears, and dancing feet, to take on, one by one, its hundreds of names and all the obstacles that stood between them. In this process, archive and experience often came together. One might, upon hearing the text describing the descent of the ancestor Ts'òzällüghügh from the sky, let one's eyes wander to the white, rocky spur on the other side of the valley, called Dzadzambù, over which the ancestor passed on his way into the Lijiang basin. One might, singing of the "origin of sorrow," recall a recent journey through the alpine passes of the Yulong range to the Yangtze river ford at Ndawgv, the path

Figure 15.   Section of a "gods' road" *(hàzhip'ì)* scroll: realm of demons, from Joseph Rock, "Studies in Na-khi Literature, II: The Na-khi Ha zhi p'i; Or the Road the Gods Decide."

Ssússä took as he went in search of years to buy. Or one might, after cleaning the wax from one's ears, simply sit and let the "good voice" of the dongba, chanting a *hóalü* formula to overcome a demon on the road, envelop one in the simple, repetitive, and incomprehensible rhythms of a walking journey. One would, in each case, find the landscapes of archive and experience to be interdependent, experience lending archive sensuous shape, archive linking experience into continuities of history, descent, and return. As these routes of text and voice pushed further north, the interpenetration of archive and experience thinned out. Long lists of place names stretched out north of the Yangtze, some comprehensible, many not. Eventually the possibility of finding experiential elements in the archival geocosmos disappeared altogether, as the routes passed along silver bridges and golden chains into the sky.

WARS

In dongba texts, rhododendrons reek of primordial battle. Plants inhabit nearly every page, but the species are few, especially compared to the hundreds of gods and demons.[45] The most common are trees *(ndzèr)*. These include pine, fir, spruce, cypress, hemlock, poplar, willow, chestnut, nut-pine, wormwood, mulberry, the "pig-intestine tree," and three types of *Rhododendron: mùn* or *mùnna, shwua,* and less frequently *mùnlua.* Joseph Rock identified *mùnna* as *R. decorum,* small, with glabrous leaves, rough black bark, and fragrant white flowers. *Shwua* was *R. rubiginosum* and *R. heliolepis,* larger and more treelike, with narrow leaves, glabrous on top and scaly on the bottom, and with pink, red, crimson, or lavender flowers.[46] The third species, *mùnlua,* was, again according to Rock, any one of the large, high-altitude rhododendrons resembling *R. adenogynum,* with pink flowers and brown tomentum beneath the leaves.[47] *Mùnna* and *shwua* are often paired in the texts. Since here the black bark of *mùnna* and the white bark of *shwua* are their defining features, I shall refer to them as the "black rhododendron" and the "white rhododendron," and I shall call the large *mùnlua* the "great rhododendron."

Along with birch for arrows, spruce for spears, and mulberry for quivers, rhododendrons were among a few plants fashioned into weapons for battles fought during the creation of the cosmos. Warriors carved sword sheaths from the limbs of the black rhododendron; they wove shields and armor from the strong, sinuous branches of the great rhododendron. Many texts relate tales of one such war, between the ancestor Muanllúddùndzi (also known as Ddù) and his archenemy Muanllússùndzi (a.k.a. Ssù), on Ngyùná

Shílo Ngyù, the great mountain of Shílo. Ddù eventually defeats the gruesome Ssù and his descendants, denizens of the black lands, living in black
iron villages and iron forts, with legions of ghost soldiers at their command.
In one account, Ddù's armies divide Ssù's flesh into nine portions and pour
his blood into nine buckets. The white bat carries a portion of flesh to the
clouds where the crane and eagle fight over it; he places a portion on a snowy
mountain peak, where the tiger and leopard fight over it; he places a portion in the forest, where the bear and boar battle over it. Eventually, the
weapons themselves begin to fight.

> the bat took one piece of Ssù's flesh
> placed it in the white birch on the mountain
> the birch arrow and quiver had never fought
> but they fought over Ssù's flesh
> ninety-nine battles of arrow and quiver ensued
> he placed a piece in the mulberry trees of the river valley
> battles of ninety-nine soldiers with ninety-nine mulberry bows ensued
> he placed a piece in the fir forests of the high plains
> battles of ten thousand spear shafts ensued
> he placed a piece amongst the wild black rhododendron on the highest
>   plateaus
> battles of ten thousand flowery sword sheathes ensued

In another text, a commander directs his ten thousand soldiers to fashion
weapons in preparation for battle:

> cut three spruce trees
> make ten thousand sharp spears
> cut three great rhododendrons
> make ten thousand shields and suits of armor
> crowded as the stars in the sky
> dense as the grass on the earth
> arrows like a swarm of bees
> spears like hemp stalks.[48]

As sheathes, shields, and armor, rhododendrons were the skin of swords
and warriors, and skin is also the form they took in weapons for ritual battle. Rhododendrons were prominent in a large and complex ceremony to
expel the demons of slander from bodies and households. This ritual, called
dtónà, was held at least twice in the Lijiang valley in the early twentieth
century, during which endemic violence created ideal conditions for slanderous demons to thrive.

Like most dongba rituals of purification and healing, *dtónà* recalled primordial acts that divided the cosmos into realms of sovereignty. The *ssù* rites

mentioned above describe one such apportionment. Such acts did not create lasting divides between social and natural. Instead, they established overlapping claims of rule or ownership asserted by many alliances of beings over many others. The texts describe hundreds of battles and wars over such claims. Frequently, lines are drawn between warring parties, often between black and white: black regions are claimed by evil beings (ghosts, demons, humans), white by good beings (gods, ancestors, priests). Some texts tell of lines of garrisons and forts built between white and black regions, reminiscent of the lines of watchtowers that once divided the realm of the Mu kings from barbarian enemies to the north and west. As soon as any such line is described it is broached by one force or another; a claim for redress is made, and a battle is joined.

Dongba rituals of healing and purification did not attempt to resolve the legitimacy of such claims. Contestation was perpetual by nature, and no judgment could be final. Instead, rituals traced the provenance of claims through a genealogy of ancient battles, all the way back to the violent emergence of heaven and earth from primordial chaos. The aim of ritual was to bring this conflict, in the form in which it afflicted a person or household, to temporary resolution, through negotiation or force. In *dtónà* the emphasis was on force.

Chinese scholarship on dongba texts frequently describes "Naxi religion" as "animistic." This is misleading. The tumult of battle that created the world and continues to give it life and vigor does not engage plants, animals, rocks, and mountains as living beings in themselves. Instead, armies of gods, ancestors, priests, demons, and ghosts marshal the world's inanimate matter to their causes by assembling it into beings with limited capacities. A tree or plant is never fully a being: it is a potential part-being which can be combined with others into a loose assemblage of parts, organs, and abilities, to make a temporary material body for a god, ghost, ancestor, demon, or living human. This body can then be ritually manipulated—perhaps executed as an enemy, perhaps launched against enemies as an ally—before being disassembled again. Many texts describe how this was done in ancient battles and give recipes for how to do it again in the present.

As potential part-beings, black and white rhododendrons emerged from the burning corpse of a vanquished enemy during the great war between Ddù and Ssù. A text read at *dtónà* describes their creation:

> [Ddù's son] Ddùzoalú was never shot
> he shot [Ssù's son] Asémiuè to death
> at Asémiuè's cremation, he was changed
> into three clouds of smoke

the clouds of smoke changed
into the white rhododendron, the black rhododendron,
and the dragon spruce; five branches emerged
on the mountain; they fought with these torches
this is the origin of purifying torches.[49]

As this excerpt implies, white and black rhododendrons were components of purifying torches. Such torches, called *soshwuà*, were exceedingly important ritual instruments, used very often to cleanse ritual grounds, houses, courtyards, and bodies. They were central in *muanbpò*, the ceremony in which, each year, the senior men of every *coqo* in Nvlvk'ö tended to the ritual health of their lineage and community. And they were central in *dtónà* as well. They were typical dongba ritual instruments: assembled of inanimate materials with diverse properties, they became living beings with limited capacities, to be employed in ritual battle. I say limited capacities because ritual technique emphasized only the organs or abilities that were must prominent or useful. For some beings it was their seeing eyes or their speaking mouths: for the *soshwuà*, it was its fiery breath.

After describing the origin of the purifying torch, this text gives a householder instructions on how to assemble one. Branches of nine trees are to be used: cypress, hemlock, bamboo, wormwood, willow, fir, spruce, white rhododendron, and black rhododendron. As the text makes clear, black rhododendron is central.

The master of this house
sends a boy fleet of foot
to cut a torch of black rhododendron from the mountain
a black nanny goat will bear it
the fleet-footed boy will lead her . . .
circling the earth and sky three times
circling the sun and moon, the stars and planets three times
circling the mountains and valleys three times
he will use black rhododendron as the torch's skin
dragon fir as the torch's hands
bamboo as the torch's bones
green wormwood as the torch's intestines
the white wind as the torch's breath
black stones as the torch's bones
the great water as the torch's blood
red fire as the torch's heart.[50]

In other texts, many of the torch's organs are assembled of other materials, but the black rhododendron, with its rough, black bark, is always the skin.[51]

Rhododendrons were also were used to assemble a variety of other be-

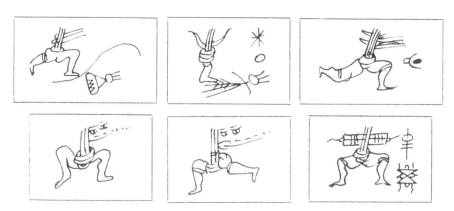

Figure 16.   Graphs for *ngawbä*, from Joseph Rock, comp., *Na-Khi Manuscripts*.

ings for *dtónà*.[52] A central act of the ritual, repeated several times, was to create beings called *ngawbä* of fir, willow, black rhododendron, and *Myricaria germania*. Dongba texts contain graphs for at least twenty kinds of *ngawbä*: headless beings with twigs for torsos and powerful running legs and feet. Some have the horns of yaks, cattle, or goats sprouting from their twig bodies, others the beaks and claws of chickens or the teeth of dogs. Some can see and call; others can laugh and spit at demons.[53] They were, one text says, "fast as the clouds and winds, swift as the fish in water, ferocious as the tiger, able to devour an ox in the forest . . . the trees of their enemies are smashed, the gates to their cliff-dwellings destroyed, their animals killed and their houses burned."[54]

Ngawbä were allies against the demons of slander and substitutes for slanderous enemies. As allies, they were launched against the enemy demons; as substitutes, they were burned and beheaded. The texts emphasize this dual character by repeatedly calling attention to their white feet and black heads. In the *dtónà* ritual, their heads were set on fire and chopped off, dividing black from white, demon from human, slanderous speech from civil speech. This act made a deep impression on Joseph Rock when the ceremony was performed in his courtyard in Nvlvk'ö: "In the northeast corner of the court . . . there is an iron pan with pine wood set on fire. The Ngawbä is held into the fire; the heads are burned, and then under a terrific beating of a drum the Ngawbä is chopped into pieces. When it is performed for the last time, the Dtombas chase the Ngawbä with swords, and the demons are put on a stretcher with pine needles; and under the din of gongs they are burned."[55]

As they burned the *ngawbä*, the dongba recited texts about burning and beheading slanderous enemies:

burn the heads of enemies
burn the heads of evil men
burn the demons with unhappy hearts
and unsmiling faces; burn their heads . . .
in the next crowded village
in this crowded village
evil-tongued men in that village
big-mouthed women in this village
burn those demons . . .

And they called upon the *ngawbä* to drive the demons of slander back to the lands of enemies.

like the clouds and wind
the ngawbä leads the demons to the enemy's house
like the storms and the snow, to the enemy's lands
like the ice and the hail, to the enemy's country[56]

And of course the texts told the story of the *ngawbä*'s origins. As with the purifying torch, one parent of the *ngawbä* was the black rhododendron. But this was a specific rhododendron tree. It grew on Ngyùná Shílo Ngyù, the vast mountain of Shílo, the gateway through which gods and ancestors passed on their way to earth and dead forbears passed on their way to the lands of the gods.

on the peak of Ngyùná Shílo Ngyù
three black rhododendrons
became the ngawbä's father
at the foot of Ngyùná Shílo Ngyù
three thin-leaved willows
became the ngawbä's mother[57]

The father of the *ngawbä*, named Gyvdtv̀nàbpú, was depicted as the symbol for the black rhododendron sprouting out of the symbol for Shílo's great mountain.[58]

Rituals like *dtónà* articulated humble daily concerns, such as physical or psychic affliction or the slanders of neighbors with a vast tumultuous world of gods, spirits, ancestors, and demons. Even more than the charismatic graphs of dongba writing, the intricate beings used in these rituals were media of communication between worlds. To most people, the texts were often mysterious and inaccessible. Most could not read them, and dongba often recited them so rapidly that only experienced listeners could follow. But many helped gather plants from mountains and ravines to assemble into torches and effigies. Those who took any interest at all knew the names of beings assembled, their attributes and uses, and the stories of their origins.

Figure 17. Gyvdtv̀nàbpú, father of the *ngawbä*. The first
graph is the symbol for *mùnna*, the black rhododendron,
growing out of the symbol for Ngyùná Shílo Ngyù, the
great mountain of Shílo. From Joseph Rock, comp., *Na-Khi
Manuscripts.*

Many of the Nvlvk'ö collectors were experts at specimen preparation.
They learned how to collect representative portions of plants and arrange
them between sheets of paper so they would dry in attractive ways. They
learned to tag specimens with numbers corresponding to Forrest's notes and
to make their own notes, mental or written, on the location and situation of
the plants. And they learned how specimens were inserted into chains of reference connecting them, on one end, to specific locations on the earth and,
on the other, to the mysterious and enormously authoritative abstraction of
scientific botanical taxonomy. Their participation in dongba ritual practice
may have provided them with tools for making sense of all this. There too,
individual plant specimens, gathered from the mountain and placed in a
courtyard, were put together with words and inserted into chains of reference. They were made tokens of species types, linked to specific passages of
text, and given genealogies and points of origin on the earth's surface. And
all this was undergirded by the mysterious authority of dongba textual practice, with its complex taxonomies of gods, species, stories, and techniques.

It should hardly seem surprising if these men found clues to what Forrest so avidly sought in this other authority, if they mapped his obsessions
onto this other cartography. And in this other cartography, rhododendrons
always pointed in a specific direction. As sheathes, shields, and armor, they
were borne into battle between black and white forces on the great mountain at the center of the world; as the skin of torches, they emerged from
the burned flesh of an enemy vanquished in battle on that mountain; as the

father of *ngawbä,* they sprang out of that mountain's peak. In the ritual language of plants—a language that articulated affect and bodily sensation with the properties of plants—rhododendrons were closely associated with the route north, which all souls walked after their deaths, towards the great mountain of Shílo.

.   .   .

It is not possible to know with any specificity how Zhao Chengzhang, Zhao Tangguan, He Nüli, Lu Wanyu, and all the others sensed this landscape as they walked and collected. What did they see when they examined a flower? What did they feel when they climbed to the top of a mountain pass and glimpsed an enormous, snow-covered mountain peak with the clouds rolling about its shoulders? These questions cannot be answered. Yet the bare fact that their home was at one of the epicenters of the dongba cult in the early twentieth century can provide some clues. If they had traveled east or south, the limited formal Chinese education that some had attained would have proved a most useful guide. But they went north and west. The dongba cult provided detailed instructions on how to approach this fractured landscape, with its deep and delicately bridged chasms between Naxi settlers and their neighbors, and between the houses and fields of humans and the vibrant, mobile world of nonhuman beings. It was the only guide available to them. It is clear that they knew this, if only because, whenever possible, they avoided traveling in regions that lay outside its detailed cartography. The most fundamental ordering principle it had to give them was that walking, reading, and vocalizing were all intertwined. The landscape was written. It was layered with prior journeys, it was laced with histories and narratives; it teemed with names no longer remembered, villages long since disappeared, and places of unknown special significance. Most of this was hidden from their eyes: they could see and admire the outward forms, as they could admire the intricate figures in a dongba book, but what those forms concealed could be divined only by experts. Thrown like a frayed net over all this potentially comprehensible if not fully comprehended significance were the paths of humans. Everywhere they walked, they trespassed on places over which nonhuman others were the titulars; everywhere they collected, they accumulated debts to the world's nonhuman inhabitants.

How might such principles have manifested in practice? Another impossible question. Yet I am reminded of a passage from Joseph Rock's magisterial dictionary of dongba script, in which he analyzes the graph for the verb *ndzer,* to recite, chant, or sing, which appears in the funeral song "Origin of Sorrow." The song names the ability to sing or chant as one of the powers the dead soul is to leave behind with her descendants. The graph shows a man with an enlarged foot to indicate that he is walking. His mouth is open, and a tree grows from his tongue: the tree is used for its phonetic value, *ndzèr*. Explaining the graph, Rock noted that the men who walked with him, sons and nephews of Zhao Chengzhang and his fellow travelers, were always singing as they walked: songs of love, travel, and the bitterness of life.[59] Forrest, deeply uninterested in his employees' lives, made no such observations. Still, it is very likely that Zhao and his companions also sang as they walked. These would have been textual songs, not made up on the spot but memorized—and taught, elaborated, and recombined during the journey. For Forrest, walking inspired writing; for travelers from Nvlvk'ö, walking very likely instigated reading—or rather the vocal recollection of texts once sung or recited in their presence.

If this intertwining of landscape, text, and voice was one fundamental value of the dongba archive, another was laid onto the landscape, giving it a powerful orientation toward the north. Dongba texts use the same word, *ggo*, to mean both "northward" and "upward," following the marked trend of the mountains of northwest Yunnan. This orientation is built into the very bedrock of this landscape, scored by its four great parallel rivers, running down from the Tibetan plateau. In the archive, the ancestors journeyed down from the north, and souls journeyed back up that way, toward the places where all beings originated, the gateways where the mountains opened up to the sky. It is likely that, as they worked their way further north, their journeys may have given the Nvlvk'ö botanists opportunities to bring archive and experience together. Opportunities to rediscover old place names in conversation with Naxi residents along their routes, to tread paths the souls of their parents and grandparents might have trod, to push up the incline in the direction that all those texts, songs, and dances pushed, towards that high mountain gate through which ancestors had once descended and towards which the dead returned.

4.   The Golden Mountain Gate

Forrest was accustomed to a dialectic of what one might call practices of beauty and practices of sublimity. He devoted enormous labor and attention to creating the most precise possible assemblages of words and things. Scouring Yunnan for material to assemble into his exquisitely crafted and meticulously documented creations, he displayed to excess qualities that Lorraine Daston argues natural scientists in the late nineteenth and early twentieth century demanded of themselves: "painstaking care and exactitude, infinite patience, unflagging perseverance, preternatural sensory activity, and an insatiable appetite for work."[1] This ethos of heroic, disciplined objectivity was the foundation for his class aspirations—his ticket into the bourgeois scientific social world of his mentors at Edinburgh. These labors were tempered by searches for moments of pure, unlimited vision, usually of mountain peaks, during which the mediating flows of time and words simply halted: moments which were, by definition, impossible to share with others. About thirteen years after he started exploring Yunnan, however, he began to reach beyond this heavily scripted dialectic of immanent social beauty and solitary transcendent subjectivity. He began longing for a place where sublime transcendence would come down to earth in an exuberant mass of species that would overwhelm his senses yet where each species might also be fashioned into specimens and seed packets and circulated among patrons and mentors, rendering that solitary sublime experience into the matter of social relations.

He believed that any knowledge his small army of explorers might possess that could not be directly transformed into Linnaean taxonomy was a distraction from his goals. He remained resolutely ignorant of their prior knowledge of plants; he took no interest in their ideas about the geographical form of the region; he viewed the ceremonies that brought them names, routes, and stories about this landscape as an irritating waste of time. Nevertheless, their awareness of that prior archive of this landscape was present at the creation of every specimen, every correlation of specimen to place, and every speculation about possible patterns of distribution. It was likely, even inevitable, that this archive would begin to inflect these creations. If it did, it was in this quest for the "center of origin" of the genus *Rhododendron*. For in that archive, transcendent visions of gods, ancestors, and demons, and immanent practices like naming routes and fashioning assemblages of words and things were always intimately intertwined.

By his third expedition, in 1913–1914, Forrest was making plans to "sweep Northwest Yunnan clear" of every genus in which his sponsors might take an interest. The herbarium in Edinburgh would be a full record of the region, with specimens of every species indexed to their area of distribution. By 1917, this idea of an evenly distributed archive had begun to take on a specific focus and orientation. Rhododendrons appeared to show a pattern of distribution in which species were concentrated in the northwest and thinned out towards the southeast. Encouraged by his mentor, Isaac Bailey Balfour, he followed this pattern further and further northwest. If he followed it far enough, he thought, he would find a place with more species than any other—the center of a global *"Rhododendron* whirlpool," which would also be the center of origin of the genus.[2] He became obsessed with finding this "promised land" and experiencing the "revelation" it would offer.

For the Nvlvk'ö botanists, the far northwest, where the great ranges that divided the Jinsha (Yangtze), Lancang (Mekong), and Qiu (Taron) rose up to meet the Tibetan plateau, was also a center of origin. Somewhere beyond the Qiu lay Ngyùná Shílo Ngyù, the vast mountain of Shílo, where three black rhododendrons became Gyvdtvnàbpú, the father of the *ngawbä* torch used to destroy the demons of slander; where Ddibba Shílo had descended to earth with the first dongba books, and where ancestors passed through the "golden mountain gate" towards the realms of the gods. This chapter follows Forrest as he developed, in consultation with his intrepid bands of explorers, his ideas of where the center of origin of *Rhododendron* lay. It shows how their search for this place, after working further and further north, bent southwest to the high remote range behind the Qiu river, called, in Naxi, T'khüdű Zhernvlv, the Great Snow Mountain of the Qiu, where connections to the Lijiang region, textual and political, historical and contemporary, were thick. I read Forrest's letters and some bits of the dongba archive side by side, hoping to illuminate possible lines of connection. My arguments are speculative and inferential. But they serve as a lens through which to examine and compare these two regimes of walking, seeing, writing, voicing, remembering, and gathering the earth that participated in this extravagantly productive engagement with southwest China.

Forrest's mentor and patron, Isaac Bailey Balfour, was prominent among those who worked to bring to Britain the revolution in botanical science that had occurred on the continent late in the prior century. The "new botany" had expanded the object of observation beyond classification to generation, development, physiology, morphology, and ecological relations. Emphasizing precision of observation, particularly in microscopy, it promoted what Daston has called a "mechanical" style of objectivity, which effaced the observer's mediating presence and emphasized his heroic restraint and self discipline. Though Forrest carried no microscope with him in the field, he did carry the sensibilities introduced to him in his training in the Royal Botanic Garden's herbarium. He strove to live up to the moral demands of science while producing precise scientific working objects for the eyes of his patrons.

In the field, he was diligent, working for twelve to fifteen hours a day, writing descriptions, developing photographs, and preparing specimens and seeds. Balfour repeatedly praised his specimens, the excellence of his descriptive notes, and his exact descriptions of locations and situations, without noting that these were all collaborative creations, which depended upon the sharp memories of the Nvlvk'ö collectors. As Balfour's work on *Rhododendron* taxonomy progressed, he began to regard geography as a key to the genus's knottiest problems. He pored over the India Survey maps of Yunnan, linking specimens, place names, and geographical features, and speculating about paths of dispersion and differentiation. The exactness with which Forrest pinpointed location allowed Balfour to create linkages between structures at vastly different scales: microscopic hair cells, visible reproductive organs, and regional maps of distribution.

Forrest did most of his writing in the days after a party had brought in a new haul, sitting with rows of specimens before him on a desk or table in Nvlvk'ö or Tengyue. This was the mode of exploration most congenial to him: where, to borrow Kant's formula, the universe best answered to his conception of it—where his senses and words seemed to achieve the most harmonious coexistence with things. He produced many thousands of pages in this mode. Take, as an example selected nearly at random, a passage of a letter to J. C. Williams, the chief sponsor of several of his later expeditions. It was December 1912. Forrest was in Tengyue; three parties of men from

Nvlvk'ö were combing the Gaoligong range where it divides the Salween (Nu) from the Shweli (Long). One party had just brought in many new tree species, each of which Forrest described in great detail to Williams.

> Amongst a host of species is a very curious and beautiful tree of 30–40 feet. The foliage is large, 4–6 inches broadly ovate-lanceolate. . . . The pads are about three or four inches in length, by about $\frac{1}{2}$ to 1 [inch] in depth, thick and leathery of a soft reddish peach shade externally. They are arranged on the apices of the fruiting branches in threes or fours at right angles to the apex. When ripe they curl and spread right out so as to expose the whole surface of the interior, and as this is of the most gleaming shade of scarlet imaginable, you may be able to conceive what a wonderful effect is produced. At even a very short distance the impression one receives is that the tree is laden with bright scarlet blossoms. I have secured about three pounds of seed of it, but it is proving a difficult thing to dry.[3]

He often folded seeds and petals into his letters, aware always that composites of words and things were more satisfactory than words alone. He was deeply wounded when, as often happened, his steady outpouring of letters received inadequate response. His letters and field notes, written in the same mode, were among the rewards that his sponsors came to expect of his expeditions.

He was advanced funds on evidence of his skill as a collector; he returned skillful assemblages of things and words. A class differential was the motor for this exchange. Was the value of his seeds, specimens, notes, and letters equal to the value of the funds advanced? It was not possible to know, especially when, after he lost Bees Ltd. as a sponsor, there was no longer any attempt to convert his products to commercial property. To retain his reputation for mastery, he tried, often with great success, to flood his sponsors with more seeds, words, and specimens than they knew what to do with. This was the source of his enormous energy and equally enormous anxiety. Were he to return an equal value for his wealthy sponsors' patronage, he might stand a chance of putting himself on an equal footing with them, morally if not socially. Yet he could never know when this balance was reached.

Still, the process of creating his assemblages was some compensation in itself. He had likely been exposed to Kant at the Kilmarnock Academy.[4] Even the shallowest acquaintance would have left an impression of the philosopher's famous definitions of beauty and sublimity, topics of many a schoolboy essay. He assumed that flowers gave others, as well as himself, the reassurances that Kant had assigned to the beautiful. Beauty was evidence of a contingent accord between nature and one's own senses. For this reason it was eminently social: if the beauty of a flower spoke to Forrest, then it would also speak to his correspondents, despite the distances in space, class, and culture that lay between them. In April 1913, Forrest wrote J. C. Williams for the second time of a wonderful find, a new *Rhododendron* species with unusually large foliage: "Some of the trusses are a great size, carrying from 40–50 blooms which are thick and fleshy, greenish-white, almost pure white, with a blotch of deep crimson at the base and fragrant. I have just received your letter of March 22nd in which you mention the skepticism of a friend regarding the above species, but if he saw the specimens I have in hand he would be lost in admiration. The young foliage is even more handsome than the mature."[5] The flower's beauty was self evident, admitting no skepticism. Describing a great many plants in great detail was a way of making his own world habitable: it provided him a deeply felt resonance with those bits of the world on which he lavished his care and attention. This resonance brought with it the certain knowledge that it could not but be the same for others. And this certainty was the ground upon which his sociality was built, and along which it reached across class divisions into scientific societies and countryside estates. His careful practice of crafting close joints between words and things was his way of acknowledging the existence of others as essential to his own.

Often, however, language failed. This failure, or rather his assertion of it, was at the core of his attempts to find in the heart of things an accord with others. He found ways to deploy this failure very early on, in his letters home to his mother, which she shared with Clementina. It was his first trip to northwest Yunnan, with G. Litton. After a long journey, they had climbed to the top of a pass in the great range that divides the Yangtze (Jinsha) and the Mekong (Lancang).

> We were again at an elevation of nearly 16,000 ft., and the view from the top is entirely beyond my powers of description. The morning was wonderfully clear, and we could see for hundreds of miles on all sides. Nothing but range after range of tremendous mountains, many of the peaks covered with eternal snow, and all glistening in the early morning sunlight like gems, lay before our eyes; add to this billows of vapour rolling about in ceaseless movement in all the valleys, and, above all, the intense stillness there was at this elevation, not even the rustle of a leaf, or blade of grass, and you perhaps form a faint idea of what the scene was like. I cannot tell you what my feelings were as I sat and gazed at it all. I could have sat and stared and dreamed all day, but we still had a long distance to go.[6]

In some places, words were arrested. It was not merely that they were inadequate. It was rather that the mediation of vision with language was no longer appropriate, even if, after the fact, language was all he was left with. He "could have stared and dreamed all day." The clarity of the vision and the "intense stillness" that accompanied it gave him vision purified: immediate, limitless, and unfiltered by subjectivity. He employed this formula repeatedly. In every journey, at the summit of a mountain pass or at the sight, suddenly revealed, of a immense range, he would find himself in that dazed state, without words, wishing only to sit still and gape forever.

Forrest's account of his experience might have been drawn straight from Kant's essay about the mathematical sublime, the feeling of enjoyment mixed with awe when imagination finds itself faced with immensity in nature: "The sight of a mountain whose snow-covered peak rises above the clouds" was one of Kant's examples.[7] Such sights, Kant argued, force the imagination to the extremity of its power, confronting it with its own limits. "A view beyond all conception" was Forrest's favorite phrase about such scenes. Even though there was nothing beyond the sensible world for imagination to take hold of, this "thrusting aside of sensible barriers" gave it the feeling of being unbounded, presented it with the infinite, and expanded the soul.[8] He bemoaned his failure to give words to this expansion, but he consoled himself that his failure demonstrated his capacity for intense feeling.

Perhaps he had read Kant at Kilmarnock; perhaps he had merely absorbed the ideas about beauty and sublimity that had worked their way into European cultures of nature writing. In any case, in practice, he would reject Kant's central point: that in reality this apparently unmediated contact with the ineffable is not an experience of the true life of things; that nature's apparent grandeur merely acts to bring forth our own capacities for feelings of expansion and unboundedness; that we cannot find an absolute or transcendent truth in the world: we can find it only within ourselves. In practice, he nurtured a conviction that the places where vision transcended language and thought were real places, embedded in the real world. In the meadows of "numberless" *Primula* hybrids, he had gone looking for the yellow hybrid that would be the key to understanding the entire range of hybrid shades. Soon he would begin to search for a meadow where the secret to the genus *Rhododendron* was hidden. He set his heart on places of transcendence, places of origin, places that held secrets that would unlock the mysteries of this entire, difficult landscape.

Kant's settlement would have made little sense to the botanists from Nvlvko. They took it for granted that the world was full of beings and places that transcended ordinary experience without, however, transcending language and sociality. The earth was already and unquestionably social through and through: beings that could be neither seen nor heard could be spoken to and exchanged with; places that let out onto other worlds could be reached with words and perhaps even with living bodies. Forrest feared his patrons would find his search for the yellow *Primula* "silly" and his quest for the center of *Rhododendron* unscientific: for him, these projects pushed at the edge of reason. But to the botanists from Nvlvk'ö the idea that the landscape held places of origin and transcendence was perfectly reasonable. Unlike Forrest, however, they knew already in which direction they lay.

ORIGINS

He began a fourth expedition in March 1917 as war raged in Europe. He had been in Nvlvk'ö when the war began, playing host to two Europeans of distinguished class, Heinrich Handel-Mazzetti of the Natural History Museum in Vienna, and Camillo Schneider, general secretary of the Austro-Hungarian Dendrological Society. Handel-Mazzetti stayed with him nine days then, stranded by the war, wandered southwest China alone for five years. Schneider walked the Yulong range with him for two months. Forrest stayed on through the autumn, packing and harvesting seeds. In December, he shipped over six thousand specimens and several crates of seed from Tengyue; in January he followed them home. All the gardeners at Edinburgh had gone to war, more than sixty in all.[9] Balfour's son had joined: he was killed in June 1915. J. C. Williams' four sons had joined and his nephew had already been killed.[10] Too old to be conscripted, Forrest lived with Clementina and their two young sons at Peebles, south of Edinburgh. He worked at the Garden without pay; he moved the family back to Lassawade to be closer to the city; he planted daffodils in his garden and planned another escape to Yunnan.[11]

Even before the war, access to the new botanical riches from China had largely moved out of the commercial realm. Veitch and Sons had given up sponsoring expeditions to exotic regions in 1906. Bees Ltd. would soon leave the business as well. Now, as the entrenched and profoundly disparate class order of Edwardian England began an overwhelming transformation, wealthy landowners arranged exclusive access to this flow of floral spoils, expensive decorations for their estates and gardens. The Royal Horticultural Society was the core of the island's network of elite gardening enthusiasts. One of its luminaries, Henry J. Elwes, gathered a syndicate to sponsor Forrest for another expedition.[12] The seven members included "one duke, one baron-to-be, three knights, at least two other ex-army officers, and at least two company directors."[13] Testy Forrest quarreled over the terms of his contract. They suggested he provide receipts for his expenses; he responded with fury: "Am I to trusted or not? If the latter, I would three thousand times rather keep out of it." Balfour explained that his charge did not like to be bullied. The syndicate agreed to his terms: a five hundred pound advance, five hundred pounds per year salary, and seven hundred pounds in expenses.[14]

He had spent his life balancing on the knife edge of class difference, this touchy son of a draper who had adopted a member of the Royal Society as another father. Now, when all that seemed solid about class was beginning to melt and shift, that edge seemed sharper than ever: it glittered more brilliantly and wounded more deeply. Flowers had brought him to this edge; flowers held out the promise of transporting him across it. He knew it was a false promise, but even so, it tempted him. Perhaps this is why he would turn to flowers for their promise of other kinds of transport.

Zhao Chengzhang and fifteen others walked to across Yunnan to Bhamo, in Burma, to meet their patron, a journey of more than a month. Forrest's Chinese was not up to the task of writing them directly: he likely wrote to the China Inland Mission station in Dali, which relayed the message to Lijiang and Nvlvk'ö. In April, he arrived in Bhamo to find them waiting for him.[15] "I have already arranged through my head man, whom I have with me, for from 40 to 50 men for the season!" he wrote the head of the syndicate. "We ought to get something surely!"[16] They walked to Dali, collecting trees and shrubs. Forrest had passed over this route twelve times; the Nvlvk'ö collectors had traveled it many more. Four men stayed to collect on the western flank of the Cangshan range. The rest walked north to Nvlvk'ö, picked up more men, and continued north over the Zhongdian plateau to Adunzi (Deqin).[17] In July, they settled in for the season in the far northwest corner of the province in a small Tibetan village on the west bank of the Mekong called Tsiriting: a few houses, a Buddhist shrine, and a small Catholic church, a branch of the French Catholic mission at Cigu a few miles south.[18]

The frontier between China and the Tibetan region of Kham had been volatile for many years. In 1904 and 1905, in reaction to the British invasion of Tibet under Francis Younghusband, the Qing state had begun a campaign to bring Kham under direct administration. In Batang, Qing officials had announced a decree reducing the numbers of monks in monasteries, forbidding the recruitment of new monks, and granting land to the French Catholic mission. This had sparked the uprising from which Forrest had narrowly escaped in 1905. Armies of monks from Batang and Tshakhalo (Yanjing) had killed the architect of these policies, Feng Quan, along with his escort, then swept down the Mekong, sacking and burning Catholic missions. The Qing response was forceful. An army of two thousand punished Batang, burned its monastery, and executed its Tibetan *tusi* and many other officials and monks. A Sichuan magistrate named Zhao Erfeng was directed to pacify Kham. He carried out the task with considerable ferocity, earning the name Butcher Zhao from the Tibetans. In 1909–1910, the Qing government sent several thousand troops from Kham into central Tibet. When the Qing was overthrown, the army withdrew, and Tibetans took control of most of Kham. Another military expedition in 1912 reestablished Chinese dominance over the area's most important towns, with much violence on both sides. In 1914, Britain and Tibet agreed to a border between "Outer Tibet" to be administered from Lhasa and "Inner Tibet," to be incorporated into the Chinese administrative system. In Kham, this boundary ran along the Mekong-Salween (Lancang-Nu) divide, swinging west to cross the Salween a few miles north of Tsiriting, where the great sacred mountain Khawa Karpo formed a natural barrier. Forrest had long desired to return to the Tibetan parts of far northwest Yunnan, but fighting had kept him away. Now, a fragile peace had settled over the area.[19]

From Tsiriting, small parties ranged up and down the Mekong-Salween divide. They found rhododendrons everywhere, many of them new. There were so many species and varieties that distinctions seemed to melt away.[20] "The wealth of Rhododendrons is almost incredible," Forrest wrote the *Gardeners' Chronicle*, "and the number of new species and forms more than confusing. I have really given up attempting to define the limits of species; each individual seems to have a form, or an affinity, on every range and divide differing essentially from the type."[21] Again, the difficulty of creating precise correlations between language and species pushed him toward the ecstatic effect he associated with linguistic disintegration and unmediated vision, even as he did, for the sake of communication, name species.

> A wilderness of Rhododendrons and many species of Primulas enclosed by jagged limestone spurs some 2,000 feet higher . . . All the Rhododendrons were in full flower and I have seldom seen anything to equal the display of colour, the masses of brilliant scarlet blooms of *R. sanguineum,* the flaring magentas and pinks of *R. saluenense,* the dark plum shades of *R. campylogynuim,* the cherry red of the new form of *R. Forrestii,* and the greeny yellow of *R. trichocladum* all distinct yet all blending most delightfully . . . Even in heavy rain and mist it was a wonderful sight, how it would appear in sunshine I can well imagine.

It was an unusual mood for him when it came to flowers, overcoming him only where the sheer variety of species overwhelmed his capacities for classification. On the Yulong range, this mood had inspired him to search for a particular *Primula* that would bring back the correlation between word and thing. Here it moved him to a thought of a similar key: a place where the tangled puzzle of *Rhododendron* would naturally unravel. "What I have seen," he wrote his scientific collaborator at Edinburgh, "points to the fact that we are approaching very close to the optimum of the genus, which I reckon is not very far from here, probably some short distance n[orth] and w[est] of the mountains of Sarong [Tsarong, now Dzayül County, Tibetan Autonomous Region]."[22]

His understanding of species diffusion was based on his interpretation of a hypothesis called "age and area" proposed two years before by the American ecologist J. C. Willis. In a region with no well-marked barriers to diffusion, Willis proposed, the area occupied by a given species would depend on the age of the species: the older the species the wider its range.[23] Given this, Forrest thought, the more species in an area, the closer that area to a place where most or all of the species of that genus would originally have evolved. In the past, he had searched for sites where individuals of particular species congregated most densely, assuming these to be the "centers of origin" of those species. Now, after exposure to Willis just prior to the present expedition, he assumed that all or most species of a genus would have originated in a single small area, where the genus itself had evolved from an ancestral form. This assumption of an infinitely generative place where species were continually coming into existence had deep roots in myth and natural history in the West—as varied as Renaissance ideas about the diffusion of species from Mt. Ararat where Noah had landed his Ark to Humboldt's location of the Garden of Eden on a mountain in the center of the South American continent.[24] The scientific literature on diffusion was suggesting that even Willis's hypothesis, which did not assert that species differentiation in a genus was concentrated in one area, was far too simple, and Forrest's mentors could have disabused him of his ideas. Yet Balfour, always mindful of his fragile ego, encouraged him, and, after Balfour's death, William Wright Smith followed his lead, waiting until he wrote Forrest's obituary to say that he gravely doubted such a place as the "center of the *Rhododendron* world" could exist.[25]

By the end of the season, having collected two hundred new species and varieties of *Rhododendron*, Forrest had located the generative center of the genus more precisely. He could find it on the map—or rather just off the map's edge—in territory the India Survey had not yet fully filled in. Writing from Tsiriting, he asked Balfour to trace with him a route towards this place. "If you are interested in my plans, procure sheet no 90 of India and Adjacent Countries, which gives the Tarong [Tsarong] Yunnan frontier. Follow this, the Mekong Salwin divide, from Tzekou [Cigu], which you will find almost on lat 28° N. long 98° 50′ E., to a point at 30° N. lat 97° 40′ E. long, where it breaks more northwards, forming the eastern confines of a basin in the shape of an irregular elongated ellipse extending over almost 2 degrees of latitude from 29° to 30° N."[26] He could not, he wrote, reach this "promised land" that season. It was north of the Yunnan border in southern Kham, which had been closed to Europeans since the Simla convention. He could have "made a dash" for it, but he had promised both the British Consul and Chinese officials that he would not trespass in Tibet. On the other hand, if Balfour would exercise his influence with the Foreign Office to pressure Lhasa for written permission, he could travel there at his ease, with mules and porters provided by local Tibetan officials.[27]

A team of collectors from Nvlvk'ö worked as far as the northern face of Khawa Karpo, which formed the border between Yunnan and Kham on the Mekong-Salween divide.[28] But plans to push further north were spoiled when Forrest quarreled with Zhao Chengzhang. His invaluable head man had "taken to drink and other evil ways," Forrest complained. In late July, Zhao deserted, leaving for Nvlvk'ö with six others. Now a respected elder with extensive networks of patronage and loyalty, Zhao must have often found Forrest's stubborn and overbearing personality difficult to bear. "You know what the average Asiatic is when he gets haggish!" Forrest wrote Smith spitefully.[29] The party was left shorthanded just as the seed harvest was coming on, making an excursion to Tsarong impossible. He hired three new collectors, Tibetans recommended by Père Valentin, the priest who ran the mission at Cigu. Their leader was Ganton (sometimes Anton to Forrest). This able Tibetan Catholic was an accomplished linguist who spoke Chinese, Tibetan, and Nu (spoken on the upper Salween) fluently, along with some Naxi and Lisu.[30] He had collected for Forrest on his previous journey to this region in 1905, protecting him as he fled the attack on the Cigu mission; and he had guided Francis Kingdon Ward on his first, extensive, journey through northwest Yunnan, in 1911.[31] When Forrest returned to Tengyue for the season, Ganton and his two companions kept working through the winter months.

Forrest could not stop thinking of the "revelation" he would experience were he to journey a hundred miles north. "I have dreams of mountain ranges clothed in Rhododendrons and nought else, and though it may seem absurd, in parts even here it amounts to that. I have the feeling that if I get further north next season I shall double my catch."[32] In the spring, after spending the winter writing up the twenty-five hundred species his party had gathered, he schemed about making Ganton and his two fellows the nucleus of an expedition to Tsarong. They would travel north up the huge ranges on the Mekong-Salween divide toward the elliptical basin he had pointed out to Balfour on the map. "As you know," he wrote Balfour again, "this is the spot I have been trying to tap for many years. If I or my men do not reach it I shall not die happy!"[33]

In 1916, Yunnan and Sichuan had both declared independence from Beijing. Their warlord governors fought over their mutual border. West Yunnan remained relatively peaceful, but early in 1917 bandits and disbanded soldiers began attacking travelers on the road from the capital, Yunnanfu, to Dali. That winter, as Forrest worked on his collections in Tengyue, the commander of the Chinese garrison in Chamdo, an important town on the border between "Inner" and "Outer" Tibet, began, on his own authority, to advance towards Lhasa, declaring that he would punish the Tibetans, who were "as servants revolting against their masters." In the mean time, the British, hoping to contain China during the Great War, had given the small, new Tibetan army five thousand Lee-Enfield rifles and several million rounds of ammunition. Assisted by Khampa volunteers, the Tibetan army pushed the Chinese forces back to Chamdo, overran the garrison there, and advanced east and south, preparing to retake all of Kham. Zhao Erfeng had encouraged soldiers from his armies to settle in Kham as farmers; the advancing Tibetan army now expelled several thousand, hundreds of whom made their way through Tengyue into British Burma.[34] In mid-1918, the British consular agent at Daqianlu (Kangding/Dartsendo) in eastern Kham negotiated a cease-fire, and a new de facto border was established along the Mekong-Yangtze divide, substantially east of the former frontier.[35]

In April 1918, Père Valentin wrote Forrest from Tsiriting of serious fighting at Tshakhalo (Yanjing), the gateway to Tsarong.[36] The Chinese garrison there had been defeated, and the area was in the hands of the Tibetan army. Forrest abandoned his plans to go to Tsarong that season, returning to Nvlvk'ö instead. There, he made up with Zhao Chengzhang and sent him and two other experienced collectors to Tsiriting. They were to work with Ganton to explore Tsarong despite the fighting. They made two journeys into the region, each lasting more than a month, to the northwest face of Khawa Karpo and beyond.[37] Other parties explored the Salween-Shweli (Nu-Long) divide north and east of Tengyue. Still others went to Muli, a small Tibetan principality in southern Sichuan, remote from the fighting. The Mu kings had conquered Muli by 1433 and held it until 1647, and communities of Naxi descendants of Mu garrisons were scattered in its southern mountains.[38] The collectors' base was the small town of Muli, near the monastery, where families of Naxi from the Baisha region had established leather shops. Thirteen of their sons and nephews would travel and reside in Muli with Joseph Rock for long periods between 1924 and 1929.

Forrest began to think of Tsarong, most of which still lay beyond the explorations of Ganton, Zhao, and their party, as a key that would make legible the entire, difficult, disarticulated geography of the region: "The whole lesson of my nine years of exploration of this region is told in a very few words, when speaking of that genus. Travel northwestwards and the species are ever on the increase; break eastwards or south and there is a marked decrease in numbers immediately! From some point northwest of Tsarong, the genus spreads out in a fan-shaped drift southeast, gradually thinning off in numbers as the lowlands or plains are reached!"[39] Crates of specimens from the last two seasons had begun to arrive in Edinburgh in September 1918. "All your specimens confirm the opinion you express about the distribution of Rhododendrons," Balfour wrote. "You have shown that the area which will probably prove richest in species is that of the high hills Northwest from Tseku [Cigu]. . . . It will be the creaming of the richest area in the world—the last of the best to be explored—when someone gets into those high hills above where Tibet, Burma, Szechwan and Yunnan meet. I hope this will be your work."[40]

He made up his mind to stay another year. In January, he and a party of Nvlvk'ö collectors returned to Tengyue, traveling through a fierce outbreak of influenza that claimed victims in every town they passed, in some places as much as fifty percent of the population. Two chair bearers and three muleteers in Forrest's party died. Four botanists from Nvlvk'ö fell ill, and two nearly died. The party arrived in Tengyue with the corpse of Yang, whom Forrest had employed as a cook since 1910 (never learning his full name), lashed to a mule. Forrest was sick too; the doctor told him that limestone in the water in Nvlvk'ö had damaged his kidneys. He resolved to spend the next season in Tengyue, sending parties north and west into Tsarong. The region was, he began to think, the center not only of *Rhododendron* but also of the flora of the entire continent. "The party which first breaks ground there will reap a rich harvest as, in my opinion, the provinces of Tsarong and Chamdo hold the cream of the flora of Asia. I have nibbled here and there on their eastern frontiers without ever having the luck to get in, and what I have secured is nothing in comparison to what will be found when the country is once opened for systematic exploration."[41]

He lay ill in Tengyue most of the season. But some fifteen collectors from Nvlvk'ö spread out over nearly the entire western frontier of Yunnan north of Tengyue. They walked up the middle reaches of the Mekong-Salween divide where, to Forrest's puzzled disappointment, the flora was much the same as that of the lower levels of the Yulong range, despite the gigantic gorges and ranges between. They traveled across the disputed border with Burma to Pianma, where their presence infuriated the botanist Reginald Farrer, attempting to make that region his own (he would die of dysentery or drink soon after). They worked the Nmai Kha-Salween (Enmaikai-Nu) divide north of Pianma, where the flora of the alpine meadows had been crowded out by dense thickets of dwarf bamboo—another severe disappointment. One party was trapped by flooding rivers for three weeks and came close to starvation, returning sick with fever.

From Cigu, Ganton and two other Tibetan Catholics walked to central Tsarong. They had a map, two mules, and a load of paper Ganton's son had hauled from Weixi.[42] They were "everywhere received kindly by the natives," Père Valentin wrote Forrest. But no precipitation had fallen on the mountains for more than seven months; the ground was dry, the flowers few, the alpine meadows burned up. The party returned to Tsiriting with only eighty-two species. Valentin directed them on another expedition west of the Salween (Nu) to the range that divides that river from the Qiu (now Dulong), a northern tributary of the Nmai Kha. They stayed for three months. Rather than giving up his hopes, Forrest shifted them. "Strange how they should decide to prospect there," he wrote, "for that is the country I should try to reach could I get north from here . . . !"[43] Ganton and his party sent two mule loads of specimens and seed to Edinburgh.

In early 1920, back in Edinburgh, Forrest dazzled the Rhododendron Society with lantern slides. Balfour was in the audience. The "nucleous and radiating point" of *Rhododendron,* Forrest declared, was somewhere north of Ganton's first failed journey to Tsarong. As one travels north up the Mekong-Salween divide, he said, "there is scarcely any ligneous vegetation but Rhododendron, the flowering season a riot of colour, gorgeous beyond any description; mountain sides splashed with colour like a giant palette." Yet all this "is but a tithe of what is yet to come, not only of *Rhododendrons,* but of many other genera—*Primula, Gentiana,* etc." He had heard from Baptist missionaries who had traveled in Kham, north of Tsarong, of an "undulating plateau, of vast extent and high altitude, enclosed by higher ranges, every where clothed in *Rhododendrons,* great and small, to the exclusion of almost all else."[44]

By March 1921, he was in Bhamo again. His syndicate had agreed to double his salary to one thousand pounds per year, of which J. C. Williams would pay half, in return for all the *Rhododendron* seeds. For the first time, he would collect birds, for a wealthy member of the British Ornithological Society, Colonel Stephenson Clarke.[45] Zhao Chengzhang and his core party of Nvlvk'ö collectors traveled through influenza-ridden Lijiang and Dali to meet him in Bhamo. After the last season, another party had remained in

Tengyue, on their own initiative, working the Salween-Shweli (Nu-Long) divide, searching for flowering specimens of *Rhododendron giganteum*, a species of great size, recently discovered. This party too joined Forrest to walk back to Dali and Lijiang.

To the east, Yunnan and Sichuan were skirmishing across the border again. To the northwest, about a thousand Tibetans from Konka Ling (Xiangcheng) had raided Zhongdian, driving out the Chinese garrison and magistrate, in protest over a new "death registry" accompanied by a tax on dead bodies.[46] To the north, another body of Tibetan rebels had raided Adunzi (Deqin). A thousand troops marched from Dali to reassert control. Forrest had planned on traveling to the upper Mekong via Adunzi, but his party found the road blocked by the authorities. Instead, Forrest, Zhao, and twenty-one other Nvlvk'ö collectors turned around to walk back to Dali and take the more southerly route through Weixi, leaving five men to work up to Muli and beyond.

In July, the party settled into a scattered village of Lisu and Tibetan dwellings on the Mekong about four miles south of Tsiriting. Forrest stayed in the house of a Tibetan farmer, "perched on the precipitous hillside like a fly stuck to a wall."[47] The Nvlvk'ö collectors worked up and down the Mekong valley, south to Yezhi and north to Khawa Karpo and beyond. But more fighting between Chinese and Tibetan forces in Tsarong made travel there too dangerous. Again, Forrest shifted his aspirations to the divide between the Nu and the Qiu, which Ganton had explored the previous season. Ganton's collection had turned out not to be the spectacular haul Forrest had hoped for, but he suspected the party had merely skirted the foothills. He arranged for Zhao Chengzhang and his "particular gang" of eight or nine to go to the range.[48] They climbed the great, glaciated Nu-Qiu divide to a pass at around sixteen thousand feet. Beyond, to the north, they glimpsed a mightier range yet. They returned to the base on the Mekong for a few days, then set off again, this time with tens of Tibetan porters carrying food. They had discovered that food was very scarce in the upper Nu (Salween) and Qiu valleys. The peoples living there, whom the Chinese called Nu (or Lu) and Qiu, after the valleys they inhabited, grew only maize, and only enough for four to five months of the year, subsisting for the rest of the year by hunting and gathering.[49]

The party worked its way over the Nu-Qiu divide and down to the timber line. Zhao Chengzhang wanted to descend to the bed of the Qiu, but Forrest, fearing fever, had forbidden it. Again, they caught sight of the great range beyond: "[They] speak of having seen, during a clear blink of the mists

which enfold these altitudes . . . a huge snow-capped range inclined trans-
versely to that on which they stood and lying several days journey to the
northwest. Consul Litton and I on our journey up the Salwin in the autumn
of 1905 at the furthest point north attained a glimpse of the same range,
and a most awe-inspiring sight it was, though distant fully 100 miles to us."[50]
On their two journeys, Zhao Chengzhang and his party gathered forty-five
species of *Rhododendron* and about a hundred of other genera. "Not bad
for what is, so far, merely scratching a new area!" Forrest exulted.[51]

Forrest had asked Balfour to apply to the viceroy of India for special pass-
ports for Tsarong and Chamdo. In the autumn, it became clear that no such
permission was forthcoming, and he walked back to Nvlvk'ö. Late in the
winter, Zhao Chengzhang and his party followed with their crates of specimens. In March, Balfour wrote that he had examined the collections of the
previous season, finding several new rhododendrons. It took no more than
this, now, to set Forrest off on the topic of his particular Eden. By now, he
had moved the precise location of the "radiating point" of the genus to the
Nu-Qiu divide—now further north than his men had yet penetrated, among
the great peaks they had glimpsed through the mists: "I can never hope to
get there, I think. However, I may safely make an addition to my prophesy
of past years that the northern spurs and peaks of the Kiu chiang–Salween
[Qiu–Nu] divide, where it breaks up and loses itself in the higher Zayul and
other cross ranges of Southwest Central Tsarong, is going to prove the richest ground of all. How much I now regret the lack of knowledge which confined me to Central and West Yunnan in the early years of my work!"[52]

He sent Zhao and his party of eight experienced collectors back to the
Nu-Qiu divide in early March, with the aim of gathering in flower the species
discovered the year before. He promised them a bonus of one Mexican silver dollar for every new *Rhododendron* gathered in flower and one half
dollar for every new *Primula*. They attempted to cross the Lancang-Nu
(Mekong-Salween) divide into the Nu gorge, but it was a heavy snow year,
and every pass was closed. Only in May were they able to cross the lowest
pass, the Londre La. They crossed the Nu and worked the giant mountains
between the Qiu and the Nu for two months. In late July, they met Forrest
in Tengyue with some two hundred species of *Rhododendron*, nearly all in
flower. With so many species flowing in over the last few years, the Edinburgh staff had fallen far behind, and the genus was a tangle. Forrest had
no way of telling how many of the party's species were new. He guessed about
twenty to forty. Presumably, though he did not mention this, he paid Zhao
and his companions a bonus of at least twenty Mexican silver dollars[53]

Figure 18.   On the road, 1920s (Zhao Chengzhang is on left). Courtesy of the Royal Botanic Garden, Edinburgh.

CONNECTIONS

To Forrest, the region of his final hope for the center of origin of *Rhododendron* was uncharted and unknown. On his India Survey map, "Burma North-Eastern Frontier, Sheet 23," the range between the upper Nu and the Qiu was represented by patches of white space.[54] To the Nvlvk'ö botanists, however, it was, though remote, well known. Its residents still participated in a tribute system that was a remnant of a structure of regional governance set up by the Mu chiefs during the Ming dynasty. This gave the eight or nine collectors from Nvlvk'ö who explored the region opportunities to converse with the representatives and village chiefs from the region, often eating and sleeping in the same courtyard and plaza. They knew of the area's thick historical connections with the Lijiang valley, and some had sat by a spring near their village in times of drought and listened to a dongba trace a route from its waters over the intervening mountains and gorges and into their valley. In the dongba archive, this remote finger of the sprawling Mu kingdom, stretching far into the northwest, was the closest place in the human realm to Ddibba Shílo's great mountain, where the dead passed from the lands of the living towards the realm of ancestors and gods.

Among Chinese administrators, geographic knowledge of the upper Nu and Qiu valleys was very thin. In 1904, G. Litton wrote that the officials in Yunnan formally charged with governing the region had only the vaguest idea where it was:

> Even the Lichiang prefect, an unusually enlightened mandarin, had no information about the Chiu Ti [Qiu territory] except that it was tributary to China, that it was a very long way off, that it was impossible to go there, and that for anyone who had the privilege of living in the Celestial land very foolish to want to go there. The prefect told me that he had twice sent men to find out where the Chiu Ti was . . . but that on one occasion his emissaries failed to get into the country and on another failed to get out.[55]

It was a very difficult place to go. Three out of four passes into the Nu valley from the Lancang (Mekong) were sealed by snow for four months out of the year; all the routes from the Nu into the Qiu were closed for even longer. And travelers could not walk up the Nu gorge from the south. In the seventeenth century, Lisu, escaping conscription into the Mu armies, had settled in the middle Nu and pushed north, taking slaves and establishing hereditary rulers in Nu villages. Acknowledging no authority outside of the gorge, their descendants frightened travelers away with their crossbows and fierce reputation. A scarcity of food was an equally significant obstacle. In the Qiu valley, a few thousand Drung (Dulong) people lived in woven bamboo longhouses in dispersed, egalitarian hamlets.[56] They raised millet and maize enough to feed their own families for about four months; they gathered seeds and hunted rodents for the rest of the year. To travel successfully there, outsiders had to have porters loaded with food.

The upper Nu and Qiu had been the most remote terminus of the Mu kings' push northwest into Tibetan territories.[57] The Mu kings had claimed it since at least the fifteenth century, but since no routes for conquest or trade passed through it, no garrisons were established there. After the Mu domain disintegrated, the Qing made the Nu and Qiu commanderie *(zhanguansi)* part of the circuit of Weixi, a town far to the south on the Mekong. Shortly after the descendants of the Mu chiefs were finally relieved of office in 1729, the Yongzheng emperor established a garrison of one thousand troops at Weixi and made that town a minor prefecture *(ting)* in the regular system of administration. At the same time, the Qing recognized as native hereditary rulers some of the Naxi lineages that the Mu overlords had established as regional administrators. In particular, the Qing granted titles to two lineages, the Nan and the Wang, both based in the village of Kangpu, a day's walk north of Weixi. The Nan, with the rank of native sergeant

Figure 19.   Wang Zanchen
(Wang Guoxiang), *mugua*
of Yezhi, 1923, from Joseph
Rock, *The Ancient Na-Khi
Kingdom of Southwest
China.*

(tubazong), controlled a territory called the Northern Circuit *(beilu)*, which included the northern portion of the Nu valley that bordered on the Tibetan region of Tsarong. The Wang, who eventually moved to Yezhi, about twelve miles further north, held the higher rank of native lieutenant *(tuqianzong)* and had jurisdiction over the Qiu and the southern portion of the Upper Nu. Both the Nan and the Wang, though formally subordinate to the Weixi subprefect *(tongpan)*, held significant power in their own fiefdoms. In the nineteenth century, as the Qing whittled away at the powers of native hereditary officials, the influence of the Nan lineage declined, and the Wang of Yezhi absorbed its territory.[58]

Their subjects knew the Wang chiefs as the *mugua* or, in Naxi, *munkwua*. In the Mu system of military rule, *munkwua*, literally "commanders of soldiers," had been district-level commanders of military and civilian affairs.[59] During the eighteenth and nineteenth centuries, the Wang *mugua* commanded military forces, adjudicated criminal cases, and owned considerable properties. As late as 1875, the *mugua* was capable of fielding two thousand Lisu warriors for a punitive expedition.[60] The *mugua* exercised their jurisdiction over the distant upper Nu and Qiu area by receiving yearly tribute. The *Yunnan beizheng zhi*, a gazetteer written in 1769, states that, beginning in 1730, the people of the Qiu river journeyed to Kangpu every year to offer tribute of thirty *jin* of beeswax, fifteen lengths of hempen cloth, and twenty skins of mountain asses.[61] By the first years of the twentieth century, this tribute was the Wang lineage's sole source of income and nearly the only remnant of its hereditary privilege.

In 1904, Forrest and G. Litton visited the *mugua*, Wang Zanchen, during the lunar new year. Litton, considerably more observant of social life than his companion, noticed the rough villagers from the Nu and Qiu streaming into the *mugua*'s spacious courtyard. Each autumn, he reported, the *mugua* sent a delegation of his relatives to the Nu and Qiu valleys. They carried a thin wooden tally (*muke*) with the *mugua*'s seal impressed on it with beeswax. Each village head they encountered cut a notch in the tally to acknowledge the *mugua*'s orders to pay tribute.[62] "Every village," he reported, "is supposed to send someone with something, a bag of grain, some gold dust, some musk, or a parcel of drugs, usually *huanglian*." On the new year, the village heads made the long journey to Yezhi. Litton watched one of the *mugua*'s brothers receive the payments at a table in the courtyard, record them in an account book, and send them to the storehouse.[63] A French Catholic missionary had witnessed the same exchange in 1877: "The Mugua has to take good care of them all for three days. Every three years he also shares out a piece of beef, which costs only a few pennies, and people have to give him back a pound of mushrooms, etc."[64] Another priest, visiting in 1908 wrote that the Nu pay "a tribute consisting of the hides of bear and antelope, musk, wax or silver. In the past each family had to give a slave. In time the slave was replaced by a ball of wax shaped like a human head. The Mugua of Yezhi has to give them salt of which the country is completely deprived."[65]

The upper Nu and Qiu were drawn into the regular system of administration in 1908. That year, Yunnan's governor ordered the border pacification commissioner at Adunzi to investigate and "pacify" the people of the Qiu. The official journeyed through both valleys with fifteen soldiers and twenty porters. Along the way, he gathered village headmen, feasted them with wine and oxen, and presented them with certificates naming them local chiefs subject to the Qing. He then named a general administrator of the Nu and Qiu *(Nu Qiu zongguan)* in Changputong, the largest village in the upper Nu.[66] After the 1911 revolution, Li Genyuan's army entered the Nu and Qiu valleys, appointed village chiefs, and reestablished an administrative post at Changputong. The post was only intermittently occupied how-

ever, and Nu and Qiu residents continued to travel to Yezhi to present tribute to the *mugua* until at least 1923.[67]

Forrest developed a close relationship with Wang Zanchen who, in 1905,
had given him food, clothing, and sympathy after his escape from the attack on the mission at Cigu.[68] During every journey up the Mekong, he made
a point of stopping in the *mugua*'s spacious and comfortable mansion. Zhao
Chengzhang and his colleagues first visited Yezhi in 1913, and they stayed
there many times thereafter: it was their gateway to the entire far northwest. The *mugua*'s courtyard, the plaza in front of his mansion, and the Naxi
homes clustered around it, were a busy crossroads through which passed
nearly every merchant, monk, official, or tribute-paying villager heading
north. Here, by keeping his ears and eyes open, G. Litton had been able to
write a substantial report on the upper Nu and Qiu region.[69] The sojourners from Nvlvk'ö, speaking the same language as the *mugua*'s many Naxi
relations and servants, could not have avoided learning a great deal about
the region in the eight years since they began visiting Yezhi. It is possible
that they developed connections or friendships among the village headmen
who repeatedly traveled to Yezhi with their offerings of grain and beeswax.

For nearly two decades, Zhao Chengzhang and his colleagues had worked
indefatigably to push further and further north in the direction toward
which the entire dongba archive was oriented, yet in 1921 the terminus of
their explorations bent south and west away from Tsarong to the range dividing the Nu and Qiu. Though easier to reach, Tsarong was just beyond
the northernmost limits of the former Mu domains. Its Tibetan district governor and Gelugpa monasteries had few connections to the Naxi hereditary
rulers and Karmapa temples to the south, and its places were not named in
the ritualized geocosmography recited by dongba. Though Forrest said he
turned away from Tsarong in 1921 because of reports of fighting coming in
from Kham, he had sent Ganton, Zhao, and two others from Nvlvk'ö into
Tsarong in 1918, as the Tibetan army battled Chinese forces throughout the
region. Zhao and his colleagues had guided Forrest's attention toward areas
described in the dongba texts many times before, sometimes subtly, sometimes taking matters more directly into their own hands. Now it is likely
that they led him away from Tsarong toward this more remote but more
familiar place.

Paper roads linked the upper Nu and Qiu with the waters of Nvlvk'ö. In
dongba texts, the people of the Salween were called Nunkhi, written with
the graphs for a yellow pea *(nun)* and a man *(khi)*. Their homeland was
Nundü, the land of the Nun. The giant range to the west, where Forrest finally
hoped to find the generative center of *Rhododendron,* was Nundü Gyìnvlv,

Figure 20. Graphs for *Nundǜ* (the upper Salween), *Nunkhi* (the people of the upper Salween), and *Nundǜ Gyìnvlv* (the great snow mountain behind the Salween). From Joseph Rock, *A Na-Khi-English Encyclopedic Dictionary.*

the "great snow mountain of the Nun." Beyond, in the Qiu valley, lived the T'khyükhi, written as a sword boring a hole (*t'khǜ*, "to bore"). Their homeland was T'khyüdǜ, land of the Qiu. Beyond that rose yet another huge range, the T'khyudǜ Zhernvlv, the "great mountain of the T'khyü.[70]

In a meadow just below Nvlvk'ö was a spring called Boashì Gkogyi. It was associated with fertility and childbirth; dongba performed a ritual there called *zämä*, "desiring children," for childless couples.[71] It was also a place that could give rain. If the rainy season had not begun by June, one of the wealthier families in the area sponsored a small ritual there, called *khǜmä*, "desiring rain." It was among the rituals for *ssù* and *llü*, the original inhabitants of the land, mentioned in chapter 3. While most of these ceremonies had last been performed decades ago, *khǜmä* was simple, practical, and common, and most people in Nvlvk'ö had probably seen it several times. At the spring, a dongba recited fourteen texts, calling upon the *ssù* and *llü* to arise from rivers, lakes, and streams throughout the former Mu realms and sweep across the land to the Lijiang valley. Only one text was specific to *khǜmä*. It was clearly written to be performed in Nvlvk'ö and nearby villages, for it named the nearby springs and cliffs of the Yulong range as "our mountain and valley spirits."

The text described the effects of drought: "When waters of silver and gold dry up, the heart of a skilled man aches; when waters of turquoise and carnelian dry up, the liver of a wise woman aches."[72] It called upon the *ssù* kings and chiefs: Muanmíbpalò, holding an eagle, Bpawuats'òbpǒ, leading a red tiger on an iron chain, Làbbut'ogkó, carrying a golden mirror, and dozens more. It implored them to call up armies of *ssù* soldiers twenty-five thousand strong from the mountains and valleys, the springs and villages, the wastelands, the clouds and peaks. It asked the *ssù* of the ponds and streams in the Yulong range for cloudbursts, naming thirty-four places in the immediate vicinity of Nvlvk'ö. It called on the *ssù* of the regions to the north-east, describing a circular path from the mountains beyond the beautiful Lugu lake of Yongning down into the Lijiang valley and Nvlvk'ö—nineteen names of mountains, lakes, and village springs. It called on the *ssù* of the southeast and southwest, tracing two more winding paths through mountains and villages along the southern borders of the former Mu realms, trending in toward the Lijiang valley and Nvlvk'ö—another thirty-four place names.

It then called upon the *ssù* of the northwest. These routes were flung much further, west to Nyinà (Weixi), then far up into Nunddǜ (the Nu) and T'khyuddǜ (the Qiu).

> alpine meadow spirits of Nyinà
> alpine meadow spirits of Dzenààhö
>   spirits of bubbling springs
> mountains of Lazhermberlèr
> mountains of Gkawgkàwssumä
> snow mountain of the Nu
> great mountain of the Qiu
>
> *Nnyi nà ghügh szi gkò*
> *Dze nà à hö ghügh szi gkò*
>   *ts'u ts'u gkò*[73]
> *La zhe mber lèr ngyù*
> *Gkaw gkàw ssu mä ngyù*
> *Nun dǜ gkyì nv lv*
> *T'khü dǜ zher nv lv*

After winding through several villages around the great mountain of the Qiu, the trail ascended to Odso Lvmänà, literally "the place of black and white rocks in Tibet." Other texts placed Odso Lvmänà at the foot of Ngyùná Shílo Ngyù: it was where Ddibba Shílo, the founder of the dongba cult, had first set foot on the earth after descending from the heavens with 360 disciples and yaks loaded with books. After Odso Lvmänà, the route descended into Nyinà again and then to Lijiang—seventeen more place names. Finally, the text called on the *ssù* of the north. It described a route along the Ssulo (Wuliang) river, in the Muli district, winding down through Bberdder (Baidi), across the Yangtze and the Yulong range, and into the vicinity of Nvlvk'ö again—another sixty-two place names.[74]

Naxi ritual brought the geocosmological archive of dongba textual practice into the intimate realms of daily life. In rituals like *khǜmä,* recitations moved through this archive, drawing listeners along looping rhythmic paths of place names. These paths connected daily experience with the inhuman substrate of the world, which could not be experienced bodily, but only spoken or read. In this archive, the great ranges above the Nu and Qiu were at the outer limit of the human, associated with the wildest of things: bear and boar, tiger and leopard, crane and eagle, spruce and rhododendron. These lands were mentioned only in texts that appealed to the *ssù* and *llü,* who commanded the forces of nonhuman life.

As they crossed the Nu and climbed into the mountains beyond for the first time, Zhao Chengzhang and his band seem to have had the sense that they were trespassing on the limits of the human. Echoes of their voices reverberate in Forrest's recitation of their report to him on their journey, as his letter to Balfour shifted into a style far more vivid and compressed than usual: "I can believe every word did they paint it even more luridly. Roads of no kind, deep jungle-choked and panther-haunted gorges separated by razor-backed spurs and bounded by break-neck precipices and dense forests at the lower altitudes; cane brakes and boulder-strewn marshy moorlands with snow drifts and eternal mists at the higher, and above all, a chaos of screes, ragged peaks, and glaciers!"[75]

This report was translated from Naxi into Chinese for Forrest's ears and again into English for Balfour's eyes. Yet in its references to leopards and panthers and its progression from low to high as it moves from gorges, to forests, to moorlands, to snow and mists, and to screes, peaks and glaciers, I cannot help but hear an echo of the style of dongba ritual recitation. In any case, in rituals of mourning, it was places like this that divided the living earth from what lay beyond. A passage from the song of the origin of sorrow quoted in chapter 3 comes to mind:

> to the dazzling white cloud gate
>   flapping, the crane will return
> all the winged creatures
>   guide the crane back to the clouds
>   his power must not pass . . .
>
> to the glittering golden mountain gate
>   dancing, the tiger will return
> all the clawed beasts
>   guide the tiger back to the mountain
>   his power must not pass
>
> we guide this ancestor to the gods' land
>   he will cross nine ridges
>   his power must not cross
>   he will ford seven rivers
>   his power must not ford[76]

Figure 21.   On the road, 1920s (Zhao Chengzhang is third from left). Courtesy of the Royal Botanic Garden, Edinburgh.

Rituals of grief tilted the world toward such places, aligning it with that great mountain at which all things originated. And as the *khùmä* rite made clear, the great snow mountains behind the Nu and the Qiu lay hard up on Odso Lvmànà, at the foot of that great mountain. As they made forays into those great ranges in search of Forrest's *Rhododendron* paradise, the nine men from Nvlvk'ö came as close as they ever would, this side of death, to the place Ddibba Shílo first set foot upon the earth, and the vast mountain that towered above it.

. . .

For Forrest too that great range came to be associated with grief and ancestral death. On their first exploration of the range, Zhao Chengzhang and his party had reported a great shining extension of the range to the north, which he took to be the range he and G. Litton had seen on their final journey together. After Litton, who died only weeks later, Forrest had no other friend in China: jealousy or impatience cut short his relationships with all the other European or American botanists and explorers that he met. He had always intended to return to see that great range again, in tribute to Litton; sending Zhao Chengzhang and his party was now as close as he would get.

After packing up the rhododendrons from the Nu-Qiu divide in 1922, Forrest headed home. In India, he learned of Isaac Bailey Balfour's death. He had always treated this man as a father, with all the complicated love, impatience, and resentment that such relations often entail. Now, he was heartsick. His search for the center of *Rhododendron* had been a partnership with Balfour, who had spent the last years of his life working out a major revision of *Rhododendron* taxonomy using the masses of material flowing in from China. He left it incomplete, and, for the time being, no one at Edinburgh had mastered the materials well enough to continue his work. Without Balfour's interest, encouragement, and scientific support, searching for the center of *Rhododendron* was pointless. Forrest never mentioned it again.

Nevertheless, Zhao Chengzhang, He Nüli, Zhao Tangguang, Lu Wanyu, and the other, unnamed men from Nvlvk'ö walked the former domains of the Mu kings for another decade. The numbers of new species they accumulated surpassed even the spectacular hauls of the previous twenty years. In 1922, they worked Muli and Yongning, both former outposts of the Mu. Forrest asked them to explore a range between the Litang and Yalong rivers, north of Muli and of the line of defensive watchtowers built to defend against the Mu armies, and beyond the pale of the dongba archive. They did not go, working instead another area new to Forrest: Bberdder, on the southeast edge of the Zhongdian plateau, the sacred center of the dongba cult. Forrest then located another "blank spot" in his maps of Yunnan, a range of mountains between the Jianchuan (now Heihuijiang) and Mekong (Lancang) rivers. A party of collectors worked the area twice, returning with some four hundred species. Though new to Forrest, the area was well known to the men from Nvlvk'ö. It was the northern hinterland of the city of Ngyudtv́ (Jianchuan), taken by Mu De in 1387 to punish the misdeeds of its native hereditary magistrate. Its highest peak was called Labà Ngyù in Naxi; its rivers and lakes were described in dongba texts, and its inhabitants spoke the Lijiang dialect of Naxi.

When Forrest left Yunnan in late 1922, the Nvlvk'o botanists continued to work all these areas. Forrest left thirty pounds with the China Inland Mission in Lijiang for their salaries. The next autumn he sent another fifty pounds for further work and to transport their specimens to Rangoon. They collected some four hundred new species that season.[77] By the spring of 1924, Forrest was back in Yunnan, sponsored by J. C. Williams and Reginald Cory. He stayed in Tengyue for two years, making short journeys up the Salween-

Shweli (Nu-Long) divide. The explorers from Nvlvk'ö ranged widely over the province and traveled for six weeks on the Burmese side of the frontier.[78] In 1929 and 1930 a small group of wealthy gardening enthusiasts who had not taken part in previous syndicates gave Forrest fifty pounds salary and two hundred pounds expenses to oversee an expedition from Edinburgh. Again, the contact was the China Inland Mission. Zhao Chengzhang quit Theodore and Kermit Roosevelt's expedition to west Sichuan to kill a giant panda and walked 250 rough miles back to Nvlvk'ö to lead the expedition. He and his men worked in the vicinities of Lijiang, Zhongdian, and Muli, gathering specimens of over 400 species and seed of some 250.

Forrest began his final journey to Yunnan in November 1930 when he was fifty-seven years old. Thirty-nine subscriptions, from twenty-five to five hundred pounds each, made it his most generously funded expedition. The syndicate's organizer, Major Lawrence Johnson, accompanied him: Forrest left him in Bhamo after a series of mishaps, quarrels, and illnesses. When he showed up in Tengyue, ill, in a sedan chair, Forrest left him there too, to return on his own to Burma. Zhao Chengzhang and his colleagues had continued to collect through the 1930 season, in Muli, Bberdder, and Haba, on the promise of salaries to be paid when Forrest arrived. They met him in Tengyue in February 1931 with specimens of over one thousand species and seed of three hundred to four hundred. In March, Zhao and his party walked to the headwaters of the Long (Shweli) river in search of the extraordinary *Rhododendron giganteum;* Forrest followed them a month later. They felled a specimen and hauled a cross-section of the trunk, nearly seven feet in circumference, back to Tengyue.[79] Forrest moved on to Nvlvk'ö, while Zhao and his fellows made sorties north of there to collect some fifteen hundred to two thousand birds and mammals as well as over one thousand botanical specimens. He then retired to Tengyue to write up and pack these extraordinary hauls, including some three hundred pounds of seed from four to five hundred species.

One January morning, he took a break from his collections to shoot snipe a few miles from Tengyue. An inquest held later that day named the three young men from Nvlvk'ö who accompanied him: Zhao Yuanbi, He Jianhua, and He Wenming. Each had worked for him only one year. They testified that their patron had laid down and died late in the morning after shooting a bird. The medical attendant to the Tengyue consulate determined that it had been heart failure.[80] Forrest was buried in the small foreign cemetery next to his friend G. Litton. The collectors stayed another month to finish packing twelve cases of specimens and five of birds. Then the consulate paid them out of Forrest's assets, and they returned home.[81]

.   .   .

Two landscapes merged in this project. Both were made of earth and paper, layered together in distinctive ways. One emerged out of an interleaving of plant specimens and their paper wrappings, visions of the earth and their paper descriptions. The processes that wrapped paper and earth together were about scientific taxonomy and the disintegration of taxonomy; they were about seeing immanent beauty and searching for transcendent sublimity. They were guided by very specific rules: about how to distinguish species, how to describe plants with precision, how to identify the limits of representation, how to think and feel about visions that seemed to transgress those limits. As this interleaving of earth and paper grew more dense, it developed an orienting horizon, an imagined place where the beauty of specific plants, described with precision, would join with the sublimity of innumerable flowers, merged in indescribable masses.

The other landscape was no less an interleaving of paper and earth. A great archive of written journeys, recited in acts of ritual power, and reexperienced through walking. This landscape too was shaped by specific rules, most of which seem to have been about memory. About how to use combinations of symbols to embed a poetic phrase in memory and then to bring that phrase back in the present in recitation. About how to voice journeys that made old names and ancient beings reemerge from the earth. And about how as one walked to let remembered fragments of texts and songs bring the earth to life as they brought back old names, old stories, unseen presences, and subtle dangers. The orienting horizon of this landscape was the origin of its interleaving of paper and earth, the place where the books, already written, had descended to earth on the backs of yaks. And it was also the end of every journey, where every soul, letting go its earthly powers to sing, recite, and walk, stepped off into another world. As the gateway between this world and what lay beyond, this horizon, too, was a place where immanent experience and the transcendent abstraction met.

Did these horizons merge in the way I have suggested? The evidence is circumstantial. And the circumstances are simple enough. First, though they explored areas outside the dongba archive with which Forrest had familiarized himself before he began collaborating with them, in nearly all other cases the men from Nvlvk'ö limited their explorations to regions described in the dongba archive, actively and passively resisting Forrest's attempts to expand them in other directions. Second, Forrest's long search for the generative center of *Rhododendron* took them further and further in the direction toward which that archive was oriented. And third, a plausible case can be made that the most specific references to the earthly location of gate-way between the lands of the living and the lands of ancestors and gods that Nvlvk'ö villagers would have heard recited put it in the precise area where Forrest last decided to search for his earthly *Rhododendron* paradise.

These circumstances are not enough to prove the case, of course. The re-sistance of Forrest's collaborators to move out beyond the boundaries of the former Mu domain could have been due merely to a combination of botan-ical knowledge and common sense. The diversity of *Rhododendron* species did increase the further north one moved along the ranges between the par-allel rivers that scored northwest Yunnan.[82] It is possible (though unlikely) that Zhao Chengzhang and his colleagues did not influence Forrest's deci-sion to shift his hopes for the center of *Rhododendron* from Tsarong to the Nu-Qiu divide. And it is also possible (though even less likely) that the so-journers from Nvlvk'ö knew and cared little about the texts that dongba re-cited in their frequent ritual performances in that village.

But proving the case is not the point. The argument has provided a path along which to explore some aspects of these two regimes. For me, the most challenging and enriching circumstance of the argument has been that their horizons did not begin to approach each other on the level of explicit dis-cursive communication. Though he spent enormous amounts of time with the Nvlvk'ö collectors, Forrest recorded very few conversations. And when he did talk to them, it was never about anything other than plants and ge-ography. He knew nothing about their domestic lives, and he openly scoffed at their ritual practices. When these regimes did communicate, it was through indexical processes, involving maps and gestures, leaves and petals, place names and orientations. The earth was a third party in all this, pro-viding the matter of reference. Everyone was always pointing at it, gestur-ing towards it, bringing it forward to the center of social life, in an ongoing process. It was in this way that the matter of the earth, and the objects and words made from it, acted as "secondary agents" in Alfred Gell's useful for-mulation, in the various social contexts through which they circulated.[83]

If these landscapes did converge, affect rather than discursive meaning formed the ground for agreement. A kind of longing, which "meant" dif-ferent things in each case, but which was laid out along the earth in the same way, through repeated acts of indexical reference, pinning it to particular plant species and then to the locations from which they were collected. Through these acts, longing took its qualities from the dramatic and unavoidable ma-terial qualities of the earth—the close alternation of parallel high mountain ranges and deep river gorges, the tilt of the land upwards and northwards—qualities that actively intruded on every act of perception.

Part II

5.   Bodies Real and Virtual

In 1934, during twenty-eight years of wandering west China, Joseph Francis Charles Rock made a brief trip to England. He clipped an obituary of an old friend from a copy of *The Times,* three years old, and pasted it in his diary. On the facing page, he pasted a photograph, two decades old, and wrote this caption: "Joseph Rock (standing) with his older friend Fred Muir at the Haw[aii] Sugar Planters' Exp[eriment] St[ation], Honolulu. Photographed in our home in Liloa Rise (Breaside), Honolulu in the spring of 1913, while our phonograph played *Spiritu Gentile,* Caruso singing."[1] Rock's parents had long been dead; he rarely saw his only sibling, Lina, who lived in Vienna, and he had no spouse or children. His significant relationships were with friends, all men. Some may have been erotic, at least in spirit. There is no evidence one way or another that his friendship with Muir was one of these.[2]

This collage invites some basic questions about the materialization of social relations. Is this a story about a friendship between two men, memorialized in a photograph and a caption? What is the material substrate of the relationship depicted here? Words? Glances? Or perhaps the memorialization itself? Does the relationship have any existence outside its materializations? Every trace of this friendship is contained in this collage. No other record of it exists—and no record existed in 1934. The friendship was called into existence for the purposes of memorialization after many years of complete absence only to melt away again after a day or two into the flux of a wandering life. What are the roles of the twin technologies of photography and phonography in defining this relationship? Two gazes oriented and aligned by the camera, two bodies harmoniously composed. I notice the jaunty slant of young Rock's arm and the tilt of Muir's head into the triangular space it protects. The evocation of domestic intimacy is heightened by Caruso's en-

Figure 22.   Page from Joseph Rock's diary, March 25, 1934.
Courtesy of the Royal Botanic Garden, Edinburgh.

ergetic voice, emanating from the unseen gramophone, and the title of his
song, suggesting harmony of spirit. Camera and gramophone create the material of memory here: they grant this friendship all the qualities of intimacy, domesticity, and harmony that this collage memorializes. Without
them, there would be no possibility of calling this old friendship back into
existence, in this remade form, for this brief moment in 1934. Can the ways
these technologies are combined here tell us anything about the status of
the bodies they compose, represent, or perhaps *replace* in this relationship?

This collage was typical of Rock's path towards engagement with experience and affect. He had begun this path in Hawai'i as he created his herbarium of indigenous Hawai'ian plants for the Division of Forestry and then the College of Hawaii (see introduction). He had wandered the islands collecting specimens, making notes, taking photographs, and in his lab room at the college, he had assembled all these traces into stacks of specimen sheets and into great volumes that traced his footsteps over the lava-strewn landscapes, embedding his relationships—intimate and distant—with botanists, ranchers, and Hawai'ian guides in a thick matrix of text and image. He was a compulsive archivist, yet his archives were not merely records of experience past. Making archives was his means of throwing himself into earth, at once drawing some part of it into social life and turning some part of himself over to its mute, fearsome, antisocial being. His archives added to the earth's accumulating social substance; but, as we shall see, at the same time, each of his photographs of single trees or mountain vastnesses deposited an empty reflection of himself at their center. His path was not merely to appropriate; it was also to be appropriated—this was the way he found to endure an existence he found, at root, unbearable.

Part 2 follows Rock as he developed his archival mode of engaging with the earth and social life. This chapter and the next investigate his use of camera and gramophone to nurture a sense of moral personhood in the face of abjection and misery, his own and others. Rock's evocation of domestic harmony with Fred Muir is of a piece with these attempts to use these archival instruments to carve out a habitable space for himself in a difficult world. He was ill, lonely, and uncomfortable during his visit to London in 1935, and he spent his time recalling domestic scenes from his past wanderings in China, scenes which often included the sounds of his gramophone. He made a lifelong habit of building real and imagined domestic spaces in which to nurture bonds of friendship or intimacy. His camera and gramophone were powerful tools in this effort: he used them to shape his own body and those of others; he used them to control the eyes of others, to silence their voices, and to give him back a relationship with the world that he could bear to live with. This chapter traces these efforts during his early wanderings in Yunnan and his adventures in Gansu; chapter 6 examines a photographic collaboration between Rock and Xiang Cicheng Zhaba, supreme religious and political authority of the multiethnic state of Muli in southwest Sichuan/ south Kham.

In 1924–1927, twelve young men from Nvlvk'ö, sons and nephews of the men who worked for Forrest, accompanied Rock on a difficult and hazardous journey from Yunnan, through war-torn Sichuan, into equally chaotic Gansu.

Chapters 7 and 8 follow them on this journey, exploring the relationship between the botanist and the young adventurers who accompanied him. In 1929, Rock turned from collecting plants to collecting texts. With the help of some skilled dongba, he created translations of many texts, and he wrote a dictionary of pictographs, several massive works of translation, and a geographical history of the Mu kingdom. Chapter 9 looks at Rock's adventures in the dongba archive. It records his own archival activities, his efforts to bring the landscape, which he now understood as thoroughly historical and as densely sedimented with texts, into a single field with his own experience of walking and writing. As a whole, part 2 is an investigation of the archival modes of engaging the earth, his own affect, and the eyes and voices of others that Rock developed, with the help of his companions from Nvlvk'ö and dongba from villages within the loop of the Yangtze, during his wanderings in China. I offer this account of Rock's archival practice because, despite its singularity, I think it has much to teach us about how, as walking, mark-making, image-making human beings, we draw the earth into our social lives.

PORTRAITS

In 1913, Rock wrote a report on his herbarium of Hawai'ian plants to the State Legislature of Hawai'i. He had already lived in Hawai'i for five years after spending his youth wandering Europe, North Africa, and America. He had taught Latin and natural history at a Honolulu middle school before convincing Hawai'i's Division of Forestry to hire him as its sole botanist, with the task of creating an herbarium of indigenous Hawai'ian plants. He had walked up every mountain on every island, accompanied by nameless, faceless Hawai'ian ranch hands, in search of plants for his herbarium, now housed in a laboratory room in the College of Hawaii. In his report, he compared a herbarium to a museum, a laboratory, and a "great illustrated volume."[3] He had learned to use photography to pluck leaves of this volume from their laboratory setting to insert between leaves of text. In his publications, he was among the first to replace botanical drawings with photographs.

At first, the photographs were beautifully rendered reproductions of herbarium specimens, complete with tape and tags, and often with rulers to show scale. They did away with the stylization of botanical drawings, rendering lucid every pit and scar, as though to place the page from the herbarium, whole and unaltered, in his text. Still, that page, with its dried, flat-

PLATE XXII.

Metrosideros collina (Forster) A. Gray subspec. polymorpha (Gaud.) var. macrophylla Rock n. v.

Figure 23. From Joseph Rock, *The Ohia Lehua Trees of Hawaii.*

tened, and faded leaves, was hardly an accurate representation. He began to photograph twigs and fruit cut recently from living plants and placed on white cards with no tags or rulers. He was trying to reproduce the specimen's style—isolated from context, arranged artfully against a paper background—while overcoming the specimen's failures of representation. In some photographs the leaves appear to be flattened with a perfectly transparent pane of glass. In others, the flattening effect is produced by the camera: he perfected a technique that did away with depth—no shadows, no foreground, no background, only surface. As one looks at these photographs, the eyes are caught up in a play of light refracting from glossy leaves and scarred fruit, a play that makes this surface vibrate with virile reality. The photographs insist upon this surface: this is all there is.

PLATE IV.

Figure 24. From
Joseph Rock,
*Notes Upon
Hawaiian Plants.*

Pittosporum halophylum Rock.
About one half the natural size.

After his death, the sole student in his 1919 systematic botany class at the College of Hawaii described his photography of herbarium specimens:

> If he were singing snatches of grand opera in French, German, or Italian when he arrived in the morning, the day would be a pleasant one. . . . In making pictures of herbarium specimens, he used a very slow emulsion plate. He would cut down the diaphragm, open the shutter, and walk off to attend to other matters; in a minute or so he would return, announce to himself that the shutter had been open long enough, and close it.[4]

The student doesn't actually say that Rock sang as he photographed, but he surely did. He had learned to love opera as a child in Vienna. With his teaching salary, he collected the Victor "Red Seal" recordings of Enrico Caruso and Dame Nellie Melba, which had begun appearing in 1902. Recordings of voices, absorbed into his body and memory, gave him the pacing with which to time the shutter. He would return to this fusion of photography and phonography in later years as he searched for ways to see his own body through alien eyes.

The history of photography is often narrated as a gradual accumulation of scientific mastery of time and space. The chronophotography of Étienne-Jules Marey and Eadweard Muybridge is often seen to mark a decisive moment in this process. Marey developed instruments to chart human movement photographically. He used a rotating shutter to make, on a single glass plate, multiple instantaneous images of a subject moving before a black background. His devices tended to directly detect and record motion, eliminating human mediation: he "wished to exceed the human senses, to correct them by detouring around them, thus eliminating human intervention, observation, and error."[5] Muybridge achieved his motion studies of racehorses by removing the human body from contact with the photographic apparatus. He placed fifty cameras along a racetrack and used trip wires to trigger electronic shutters as the horse galloped past. Later, he used similar automated process to do studies of human bodies, nude and clothed, walking and working.

Scholarship on Marey and Muybridge often sees their work as beginning a long process in which the photograph apparatus gradually freed itself from mediation by the human body, while subjecting the body to mechanical measures of time and motion. Photographic technology is seen to produce an extensive, metrical time which becomes the measure of movement, breaking movement into segments, each of which has the same form and speed as every other. This is then reflected back to perception through pictures and moving pictures.[6] For Rock, to ignore the shutter timer and time his photographs "by feel"—or, as I have claimed, by voice—was to sidestep this narrative of progressive mechanical abstraction, already, by the end of the nineteenth century, attached to the camera. It raises the possibility of a different history, a history of human bodies, voices, and temporalities becoming intimate with the technologies of mechanical reproduction, fusing with them, and shaping their rhythms.

After a scant five years in Hawai'i, he published a masterpiece: *The Indigenous Trees of the Hawaiian Islands* (1913). It described more than three hundred native Hawai'ian trees, including seventy-two new species, varieties, or forms, and it included 215 full-page photographs. He tried to present three photographs of each tree.[7] The first were identical in style to photo-

PLATE 66.

ACACIA KOA Gray.
Koa.
Showing trunk, bark and flowering branch; near tree-molds, Kilauea, Hawaii; elevation
4000 feet.

Figure 25.
Acacia koa,
with hat to
mark scale. From
Joseph Rock,
Indigenous Trees
of the Hawaiian
Islands.

graphs in earlier publications: twigs, leaves and fruits arranged on white cards, sometimes with rulers along the bottom or side, photographed with rigorous clarity. The second showed a section of the tree trunk, usually with a flowering or fruiting branch pinned to it. Each included a measure of scale: his knife, used to pin the cut branch against the trunk, or simply stuck horizontally into the bare bark. It was always the same knife, about eight inches long, with a sharp tip and a cross-hatched handle. The third set of photographs, of the entire tree, included a similar measure, usually his soft, broadbrimmed hat lying against the base of the trunk, sometimes his horse and empty saddle. It is significant that these marks—knife, hat, horse—were extensions of his body. It was as though he was explicitly, if intuitively, rejecting the idea that a photograph of a plant specimen is merely "objective," merely a portrait of a plant, and suggesting that it is also a self-portrait.

In some photographs of the full tree, his horse is held by a guide, face obscured by his hat. In many, the measure is a Hawai'ian guide alone. The guides—native Hawai'ian ranch hands whom Rock borrowed from their employers one day at a time—crouch or stand in rumpled work clothing, facing the camera in formal poses. Their faces are always difficult to discern—darkened, muted, or obscured. These photos contrast markedly with the few where a Euro-American replaces the Hawai'ian. In these, the features are clear, and the men stand in attitudes that place them in relationship with the tree: looking up at it or leaning against it. In plate 68, a white man stands in jacket and tie by a gigantic *Acacia koa,* his right hand resting on the tree trunk. A shaft of sunlight dapples the trunk of the tree and illuminates the man's face and white shirt. Behind him, up the hill, is a Hawai'ian man, his body melting into the undergrowth, his face in deep shade. The photo was taken by a rancher named R. S. Homer: none of Rock's own photographs in *Indigenous Trees* include more than one man. Still, it illustrates the minimal conventions of race within which Rock was working, in which it was natural, even obligatory, to render a Hawai'ian worker anonymous. In Rock's photographs, his faceless guides are placed in a frontal relationship to the camera but no relationship to the tree; their obstensible function is to show scale, replacing ruler, knife, hat, or riderless horse. But emptied of features and personality, they act in these photos exactly the same way as knife or hat, as placeholders. They mark Rock's absence from the photograph while mirroring, with their straightforward stares, his presence behind the camera.

He regarded his publications as extensions of the herbarium. If recordings of arias offered him a means to integrate his bodily sense of time with the photographic apparatus, mediating specimen and photograph, the spatial resources of the herbarium gave him ways to align body and memory to the island landscape, mediating landscape and text. His student in 1919 described the herbarium room that doubled as his classroom: "Because there were few cabinets, bookshelves, and other working facilities, herbarium sheets and books were in piles all over the room. A particular specimen of *Plantago princeps* from the bog on the summit of Mt. Eeke, Maui, was the third from the top in the pile under the table in the corner. He gave the librarian strict instructions not to rearrange the books when she came in to take inventory; if she did, he wouldn't be able to find what he needed."[8] Just as he eschewed the shutter-speed timer, he did without the filing technology of a conventional herbarium. The room was an extension of bodily space and bodily memory, allowing him to coordinate this felt space with the com-

PLATE 68.

ACACIA KOA Gray.
Koa.
Showing straight growth of bole in wet or fern forest, near Volcano Kilauea, Hawaii;
elevation 4000 feet.

Figure 26.
Acacia koa, with white companion and Hawai'ian guide. From Joseph Rock, *Indigenous Trees of the Hawaiian Islands*.

plex double map of landscape and taxonomy that was the topic of his botanical publications.

FILTH

In 1922, having left Hawai'i and traveled to Burma and Assam in search of chaulmoogra trees, he walked from Siam into the Tai Lü (Shan) state of Sipsong Panna in southern Yunnan, in the employ of the U.S. Department of Agriculture. With two medical missionaries named Mason and Baer, a Lao interpreter, six horses, five porters, and Boomah, his "boy" from Siam, he walked to the tea-trading towns of Simao and Pu'er and up the valley of the

Black River (Heishui He). Everywhere he looked, he saw the dead, recording each corpse, coffin, or graveyard in his diary. Often the landscape seemed divided into parallel zones for the living and for the dead: "Left Meng Hua Ting early . . . To our left is a huge graveyard, the population of Meng Hua of bygone days."[9] "To my left a bleached skeleton of a man, a robber shot and left to rot. On the slopes to the right is the village of Wuen Yaul."[10] He complained about the towns incessantly: "flies by the millions; one can't eat properly; it is to me a great plague, these flies."[11] He much preferred to live, lunch, and sleep amongst the dead. Before arriving at any substantial town he would send a soldier ahead of him to the official there with a letter asking him to prepare a temple for him to stay in. The temples he preferred were on the outskirts of towns, surrounded by graves.

In Dali, he met Forrest who took him to Nvlvk'ö and helped him rent a house there, deciding that, since Rock was after trees, their interests did not clash, "though," he commented, "I cannot say I have any particular love for Americans."[12] After this, whenever Rock was in Nvlvk'ö, Forrest avoided going there. In September, a party of men from the village took Rock to Tengyue and the Tai polity of Zhanda. They went to Sadon in Burma, then returned to Nvlvk'ö on a northerly route, crossing the great gorges of the Salween and Mekong. He spent the summer of 1923 camping on the Yulong range with Boomah and eight men from Nvlvk'ö. That October, having fired Boomah, he headed west again, with a party of Nvlvk'ö men, to the Mekong valley. At the Catholic mission at Cigu he picked up Ganton, who, three years before, had explored Tsarong for Forrest. The Tibetan Catholic botanist guided Rock's party past the great sacred mountain Khawa Karpo to Forrest's former camp at Tsiriting, over the Dokar La, to Mengong (sMan-khang), the capital of the Tsarong region, and over the Si La to the Salween. This was all Forrest's dreamland, which Ganton, Zhao Chengzhang, and the Nvlvk'ö collectors had explored in his stead.

Having a body, moving through the world, seeing it, being seen in it, he often found almost unbearable. This is what I find most appealing about him. He had terrible faults, but he was sensitive to the abjection of bodily being—his own and others'. His desire to find solutions and escapes, practical and fantastical, to the abject conditions of bodily being were his furies, compelling him relentlessly through the Sino-Tibetan borderlands. He came to this engagement with abjection gradually. At first, it was merely by observing "filth" and being disgusted with it. From the moment he crossed the border from Siam, nearly every page of his diary takes up the topic of filth. Here is a typical entry, made in the town of Nanjian, en route from Sipsong Panna to Dali in April 1922: "Today was market and crowds were on the open spaces outside of the town, peas, beans, corn and sugar cane were the only articles of which there were quantities, pigs wallowed in the large stinking pools surrounding the town, myriads of flies hovered and buzzed about quagmire, Chinese dishes, sugar cane, pigs, and the dirty faces of the children encrusted with filth and skin diseases. We bought some sugar cane to be sent to Washington."[13] He tried to hold himself apart from filth, avoiding markets, skirting towns, sending little novice monks away while eating lunch at a monastery, nearly thrashing a "dirty Chinaman" who "wiped his hands on my nice American flag."[14] But of course it proved impossible. For one thing, the very roads beneath his feet were covered in filth. In the town of Shuizhai, on the road from Dali to Tengyue in October 1922: " . . . like every other Chinese village no more than a filthy piggery; pig stables in America are a beautiful parlor in comparison to a Chinese place of abode; instead of cleaning their roads they throw all the rubbish out, hold babies out the door over the roadside to relieve themselves and let the attending dogs clean them, this instead of "Onliwon" [a brand of toilet paper]. In the middle of the road a pig is cleaned and all the dirt and filth remain in front of the house, people stepping over or in it."[15] More often, it was not immediate physical contact that immersed him in filth; it was the exchange of glances. His own gaze he armed and protected with his camera. This was the case in his very first diary entry about filth, written before he even crossed into China, in a small Hmong village in northern Siam: "The dirt and filth is indescribable. The people wear dark blue home woven cloth with long narrow blue aprons in front. . . . they never wash their faces or bodies . . . several of them were idiots. . . . Took a picture of a spinning wheel; the girl working on the wheel ran away at our approach as did all the women, they hid themselves in their houses and refused to come out."[16] But not everyone hid; most people looked back, and often it was he who had to run away. His instinctive response, sometimes after running, was to battle it out with the most potent weapon he had: his camera. In the spring of 1923, in an Yi ("Lolo") region of north Yunnan, he visited the market town of Wannian Jie. His diary details a complex, running battle of gaze and counter-gaze:

> . . . the dirtiest wares as brown sugar exposed on the dirty ground, tobacco leaves, pigs, chickens, silver ornaments for the tribes-women, rice, blue cotton cloth, . . . strings of cash . . . mules and horses standing in the middle of the filthy street, pigs squealing as they are stepped upon by the pedestrians; beggars, their clothing made of tiny bits of rags picked up on the street and stitched together haphazardly so that every square inch of the surface exposes some part of the skin; these poor wretches with hair like a witch's . . . matted with filth . . . their finger-

nails like claws . . . barked at by every mangy dog . . . such rub shoulder to shoulder with the gentry, such as, it is in their abode of filth.

This place boasts a small temple and hither we went . . . and clouds of dust announced that the whole of the market behind us is descending on us like a mad riot. . . . Women and children shrieked for fear at the sight of me and the men fought for the small space in front of the temple. I took three photos of the mob . . . I went to the dark narrow room of the temple and closed the gates . . . . in every hole . . . [in the latticed door] . . . there poked a nose, or some eye strained sideways to get a glimpse of me.[17]

Seeing and, especially, being seen drew him closer into the world around him, too close to be comfortable, close enough to be *visceral*. In *The Visible and the Invisible* (1968 [1964]), Merleau-Ponty elaborated a vocabulary to describe the lived body as a worldly object among worldly objects. The body, in his words, is "a being of two leaves, from one side a thing among things and [from the other] what sees those things and touches them." A body is a worldly thing "in a stronger and deeper sense" than are insensible things: seeing is an encounter in which one's own body becomes visible to oneself as though from the position of things seen, things that are outside the body, but that may also be a part of it, "even entering into its enclosures."[18] He quoted a painter who felt as though the trees were looking at him. It is not that the trees have eyes; it is that vision places one in the midst of things; one always sees oneself looking as though from the place of the visible. There is, he wrote, a "fundamental narcissism to all vision." This is yet more the case when the visible is not trees but other bodies, with their own eyes: "As soon as we see other seers we no longer have before us only the look without a pupil, the plate of glass of the things with that feeble reflection, that phantom of ourselves they evoke by designating a place among themselves from whence we see them: henceforth through other eyes we are ourselves fully visible; that lacuna, where our eyes, our back, lie, is filled, filled still by the visible, of which we are not the titulars. . . . I appear to myself completely turned inside out under my own eyes."[19] Merleau-Ponty made these observations in his efforts to understand the carnal body as involved in the "flesh of the world." Sensible bodies are, according to him, "exemplary prototypes," or "remarkable variants" in a "massive corporeality," in which the sensed always coils back upon the sensing body.

Seeing and being seen made Rock aware of his deep involvement in the flesh of the world—and its viscera, its filth. As Javier Sanjines remarks in a reading of Dario Antezana's painting *Complicidad*, of two massive, slimy bureaucrats, with satiated expressions and extended, semitransparent guts,

Figure 27.   Market crowd in Dali, Yunnan, 1922. Courtesy of the Arnold Arboretum.

Merleau-Ponty's analysis of the "flesh of the world" is hardly carnal enough. For all its virtues, Merleau-Ponty's extended descriptions of the lived body stay with its visible and palpable surfaces, neglecting its innards, and the flows that pass in and out of its viscera. "Flesh," writes Sanjines, "is mass . . . the stomach, buttocks, cheeks, tongues, palms, eyes. . . . Bodies are slimy surfaces that sweat, the hairs that grow in total disorder, the stomach fluids that exude gases." Bodies exhale shit, piss, mucus, fluids, microbes; they spread viruses, toxins.[20] This viscerality involves the lived body as a thing among worldly things as much as does seeing and being seen or touching and being touched. For Rock, being seen drew him into the "massive corporeality" of the world. But this was not merely the corporeality of his sensing gaze coiling back upon itself through the eyes of others. It was a visceral corporeality, of bodily flows and wastes, in which bodies shared their substance not only through touching and seeing but also through ingestion and infection.

Rock experienced the unveiling and projection into the world's flesh that the eyes of others exercised upon his body as an assault rather than as the joyful opening up into the world that Merleau-Ponty imagined it to be. At bay before that little temple at Wannian Jie, he turned on the crowd and fired back three times with his camera. The camera did not effect an "objectification" in the usual sense. At the moment, there was no object—no photograph: he was looking through the shutter. Instead, the camera acted on his own body. For the moment, it replaced his body, forcing the crowd to see it only through the lens. In this way, it opened up again those dark "lacunae" that had been filled in by all those eyes. If he had been "completely turned inside out" under those eyes, the camera let the portions of his body previously unseen be unseen again, allowing him to cast off the weight of eyes on his back, his legs, his torso. It made him a thing in a world of objects again, gave him back that "plate of glass of things," that "feeble reflection" or "phantom" of himself that those things allowed him. In short, it was an attempt to reestablish the relationship with himself as a visible being with which he felt most at home, displayed in those photographs of Hawai'ian trees, for instance, in which he appeared as a mark—a knife or hat—in a world of surfaces.

Back in the market, in that tumultuous battlefield of gazes, were the beggars, their skin stitched over in a mosaic of bits of cloth, their hair like witchs' hair, their fingernails like claws. In this field of visibility, in Merleau-Ponty's account, other bodies, like mine, are both worldly things and sensible beings; they too have the inverse and obverse sides of the seen and unseen. But not for me. Though my involvement in the flesh of the world may give me full access to their sensed side, the side of them that is "object," my access to their sensing side is partial and variable. I can discover it only to the extent that their senses touch upon me—that I feel the palpations of their eyes—and even then imperfectly. For Rock, the beggars had no eyes. In great contrast to everyone else, they seemed insensible of him. They were a kind of a hole in the field of visibility that enmeshed and tortured him. But at the same time, they were full participants in the field of corporeal flows—as they scrabbled with the dogs for grains of spilled rice. They were bodies, with all the mass and all the flowing viscera of bodies, yet withdrawn from that network of gazes that make bodies social. This was the status of the abject for Rock: sensing flesh reduced to insensibility—visceral meat. His attitude towards abjection would come to resemble that which Deleuze ascribes to the artist Francis Bacon—"an immense pity for meat."[21]

He soon learned to focus his revulsion on a particular form of excreta—the stickiest and slimiest kind. In December, 1924, he was on his way from Yunnan Sichuan and Gansu. An assortment of travelers met along the way had sought the protection of his large military escort: "We stopped for lunch

Figure 28.   Crowd peering into temple in Gatoubu, Gansu, 1925. Courtesy of the Arnold Arboretum.

among some trees near the streambed while the soldiers and Chinese travelers regaled themselves in the fly-ridden food stalls with their mangy dogs and the usual filth accompanying such impossible places. The continuous spitting, cleaning of the nose with the fingers and wiping them either on rocks, trees, shoes or clothing or still better table and chairs or the nearest thing handy, the filthy muddy road, the squealing pigs, and one has a picture of the lunch stops on a caravan road in Yunnan."[22] It is difficult to show just how central mucus was to Rock's diaries. He wrote of it frequently, intensely, and at length. These diary entries were not merely the occasional expressions of disgust of a man fed up with traveling. They were a deep and sustained engagement with filth, a lingering, meticulous exploration of its trajectories from body to body, and around to the proximate world. In February 1931, he was traveling with the journalist Edgar Snow, whom he disliked. He was in a very bad temper, since Snow had awakened him at two in the morning, "scrabbling for something":

> We reached the rotten hole of a place called Hsia Kwan [Xiaguan] with its indescribably black oozy filth in which pigs wallowed, with the walls of the houses cracked and leaning every direction. The filthy populace is lousy and encrusted with the dirt of years, the so called gentlemen leaning forward and, taking their nose between two fingers, wringing it and then throwing the thus recovered contents of their filthy noses in a wide circle from them, and wiping the remaining slime with the palm of their hand in an upward direction or movement over their face and the hand on the nearest object, post, wall, chair, table, or in the absence of all the above on their shoes or clothing or rubbing both hands together until all the filth has been thoroughly rubbed into the skin; such is a mild description of unspeakable Hsia Kwan and its lousy degenerated populace.[23]

It appears that at times this investment seemed excessive even to him. He sometimes struggled to domesticate it, to frame it within more acceptable discourses. Of course the most readily available of these, fashioned with great energy to be adaptable for just such purposes, was the discourse of racism. He made, at times, ample, if anguished, use of it. His diary entry about Xiaguan continues.

> Missing links exist here by the hundreds, not single ones, the faces of the people remaining of the lowest type of apes, of a degenerated and idiotic type at that, with goiters which hang down their chests. They are . . . lower in mentality than a cow or water buffalo; . . . their lips cannot even hold slobber, and the slime and saliva runs out of their mouths like babes in arms; what a wretched race. . . . White people are depraved enough and silly to want to save the souls of such a despicable people. They have no souls, no more than lousy flea ridden dogs have. Why not let them steep in their filth and ignorance and leave them alone to let them rot in their bandit-ridden country. The best thing that could happen to the world would be a huge catastrophe which would annihilate that miserable selfish mean filthy degenerate race. O how I hate and despise them.[24]

His was a tormented racism, miserable, often hysterical. But at least it wasn't "scientific." He played at random with a few concepts—"degenerated type," "yellow peril"—but always in the most unsystematic way. He had no theories at all. He was contemptuous of missionaries who thought they were bringing the light of civilization to Chinese or Tibetans; he believed them to be moral worms. Though he sometimes thought of peoples such as Kachins or Tibetan nomads as "primitives," he erected no hierarchies. Instead, he kept his racism close to its roots—his revulsion at the "filth" of bodily emissions.

STAGING ABJECTION

On his journey to the Tai state of Zhanda in the autumn of 1922, Rock was riding his horse through a forest after finding a chestnut: "one of the loveliest trees I know . . . small leaves, a rich deep green, golden brown beneath, silvery when young." His horse nearly trod on a man lying on the forest path. He could not see the man's face from his horse, but when he dismounted, "death looked up at me from his staring eyes." As always, the abjection that compelled him was that in which a body became an unsensing object—meat: "From his neck to the center of his breast he was laid open, ending in a five-inch-broad wound, the skin and fat hanging in rags from him. He was ice cold and shivering. I felt his pulse and, when I touched his hand, he was terrified. He could not speak. He was out of his mind. . . . His terrible wound was running with pus, and in such a condition he was lying in the cold and wet 2 days and 2 nights."[25] The men from Nvlvk'ö accompanying Rock made a litter of bamboo and carried the man down the trail. They met some laborers in a rice field, and Rock paid them to carry the man five miles to the town of Longling. In a temple there, he cleaned and bandaged the wound and gave the man a clean shirt.

The laborers stopped in the market to eat, and there they ran into their employer. "He raised a terrible howl and cursed me up and down. What right had I to take his men from the field *wasting their time carrying that man to Lung Ling*." Enraged, Rock grabbed a big stick and, trailed by a crowd, marched to the man's shop in the market. "He was leaning against the wall, surrounded by friends, laughing, filthy as a pig and as fat. A soldier pointed him out. I made one dash for him, grabbed him by the throat, and pulled him out into the street, a multitude of people watching me." Rock pushed the man against the wall and screamed at him until he was hoarse. He then dragged him before the magistrate, "followed by the whole of Lung Ling." He said that the man had insulted him, and he insisted that he be punished. The magistrate sentenced the man to six weeks in prison. "Such are the people of this province," commented Rock. "God forbid that a similar fate should befall *me*."[26]

All the tumbling affects of compassion, self-righteous indignation, and racist disgust in this incident were subtended by a precise organization of gazes. On the forest path, Rock was captured by "dead eyes" that swallowed

Figure 29. *Castanopsis,* from near Longling, Yunnan, 1922. Courtesy of the Arnold Arboretum.

up his own gaze rather than reflecting it as objects did, a deep pit in the happy visibility of the world that he had just been enjoying as he collected his lovely chestnut. For all his horror of the man's terrible wound, he felt a certain competence in this form of relationship to another body. Drawn up close to this body, close enough that its wounds were the central objects of his encounter, "filth" was not an issue.

But then he found himself in the market, drawn again into the viscerality of that world, where the most significant social exchange seemed to be the spread of excreta among bodies. His response was to create a theater. Many historians have remarked upon the theatrical organization of the county courts of justice in the late Qing—forms which carried over unreformed into the Republic in remote, rural counties like this one.[27] Rock played up this theatrical effect: "I thought the people should see such a lesson," he wrote. It was a deliberate realignment of that network of gazes, just as when he had turned his camera on the crowd in the market at Wannian Jie. He faced all those eyes frontally, catching them with his own. He became again, to a degree, as an object in a world of sensing objects, seen by them, to be sure, but seen as the reflection of his own gaze. He sought to empty those gazes of their inherent ambiguity as sensed bodies and sensing beings. What mattered about them to him now was their eyes: the visceral corporeality that so frightened him melted away as they were reduced to their more purely sensing side.

It is true that these are relations of sensation. They are ways of organizing perception, one's own and that of others, of arranging lines of sight and of deciding what, and whom, may be heard. But they were also *social* relations in as full and solid a sense as relations of kinship or kingship. Rock's exploration of "slime and slobber," his capacity to be drawn into the dead eyes of abject bodies, his attempts to extricate himself from visceral exchanges with camera or stage—all this was a deep and serious struggle to negotiate a place for himself in the social world, moment by moment, and to find ways of acting that would make inhabiting that place bearable.

Composing himself as a subject in relation to the social world was a matter of working on the world, not just upon himself. Whenever the opportunity arose, he arranged the world into stages—geometries for aligning and focusing gazes. In May 1922, on his way from Dali to Lijiang: "On arrival at Peisumei, we found a dirty smoky school full of dirty brats so we decided to stop at a theatrical stage. A ladder was procured, and we climbed the stage, where we are now comfortably located, the caravan being below. I attended to the horse boy, who had very sore eyes, etc."[28] He had already found, only four months into his first sojourn in China, the scene that suited him. It had each of the elements that he would come to rely upon: the filthy crowd (here of children); the stage on which he situated himself, lining up the crowd's gazes; the abject body at center stage, which he, intimately, treats, and with which he empathizes as a meaty and insensible double of his own.

When in a town, whenever possible, he put up in a temple: the alternative, public inns, he found unbearable. He often remarked upon the little theatrical stage that one found facing the courtyard of every temple in Yunnan, used for performances of *Dianxi,* Yunnan opera. At times, he pitched his tent on this stage. More often, he set up a cot at the foot of the god icons—theatrical enough. He liked the backdrop the gods provided him, especially on evenings when he felt lonely. Often, he and the gods shared their stage. The dead were frequently present, in coffins lying in the courtyard or stashed under the little stage, and there were always graves just outside. In 1925, journeying from Yunnan to Gansu, he set up his tent inside a temple near Zhaotong to wait out a storm. "It was a lonely place; the only neighbors I had were a few graves. . . . One night, it was moonlight, wolves came to the temple door and howled for hours. . . . they [dug] up a newly buried man, and devoured him. All we found the next morning were a few blue rags and the open coffin."[29] In 1930, walking through west Yunnan, he and Snow avoided the "filthy hole of a place" that was the city of Jianchuan, "and climbed to the temple among the dead on the western hillside, where we used to stop to lunch at previous times, the dead being a thousand times preferable to a living Chinaman."[30]

When the bodies were living, they were abject. In a temple near Dali: "Ten soldiers are guarding the front of my temple, not outside, but within the courtyard and in front of the doors of the shrine. They are all diseased with skin troubles and ulcers. . . . Every one I have treated every day, and they seem to be grateful."[31] In these war-wracked times, the temples were often full of wounded militia. In 1928, passing through the town of Sanying, near Dali, he repaired to the temple for the god of wealth, where he had often stayed. The town had been attacked by bandits the day before. "What a dreadful sight I beheld! Some of the soldiers had their feet chopped off, others their legs; one sat on a platform with a rag on his head, and as I removed it his brain was exposed and pulsating. He died shortly afterwards. I did not have enough bandages, so I cut up my irreplaceable bed sheets and worked all night bandaging the wounded."[32] Abject bodies moved him when they shared his stage. On the road, in a marketplace, on the streets of a town, he was rarely compelled to treat diseases or bandage wounded limbs. (Even on that early occasion when he stopped to help the wounded man in Zhanda, he repaired to the temple to treat him.) Yet, bounded in this way, compassion was absolutely at the core of the social subjectivity he was arranging for himself. Corpses dug up and devoured by wolves, soldiers abandoned and at the point of death: these bodies were at the very threshold of the social. They were in the process of being purified of all subjectivity, becoming senseless—becoming meat. As he negotiated a social geometry within which he could establish his own subjectivity, Rock found himself pressing right up against this threshold. On his stage, with his gods, when he was not finding fellowship with abject bodies, he was often contemplating his own death. It was always (and quite unrealistically) the death of a stranger—forlorn, unmourned, torn by wolves, left to rot in the wasteland.

It was a diagram: an arrangement of gazes, actual and virtual, an organization of intimacies and distances, of written voices and enforced silences. To speak of social subjectivity as merely internalized is always an error, of course: one is social always through arrangements of relationships, some materialized, others imagined, involving other beings, human and inhuman. Rock's difficulty was that he had to carry his sociality about with him, like a snail carrying its shell—or rather a hermit crab which, at each resting place, finds another shell and collects objects to arrange about it. He was not without resources in this endeavor however. He had his machines.

VIRTUALITY

Rock's most eventful journey was a three-year-long expedition to Gansu/ Amdo (described in chapters 7 and 8). It was amply funded by the Arnold Arboretum. He hired twelve young men from Nvlvk'ö, most of whom had already accompanied him on his wanderings around Yunnan. For the first time in China he bought a top-of-the-line portable Victor gramophone, invested in a collection of Victor Red Seal recordings, and assigned a mule to carry it all. In December 1924, a day after setting off from Kunming, the party stopped at a village called Yanglin. There was a harvest festival at the temple, so Rock decided to stop in a public rest house: "The window consists of iron bars; the floor is dirt; there is a large upstairs, but it was so filthy that it would have taken weeks to clean it. The place is inhabited by idiots. To cheer myself up I unpacked the phonograph and played Caruso's favorite operas. A large crowd gathered at the window outside."[33]

Three months later, the party was traveling through the dusty plains of Gansu—"nothing but stones and stones." They reached the hamlet of Goutou Ba, "a miserable conglomeration of houses built of loosely piled-up rocks." The night before, he had slept in an inn. "The smells were very bad, and as the mule men continuously opened the doors into the stable, the stench became unbearable." He had ordered a soldier to tell the landlady, "a nasty woman" who kept "chatting and chatting, true to her sex," to shut up, so he could write. To avoid a repetition of all that he pitched his tent in the village, among the scattered houses: "Today I played the phonograph in my tent; it attracted a very peaceful and well-behaved crowd which seemed very amused." After this, his irritability melted away, and he wrote that it was the countryside, *"not the people"* (Rock's emphasis) that was inhospitable.[34] In a few days, they stopped in the village of Qingshui, where the tiny temple was too small for his bed: "We go to an inn . . . I stop in a room containing the family shrine. . . . I opened the phonograph and gave the people of Chingshui a concert; they . . . stood in amazement amassed around the music box listening to Caruso's Celest Aida, the Quartet of Rigoletto, etc. I took several pictures of the mob while they listened to the phonograph. I then took the phonograph to the top of the roof and photographed the throng that gathered on the roofs of the houses across the narrow dirty street."[35] It became a habit. During this expedition, he played his Victor on village streets, in the grasslands, in monasteries—whenever he felt lonely or miserable.

In the spring of 1926, the party set off from its base in the monastery town of Chone towards the Labrang monastery and the great mountain that

Figure 30.   In Qingshui, Gansu, listening to Caruso on Rock's gramophone.
Courtesy of the Arnold Arboretum.

Rock had made the expedition's goal, the Amnye Machen. From Labrang,
they were escorted by a group of *drokwa*, nomads.

> Yesterday I played my Victor to an audience of nomads of the Sokwo
> Arik tribe. They listened and listened and still wanted more . . . I had
> taken a number of pictures of these sons of the grasslands, each with
> his scalp lock sticking up from his otherwise bare head, as they sat
> around the Victor listening to Caruso and Melba, Plancon, Titta Rufo,
> and others, screaming with laughter at the most pathetic passages—
> "Boheme" or "force del destino." They said what funny clothes I wore
> and what a funny hat. The son of the Chief (now dying of paralysis)
> looked the picture of an African negro.[36]

Rock's photographs show twelve men and boys and a little girl sitting on
the ground around the Victor, perched on a packing case. Their attention is
split: some attend to the gramophone, others to the camera. These photos
might be contrasted with his other pictures of his escort. When he was able,
he separated them into small groups, posed them against a smooth background, and shot vivid frontal portraits. An autodidact in all his enthusiasms,
he had a facility for investing in the emblematic forms at the roots of disci-
plines and exaggerating their formal properties to surreal perfection. He
coupled this with an autodidact's tone-deafness to disciplinary flux. His por-

Figure 31.   Sokwo Arik nomads with Rock's gramophone. Courtesy of the Royal Botanic Garden, Edinburgh.

traits have roots in the anthropometric photography of the nineteenth century, from which had grown the massive trunk of comparative ethnological portraits.[37] Like those, Rock's portraits are of "types" (a common label in the Royal Geographical Society archives) oriented to the camera and severed from their social or material context. But, like his specimen photographs, they exceeded the form of "type" photographs, by exploring the particularity of his subjects, insisting upon their individuality.

Much has been made of the capacity of photography, frequently observed, "to induce an imaginary *rigor mortis* through its capture of the frozen moment," as Christopher Pinney puts it.[38] The most precise reflections on this connection remain those of Barthes, in *Camera Lucida*. Looking at the image of Lewis Payne awaiting execution, Barthes reflected that the "punctum"—that which stings, or wounds—in this photograph "is: he is going to die. This will be, and this has been." The photograph is a real moment, a moment that happened, a man who is about to die, and who is now dead: the sting is the "inevitability of one's own mortality."[39] The Sokwo Arik nomads in Rock's photograph are not condemned to die, any more than you or I, but they are dead now, as we will be soon enough. Rock often obsessed about this: his portraits were all, in his view, of people and ways of living that would very soon "disappear from the surface of the earth."

Figure 32.   After the concert. Reproduced by permission of the Harvard Yenching Institute.

As Paula Carabell reminds us, such reflections depend upon an "essentialist theory of image making," the idea that referent and representation are physically linked, that the photograph "partakes of the object's identity," justified by the fact that, in photography, light physically traveled from the object to make an imprint upon the film. "Photography is an imprint, a photochemically processed trace like fingerprints or a death mask."[40] In an essay that describes how photographs are closer to fetishes than film, Christian Metz, a founder of the psychoanalytic approach to cinema, comments on the consequences of this central feature of photography, which may be called its indexicality. "What is indexical," Metz writes, "is the mode of production itself, the principle of the *taking*." Compared to film, "photography . . . remains closer to the pure index, stubbornly pointing to the print of what *was* but no longer *is*."[41]

> The snapshot, like death, is an instantaneous abduction of the object out of the world into another world, into another kind of time . . . the photographic *take* is immediate and definitive, like death and like the constitution of the fetish in the unconscious, fixed by a glance in childhood. . . . Photography is a cut inside the referent, it cuts off a piece of it, a fragment, a part object, for a long immobile travel of no return. Dubois remarks that with each photograph, a tiny piece of time bru-

Figure 33.   He Xueshan, one of Rock's assistants, with head of blue sheep.
Courtesy of the Arnold Arboretum.

tally and forever escapes its ordinary fate and thus is protected against
its own loss. . . . Not by chance the photographic act (or acting, who
knows?) has been frequently compared with shooting, and the camera
with a gun.[42]

One does not have to invest in this "essentialist" idea of the photographic
image to observe its effects on those who might. For Rock, photography
could, in nearly every way, substitute for collecting. These were short steps:
from collecting pieces of plants, to photographing those pieces as specimens,
to making photographs of people that functioned, in all important ways, like
specimens. This kind of photography was indeed "a cut inside the referent,"
liberating fragments of the world from their fate, preserving living beings
as dead specimens. Rock's portraits imitate his specimen photographs in all
the ways I have already noted: the careful arrangement of the object against
a featureless backdrop, the obsessive attention to detail, the careful flattening. The only difference is that the mark of scale, the empty mark of his
own presence, is absent from the portraits. But not really. He is present there,
as a reflection in his subjects' eyes, an empty reflection, devoid of content
as the knife or empty hat. By rendering his subjects as specimens he made
them into things again, or as close to things as was possible. And these things

Figure 34.   Concert at Dzangar monastery. Courtesy of the Royal Botanic Garden, Edinburgh.

gave him back his own body not as a complex social organism involved in the visceral flesh of the world but as that feeble phantom of himself that his vision reflected from the world of things allowed him.

This relationship was easy enough to create with individuals or groups of two or three. But it was much more difficult with crowds, even small ones. And it appears that this is when the gramophone became useful. The disembodied voice suspended the heterogenous interactions of social life that he found so irritating or threatening (the chatting landlady, the stinking latrine) turning all ears toward the box, all eyes toward the camera. On their way to the Amnye Machen, the party visited the grassland monastery of Dzangar: "I invited the lamas to my quarters and there entertained them with a concert furnished by Caruso, Melba, Titta Ruffo, Plancon, Kreisler, Heifetz, etc. It was a most peculiar scene, an old Living Buddha sitting in my camp chair and intently listening, . . . door and window packed with lamas, also the roof of the building surrounding the courtyard."[43]

It was a cliche, of course. Juxtaposing phonographs and "primitives" had become an instinctive gesture for American travelers playing out fantasies about the power of American technology to penetrate the globe's remote places. A 1910 *New York Times* article titled "America's Advance Guard" summed it up: " 'I have been all over North and South America and Africa,' remarked a New Yorker, 'and was surprised to find the phonograph and sewing machine everywhere. . . . I have heard Caruso singing in a Klondike dance hall, in the middle of the long arctic night, and Melba entertaining a crowd of naked niggers at Lagos on the west coast of Africa. I think these incidents gave me an even greater shock than the squaw with her sewing machine."[44] Such scenes were a staple of expedition films—a popular genre in the 1920s—particularly of films about Africa and the Arctic. By 1929, a *Times* film critic could write, "It would seem that to show a film of Africa the following ingredients must be incorporated: A shot of natives listening to a phonograph; native women with pickaninnies slung on their backs; . . . scenes of tribal dances with subtitles comparing them to the Black Bottom and Charleston."[45] Yet there is a distinctive difference between most cinematic phonograph scenes and Rock's photographs. Expedition films showcased a "primitive" sense of wonder at the disembodied voice; the focus of the concert audience's attention was nearly always the phonograph. But in Rock's photographs, the audience is not attending to the phonograph, and their faces do not obviously display wonder or amazement. Instead, silent listeners turn their gazes, emptied and unified, directly on the camera. Nor do Rock's *accounts* linger on the wonder of the audience, except to comment on how their astonishment drew their gazes to the box, then to the camera. And to him, the magic of the phonograph did not depend on the audience's "primitivity": nomads, ordinary villagers, or literate monks would all do. What was important was that they be rendered speechless, their gazes emptied, and their attention led to the camera lens.

Our bodies are given to us by the gazes of others, in which our own gazes have their origin; they are given to us by the voices of others, from which our own voices arise. Those gazes and voices clothe us with our own flesh. To have a body in this sense is be part of that portion of the world that is added to it by the vision of others; it is to be subsumed in that coiling of vision back upon the world's "massive corporeality." The sliminess of the world was evidence for Rock of this inevitable sociality of corporeal being. The sputum that flowed over tables and floors, the mucus stuck to posts and walls, the feces that flowed in the streets, were shared corporeal substance. They were evidence that the lines between bodies and the world were indefinable; that his own body was given to him by gazes, voices, fluids, microbes, of which he was neither the origin nor the titular. The unbearable sign of this, for Rock, was the contingency and heterogeneity of social experience. It is whenever this heterogeneity was paramount—when the market goods were jumbled together, when the crowd was looking in various directions and speaking of various things, rather than being "peaceful and well-behaved"—that he began to rave about slime.

Camera and gramophone were powerful instruments for dispelling contingency and heterogeneity. I suggested above that scholarship on photography has encouraged us to view the camera as an alienated extension of the senses. But Rock's problem was not Étienne-Jules Marey's, Eadweard Muybridge's, or Walter Benjamin's. It was not to divide time in order to perceive what the unaided eye could not. It was not to add appendages to the body to extend the reach of the senses in time or space. It was not to delve beneath the gaze or to open up an optical unconscious. It was to replace the body.

In Hawai'i, using the pacing of Caruso's arias to time his camera's shutter, he had relied upon the mechanism of the gramophone to insert his body into the mechanism of the camera, integrating body and camera more fully. In west China, he went one step further. If a body is a social relation, he was replacing the terms of that relation with gramophone and camera. He was using the voice of the gramophone to organize the voices of others, using the eye of the camera to align the gazes of others, so that those voices and gazes would give him back a virtual body, composed of nothing more than camera and gramophone. The *virtue* of this virtual body was that it could encounter gazes and voices while being raised above the filth and the slime on packing cases and tripods. His carnal body, holding the camera, was reduced to a tripod—a support for this purified eye.

·   ·   ·

Social relations do not exist apart from their materializations. Wandering west China, Rock found social relations taking material form about him at every step—along roads, in marketplaces, in temples, among bodies gathered or dispersed. These material forms caught him up bodily, inserting him into various forms of relationships, with or without his conscious participation. His efforts with camera and gramophone were attempts to reshape these processes of materialization, moment by moment, finding ways to inhabit the world, just as he did many years later, in his photograph of his younger self and Edward Muir, listening to Caruso on the gramophone. And these efforts always included reshaping his own body.

Gradually he discovered ways to use the resources available to him to reorganize the world, to arrange a social geometry that he could live with. An arrangement of gazes that allowed him to present himself frontally to a world that gazed back at him, surface reflecting surface. An arrangement of voices through which he could silence the voices of others, filling the world with a single, clear voice and placing his body at its source. The creation, through these arrangements, of a virtual body, fully frontal, fully visible, raised above the viscerality of the social world, behind which a private self could continue to exist, protected from the gazes and excreta of others. The mobile habitation he devised to produce and protect this virtual body included at its core compassionate fellowship with different, abject bodies. He made this fellowship possible by strictly controlling the conditions under which he encountered these bodies—depriving them of gaze and voice, cutting them out of those social networks through which bodily excreta spread, cultivating in himself an "immense pity for meat."

I have noted that he had terrible faults. And I have not, perhaps, treated him so far with a very sympathetic eye. But I want to suggest that his acts of caring for a few of the dying victims of this immensely violent time, though rigorously limited by a social diagram of his own devising, were difficult ethical acts, made possible by the limits set by that diagram. Rocks's diaries overflow with evidence that he had an immensely difficult time moving through the populous roads and towns of west China, where he found himself caught up in a network of gazes that drew him repeatedly into the viscera of the social world. Every step of his journeys involved negotiating a stream of affective responses to these difficulties—mainly varied shades of fear, disgust, and horror. With his camera and gramophone, Rock found ways to manage that stream of affect, distancing himself from others and forcing them to encounter him on his own terms. But he also found ways to connect with the most abject of others: placing them beside him on the stage he devised. They were certainly not his equals on this stage, but they were potentially so—for, once they were there, he could always imagine himself in their ragged clothing, divested of gaze, voice, and sociality, and slipping towards death.

6.   Lost Worlds

An archipelago of tiny states was scattered through these border regions. Most had paid tribute to China, Tibet, Burma, or Siam for centuries while governing their own affairs with little practical outside interference. Within China's twentieth-century borders, a surprising number survived into the Republic, though all were subject to new policies bringing them increasingly, if sporadically, under the supervision of regional governments. From the perspective of the Chinese state, they were what remained of the once extensive system of native hereditary chieftanships *(tusi zhidu)*, instituted during the Yuan and reformed repeatedly in the Ming and Qing, through which the Imperial state had exercised its sovereignty over the peoples of the border areas. From local perspectives, they were autochthonous principalities with rulers whose ranks and powers, sustained through deep and carefully cultivated genealogies, were semidivine. Many had existed for centuries and had developed elaborate systems of rule, rooted in ritual, intricately balanced, and unusually effective, particularly in comparison to the often stateless chaos that surrounded them in the early twentieth century.

Rock became infatuated with a few of these borderland states. Their institutions of kingship, contemporary and historical, came to center his geographic sense of the region. They became zones of heightened intensity for him—botanical, historical, and textual—where knowledge in each of these areas coalesced to form dense webs of interrelation. And they became centers of an affective geography as well: bright islands of peace in a great, flat swamp of misery. It was as though the majesty of monarchs—their capacity to focus time on their persons through their genealogies and to concentrate space on their persons through their powers over peoples, armies, funds, and territories—gave him a purchase on this landscape, emotional and intellectual. Their mechanics of rule, ritualized and theatrical, provided happy so-

lutions to his own struggle to rearrange the world. A few of these states became very important to him: the small, multiethnic state of Muli in southwest Sichuan, the smaller polity of Yongning on the Yunnan-Sichuan border, the Tibetan principality of Chone in south Gansu, the monastery state
of Labrang on the grasslands of Amdo, and, most significantly, the historical kingdom of the Mu dynasty of Lijiang.

The semi-independent polities that lasted until the twentieth century in
the southwest might be grouped into two categories. Along Yunnan's borders with Burma, Laos, and Siam were the Tai states, known to British col-
onizers of the region as the Shan states (and the subjects of a famous study
by Edmund Leach). Within China's borders, some twenty-four of these states
survived into the twentieth century, subject to increasing pressure from Qing
and Republican governments. In the Tibetan region known as Kham, consisting of the northern portions of the four valleys of the Ngül (Salween),
Dza (Mekong) Dri (Yangtze) and Nya (Yalong), some twenty-six polities
had maintained relative independence from both Lhasa and China before
1905, when Qing armies, under Zhao Erfeng, had invaded the region. Zhao
Erfeng subsumed many into China's regular system of administration, but
a few survived. One survivor was Muli, a small, multiethnic state with a
system of rule unique among Khampa polities.

Rock and his party of young men from Nvlvk'ö traveled to Muli several
times. Their first visit was in 1924, shortly before they began their adventurous journey to Gansu. They visited again in early 1928 after returning
from Gansu, attempting to secure permission from Muli's head of state to
travel to the Konka Ling. They went back to Muli upon returning from that
range, and they visited again after a second tour of the region near the range.
They visited the polity again at the end of 1928, during an unsuccessful attempt to travel north to explore the Minya Konka range. And they visited
a final time in the summer of 1929 after their successful tour of that range.

Almost immediately upon arriving in Muli for the first time, Rock immersed himself in this little state's ritualized theatrics of power. He entered
into a photographic collaboration with the state's supreme religious and political authority, Xiang Cicheng Zhaba, or Tshul khrims grags pa, charismatic
heir to a long dynasty of ruling lamas.[1] This collaboration continued and
deepened during later visits, serving the different ends of monarch and explorer, the theatrics of state power and a personalized theater of subjectivity reaching happy, if temporary, accord. Soon enough however, Rock, looking down, discovered one of the hidden circumstances of monarchical
splendor: the abject inhabitants of a dungeon beneath one of the monarch's
palaces. He entered a brief collaboration with them as well, his driving need

to bring them to the surface allowing them a slightly less painful existence and a few moments in the sunshine in front of his camera. This chapter examines Rock's collaborations with Xiang Cicheng Zhaba, the Great Lama of Muli, and with eight of his prisoners, seven accused of murder and one of harboring a thief. Perceptual technologies are very often analyzed as instruments of power, and, as we have seen, Rock used his Kodak Autographic to create power-laden diagrams to address the problems of difference and subjectivity with which he found himself confronted daily. But his camera was also an ethical instrument, and the diagrams it produced also made room for limited acts of compassion.

THE STATE OF MULI

The first botanists from Nvlvk'ö to venture into Muli, in the summer of 1918, found that they could not roam the hills unsupervised as they were used to doing in Yunnan. "The region is absolutely under the sway of the lamas of the yellow sect," Forrest fulminated after they reported to him on their return to Nvlvk'ö. "Their lamaseries are on every mountain slope and my men, being strangers, were treated with suspicion wherever they went, bullied and hindered in their work, and even threatened with death if they dared attempt explore certain portions of the ranges!"[2] Before sending the men back, Forrest prudently sought letters of introduction from Mu Yin, the heir of the Mu chiefs of Lijiang, to Muli's supreme authority, the Great Lama Xiang Jiangpu Zhaba (Byams pa grags pa). Jiangpu Zhaba allowed the next party into Muli, but he kept a close eye upon it and subsequent explorers. When three collectors from Nvlvk'ö guided Francis Kingdon Ward to Muli in the summer of 1921, Jiangpu Zhaba received him hospitably.[3] The next summer, when the Great Lama refused Ward permission to go to the Konka Ling range to the northwest of his realm, Ward attempted to steal out of Muli towards the range at night. Jiangpu Zhaba, feeling his trust abused, ordered a cantilever bridge dismantled to keep Ward from passing, declared that he "had more than sufficient of flower collectors," expelled a party of seven Nvlvk'ö botanists working for Forrest, and denied Rock's written request to enter his polity. Xiang Jiangpu Zhaba died in January 1924, and his successor, his younger brother Xiang Cicheng Zhaba, received Rock and allowed the collectors working for Forrest to continue exploring the state's remarkably rich alpine flora.

Muli was one of several polities in the Kham region that still enjoyed relative independence from both China and Lhasa. It lay in three river valleys and the surrounding mountains: the valleys of the Yalong, its tributary, the Litang, and a tributary of the Yangtze, the Shuiluo (Sholug in Tibetan, Shulo in Naxi). Tucked away in the southeastern corner of the Kham region, it was unusually diverse for a Khampa state. Its population was around 49,800 (in 1949), of which around a third was ethnically Tibetan (now considered Premi). The other two thirds included Naxi descendants of garrisoned soldiers of the Mu kings, who occupied Muli in 1605, Mongol descendants of the armies of Güshri Khan, who invaded Kham later in the seventeenth century, and large numbers of Nuosu (or Yi), who had emigrated from the Liangshan mountains in the nineteenth century. There were also small numbers of Han, Hmong, Lisu, and Zhuang immigrants.[4]

The roots of kingship in this little state lay in the Gelugpa sect's consolidation of control over Tibet. The Gelugpa emerged from the followers of the great scholar Tsongkhapa (1357–1419). The Gelugpa gained influence in the central province of Ü, but its power was opposed by two successive dynasties of kings based in the western province of Tsang. It was in this context that, in 1578, the head of the Gelugpa sect, Sonam Gyatsho, established an alliance with the Altan Khan of the Tümed Mongols and accepted from him the title of Dalai Lama for himself and his two previous incarnations. The Third Dalai Lama traveled extensively through Mongolia, Amdo, and Kham, winning the patronage of Mongol chiefs and founding monasteries, or *gompa.* In 1580, he established a Gelugpa gompa in Litang, the polity closest to Muli. From Litang, he sent an incarnate lama named Qoigyi Sanggyi (Chos-kyi Sangs-rgyas, Quejie Songzanjiacuo in Chinese) south along the Litang river to bring Gelugpa teachings to the Muli region.[5]

Over the next twenty years, Qoigyi Sanggyi and his assistants founded two large gompa in the mountains of Muli. The region was already under the influence of the Mu kings of Lijiang, who patronized the Karmapa subsect of the Kagyupa sect, and the two sects engaged in a fierce struggle for the territory. A remarkable history of Muli from 1580 through 1785, the *Muli chos 'byung,* "The History of Buddhism in Muli," written by the secretaries of the Great Lamas, tells of how the gompas became centers of military power, gradually subordinating villages in the surrounding territory.[6] Qoigyi Sanggyi worked to win over the local chiefs by absorbing their sons into the gompa as monks. When he died in 1584, the infant son of a powerful local chief from the district of Bar (hBar, Ba'er in Chinese) was recognized as his reincarnation. At the age of sixteen, the boy, Jiangyang Rongbu ('Jam dbyangs rong po) was sent to central Tibet to study. Fifty-eight years later, he would return to Muli to become the head of state, the first of nineteen generations.[7]

The Mu kings had not remained unaware of this threat on their northern frontier. In 1598, Mu Zeng became the last great military leader of the Mu kingdom. During the first year of his reign, he began to campaign aggressively in the north. The Mu chronicles relate that in 1601, after achieving victories over rebels in Zhongdian and Zuosuo on Muli's western and southern borders, he destroyed the rebels of Shuiluo (Shulo in Naxi), Muli's western district. By 1604, his armies had overrun all of Muli, where he became known as the *tian wang,* the "celestial king." His Naxi armies established garrisons, built terraced fields to supply themselves with rice, and constructed watchtowers along the western and northern borders against incursions from the unconquered regions beyond. Mu Zeng burned the Gelugpa gompa and expelled the monks. The *Muli chos 'byung* has him declaring, "Send your sons and brothers to the temples to become monks: we will sever their heads and limbs and summon their mothers and fathers to bear their corpses home."[8]

The Mu occupation of Muli lasted only forty-three years. By 1648, the Ming had disintegrated; Yunnan and Sichuan had fallen to the rebel general Wu Sangui, and Mu military forces were busy fighting internal rebellions. In Tibet, the Fifth Dalai Lama had allied with Güshri Khan of the Khoshut Mongols, who swept from the Kokenor region to conquer central Tibet and most of Kham. After his long exile in central Tibet, Jiangyang Rongbu had, with sixty exiled monks, founded a gompa in Yanyuan, south of Muli, as a military base, and he had established alliances with various local *tusi.* With the aid of Tibetan and Mongol armies, he entered Muli and expelled the disintegrating Mu forces. An assembly of monks declared him the First Great Lama of Muli and established the principle that this title should be inherited by members of his family, the Bar lineage. The succession hardly proved stable, however. The *Muli chos 'byung* records a cascade of bloody battles of succession among the important families of the Gelugpa hierarchy, lasting seventy years. In 1717, two contenders for the throne sent representatives to Lhasa to argue their cases. The Lhasa government made a clear ruling reaffirming that the position of the Great Lama must be restricted to the Bar lineage. After this, the lineage maintained a stable monopoly of military, political, and religious authority until 1953.[9]

The Qing court was slow to ratify the authority of Muli's ruling house. But in 1729, the Yongzheng emperor summoned the Great Lama to Beijing to receive the Chinese surname Xiang and the title of *anfusi,* or "pacification commissioner" of Muli.[10] In 1787, the Eleventh Great Lama was summoned to the capitol, where the Qianlong emperor himself received him. A hundred years later, the Tongzhi emperor promoted the Great Lama to the rank of *xuanweisi*, the highest rank of *tusi*. The Guomindang government ratified this rank and added to it the title of "regional military commander."

At the apex of Muli's system of government were three official positions of equal prestige, if not equal earthly power. First was the Great Lama, in whose hands was concentrated power over the military, the political administration, and the religious establishment. As a Gelugpa lama, the Great Lama was formally celibate, so his title could not pass from father to son as with the lay princes of many Khampa states. In principle, the first son of the head of the Bar lineage was initiated into the gompa as a monk and groomed for succession as head of state. The second son married and succeeded to the position of the lineage head. Younger sons also entered the monastery to be trained for the position of Great Lama in case of their elder brother's death, and in many cases royal power did pass from elder to younger brother.

As father or brother of the Great Lama, the head of the Bar lineage wielded great influence over matters of state. He lived with his family in Muli's biggest village on a landed estate with lavish residences, a jail, an armory, and a law court. In principle, if the lineage head had no sons, his sisters' sons would take their places, both as lineage head and as Great Lama.[11] The lineage's affinal relations were thus also granted great prestige as possible sources of lineage continuity. All high officials who were not Gelugpa monks who had studied in central Tibet were required to be either from the Bar lineage or from the lineages of its affines, a principle affirmed by the Lhasa government in 1730. In addition, many affines were granted landed estates, from which they could attain great wealth. In 1928, Rock stayed for a few days in the house of the Great Lama's "brother-in-law" in a village near the Kulu gompa, where he was greatly taken by his luxurious surroundings: "The room is new and clean and very bright and airy; it has four glass windows, and its walls are beautifully painted, resembling temple banners; every panel and beam is beautifully decorated. . . . Even the King [the Great Lama] in his massive palaces does not occupy such a well-lighted and appointed room."[12]

Muli's system of royal inheritance is unusual, if not unique, among Tibetan polities. Many of the two dozen states of Kham were ruled by lay princes, or *gyelpo (rgyal po)*. In others, the head of state was an incarnate lama, or *tulku (sprul sku)*, and in at least one, the power rotated among different lineages of *tulku*. Some were, at some stages of their existence, governed by officials appointed by the Lhasa government; a few were ruled by officials appointed by the *amban*, the Qing court's representative in Lhasa. Some had subordinate status within other states. Most of any size had subordinate units called *dzong (rdzong*, literally "fort") ruled by *dzongpön*, hereditary officials who enjoyed varying degrees of independence from the center. In the early years of the large monastic estate of Sakya in southwest Tibet, and in the Khampa polity of Derge, the position of head of state passed from a celibate monk to his brother or brother's son, but in neither state did this form of succession last for more than a few generations.[13] It seems that the flexibility of Muli's system, allowing succession to pass to affines when necessary, was important in creating the stable line of inheritance, which continued for nineteen generations from Jiangyang Rongbu to Xiang Peichu Zhaba (Phun tshogs grags pa), the last Great Lama, who ruled from 1950 to 1953.

The third great personage in Muli was the incarnation of the founder, Qoigyi Sanggyi (Chos-kyi Sangs-rgyas). Jiangyang Rongbu had been both Qoigyi Sanggyi's incarnation and the head of state. In the settlement that ended the battles of succession that followed his death, the two positions were separated. The incarnate lama did not usually participate in the political administration of the state, concerning himself instead with the administration of Muli's three great gompa. He maintained his own household in Wacing gompa, but his residence rotated each year among Muli's three great gompa with the residence of the Great Lama. As the incarnation of a disciple of the Third Dalai Lama, the *tulku* had great prestige in other parts of Kham and Amdo. *Tulku* were important manifestations of continuity and links with an ancient and storied past, whose presence greatly increased the potential political and economic power of any monastery. The rank and depth of the Muli *tulku* lineage were unusual: most such lineages were relatively shallow. While Muli's small size and peripheral location made it relatively unimportant among Khampa polities, the presence of a famous incarnate lama nevertheless reassured its citizens that it was a state of renown.

The Great Lama appointed a cabinet of six. Ranking highest was the minister *(mengong* or *chiaza)*, with a renewable three-year term. The cabinet also included a treasurer, a Tibetan-language secretary, a Chinese-language secretary, a personal assistant, and the chief of the Great Lama's personal guard.[14] The location of the government rotated among Muli's three great gompa: Wacing (Wan'erzhai in Chinese), founded in 1584, Kulu, founded in 1604, and Muli, founded in 1674. Each summer, the Great Lama, the *tulku*, the cabinet, and the Great Lama's personal guard of about twenty soldiers made a three-day journey to the next of the three gompa, in which they would reside for a year. The three gompa were situated in a triangle that spanned Muli's historical core. In some ways, they appear to have been an attempt to create in Muli a microcosm of central Tibet, with its three great

Gelugpa monasteries. Thus, in principle, Wacing had 330 monks as Ganden, in central Tibet, was said to have 3,300; Kulu had 550 as Sera had 5,500; and Muli had 770 as Drepung had 7,700.[15]

Parallel organizations of administrative and religious officials within each gompa filled out the state structure. The abbot of each gompa was the Great Lama's assistant when he was resident and his representative when he was not. Each gompa also had a treasurer, a quartermaster, and a host of minor political and religious officials who administered the gompa's large territory. The Muli gompa governed some 94 villages, Kulu 155, and Wacing 51. Each gompa had six smaller monasteries scattered through its territory, which served as minor administrative centers. In the villages, minor hereditary officials (dingbön) administered ten to three hundred households each.[16] Their main task was to collect taxes and corvée, assisted by lower-level village headmen (maise), appointed by the gompa.

If the three great gompa served as the central state's administrative arms, they could also, in concert, form a competing center of power. Most significantly, they held a periodic assembly called gesong of the important political and religious officials from each monastery, amounting to some thirty or forty persons. The Great Lama, the Bar lineage head, the tulku, and the central cabinet were excluded. This assembly had influence in many civil and religious affairs, including waging war, negotiating conflicts between the state and its neighbors, and deciding internal disputes, particularly if they involved the Great Lama or his relatives.

Finally there were three hereditary positions, called batsong, held by lineages with landed estates on Muli's borders. The batsong helped the central state manage relations with the outside world. They were expected to be fluent in Chinese, Tibetan, Nuosu, Naxi, and Mongol. From their border estates, they wrote the Great Lama letters apprising him of events in the surrounding territories. They could also be summoned to one of the great gompa to translate for Chinese-speaking officials or Nuosu headmen. Their estates were significant: the largest had over two hundred tenant households.

In Muli, as in other Tibetan polities, all land, in principle, belonged to the state. The state granted landed estates to hereditary nobles, high officials, local chiefs, and gompa, along with a variety of different rights to collect tax, rent, and corvée from their tenants. Most ordinary people were farmers; a few were pastoralists. In villages where Tibetans, Mongols, or Naxi predominated, most households were bonded to an estate. In the early 1950s, some 2,878 of Muli's 5,230 peasant households were found to belong to estates held directly by the state; 853 to estates of the noble classes, headed by the Bar lineage; 631 to estates of upper classes of monks and lay people;

and 614 to estates of gompa.[17] These figures exclude the third of Muli's population that was of Nuosu (Yi) ethnicity. Usually every household in a village belonged to the same estate.

The most privileged peasant households were *drungpa,* regular tax-paying households. Their estate allocated them land and received taxes, rent, and corvée from them. A similar class of households, bonded to monastic estates, received land from the estate and paid rent, taxes, and corvée to both the estate and the central state. The rank and obligations of these households were hereditary, and they were not allowed to move off their estates. Another class was tenant farmers. These were mostly immigrants of Han, Hmong, Lisu, and Zhuang ethnicity, not bonded to any estate, who rented land from *drungpa* and paid taxes and corvée to their landlords' estates. Some tenant farmers eventually attained land and the status of *drungpa*. Another class of immigrants, who lacked the means to form households, worked as tenants or casual laborers on the lands of *drungpa* or tenant farmers. Finally, nobles, high officials, and estate lords often owned slaves, who were usually people of the lower classes who had failed in their obligations or been convicted of crimes.

The few descriptions of Muli society before Liberation use documents created during the "democratic reforms" of 1956 as sources. These deserve to be read with skepticism, as they tend to exaggerate the degree to which farmers and laborers were exploited. Nevertheless, it does seem that the obligations imposed on peasants were very heavy. Land rent, paid in grain, was at least 40 percent, sometimes as high as 80 percent. In addition, nearly every productive activity was taxed. One document lists fifty-four taxes paid in kind each year by farmers in the village of Xirong, under the administration of Wacing gompa: measured amounts of grain, pork, yak meat, goat meat, deer meat, paper, clay, pine resin, gold dust, silver coins, hemp stalks, hempen cloth, white woolen cloth, and black woolen cloth for the central state; larger amounts of grain, paper, gold, silver coins, hempen cloth, white woolen cloth, and black woolen cloth for the estate lord; grain and butter for Wacing gompa; and grain, hemp stalks, and hempen cloth for the minor gompa. In addition, the village's forty-two households had forty-six collective obligations to pay cloth, meat, felt, clay, mud, pine torches, gold, and silver coins to the state and estate.[18]

Perhaps the most astute critical observer of Muli in the decades before Liberation was Ah Yunshan, the administrator *(zongguan)* of the much smaller neighboring polity of Yongning.[19] The members of Ah Yunshan's lineage had governed their little state for six hundred years. They were long-time allies of Muli; they had periodically intervened to settle its wars with neighboring polities, and they had recently lent the Great Lama soldiers to defend against bandits and Republican generals. Rock recorded several long conversations with Ah Yunshan in his diary. While Ah expressed personal respect for the Great Lama, he disapproved of the monarch's autocratic style of rule: "Any grievance the peasants might have they must bring before the lamas who eat them and squeeze them. They are the poorest peasant class existing on God's earth. . . . This present King sits upon a volcano which might go off any day. He is not liked at all either by his people nor by the lamas. The King is a big fool who believes in his divinity and allows none of his people to look upon him, still less speak to him."[20]

Outside of the estate system, a special system of land tenure applied exclusively to the ruling strata of Nuosu, or Yi, ethnicity. Muli bordered to the west, north, and northeast on core areas of the Liangshan ranges, the refuge of Nuosu societies. Beginning in the Yuan dynasty, most parts of the Liangshan were ruled by Nuosu aristocrats, recognized as *tusi* by the Chinese state. In rebellions during the late Ming and early Qing, Nuosu aristocratic clans had driven the *tusi* from most of Liangshan and established societies not subject to any state. During the nineteenth century several clans expanded into Muli, taking over mountain areas suitable only for swidden cultivation. Nuosu were not easily absorbed into the estate system. Their societies were organized into hereditary class strata, which were strictly endogamous. The aristocratic classes *(nzymo* and *nuoho)* were tightly ordered into competing, egalitarian patriclans. They held sway over much larger classes of commoners *(quho* or *quno)* and serfs *(mgajie)*. During the nineteenth century, the ruling clans assiduously accumulated firearms, which they used to feud with each other and raid peripheral areas, seizing captives who became slaves *(gaxi)*.[21]

Nuosu immigrated to Muli with their class and clan structures intact. Lords of the two aristocratic classes granted commoners land and collected rent, taxes, and corvée. The ruling clans retained strong alliances in the Liangshan, and they proved impossible for the Muli state to subordinate, despite its considerable military power. In the context of an unstable balance of power, the state negotiated a simplified system of land tenure with the Nuosu ruling classes. Nuosu lords who desired land called on the Great Lama with large gifts of silver; the monarch responded with a grant of land formalized in a document that specified an amount of silver to be paid to the state each year. This document allowed the Nuosu lord to freely grant land to his subordinates. Other than the yearly indemnity, neither the lord nor his dependents had any further economic obligations to the state.[22]

Great Lamas had considerable difficulty controlling their Nuosu subjects.

In 1928, the Great Lama accused Nuosu from Yiji in southern Muli of attacking a delegation of lamas and robbing them of thirty ounces of gold. He took "severe measures," and when Rock traveled through the region, he found that all the Nuosu inhabitants had fled. Near Muli's southern borders, Rock's party came upon the abbots of two of Muli's great gompa negotiating with several hundred Nuosu in a mountainside camp:

> They were indeed a wild bunch; all wore the black felt Lolo [Nuosu] cloak and belts of cartridges, and each carried a rifle, most of them of German make . . . which each hugged tightly like a baby. They were all lined up and were being addressed by the two Muli lamas. . . . The lamas had finally come to terms with them. The Lolos said they had not taken the gold, but still they admitted to have done wrong. The gold they did not return; this would have meant the admission of their guilt and a loss of face. So they kowtowed and paid the lamas. . . . some 700 taels of silver and five fast riding horses, mules, and cattle, in all fourteen animals.[23]

A rite of alliance followed, in which a bull was sacrificed, speeches were made, and much liquor consumed. Rock, who did not like Nuosu, strongly disapproved of this result: he felt the Great Lama had lost face by agreeing to a compromise. The Great Lama demurred: the Nuosu were not actually his subjects, being from the neighboring polity of Zuosuo, so a compromise was appropriate.[24]

For much of its history, Muli was a very warlike place. Each household was expected to maintain weapons, horses, and grain and, when ordered, to assemble its men for military service. The *Mu li chos 'byung* describes endless wars, battles, skirmishes: wars at the state's founding, wars over the succession, wars of expansion, and boundary wars with neighbors. The state also conducted wars on behalf of other powers: in 1656–1679 it invaded several neighbors for the Lhasa government; in 1691 it attacked rebellious Nuosu clans for the Qing emperor; in 1729, it sent two thousand soldiers to aid a Nuosu lord against his enemies.

The Republican era was a particularly violent time. The polity's peripheral location had spared it from the Qing troops that had subdued most Khampa states in 1905. But the Qing invasion had freed the mountains to Muli's north from the control of the Tibetan polity of Litang, and bandit armies from these mountains, numbering as many as a thousand fighters, raided Muli in 1914, 1916, 1924, and 1939. The polity also fended off Republican generals intent on seizing some of its considerable mineral resources. In 1916, when the Great Lama closed down a mine at Longsong controlled by a Republican militarist and expelled the miners, over a thousand people died in battle. In 1913, under orders from Yunnan's new governor, Long Yun, the Great Lama ambushed a division of the army of Long's rival Hu Ruoyu as it passed through Muli on its way to take refuge with the Sichuanese warlord Liu Wenhui. About half the division's one thousand men survived the attack, only to be decimated on the other side by Long Yun's Yunnan Army. The governor presented the Great Lama with sixty rifles and promises of friendship as a reward. The next year, the Great Lama defeated an attempt an attempt by another general associated with Liu Wenhui to reopen the mine at Longsong. In 1934, Liu Wenhui sent a delegation to Muli, promising to grant Cicheng Zhaba a new rank in his army. The delegation met the Great Lama in a meadow, murdered him, and took his heir Xiang Zhaba Songdian prisoner. Long Yun eventually negotiated Zhaba Songdian's release.[25]

In this dangerous context of pressure from competing warlords, Zhaba Songdian ordered each village to send one armed man to participate in a new standing army. As the People's Liberation Army passed through Yunnan towards Sichuan in 1935, the Sichuan warlords gave the Great Lama the title of military commander, presented him with ten thousand bullets, and ordered him to stop the Communists as they crossed the Yalong River. Zhaba Songdian deployed some 450 men at key points on the river, but he avoided significant battle. Taking pains to cultivate friendships with warlords in Yunnan, Sichuan, and the new province of Xikang (which incorporated most of Kham), he and his successors averted further conflicts with outside forces until 1946, when Muli began a three-year-long feud with neighboring Zuosuo. In 1950, even as the People's Liberation Army was making arrangements for Muli's peaceful "liberation," the Great Lama entered a damaging war with a Nuosu *tusi* in neighboring Yanyuan County. When the People's Liberation Army counted the weapons in Muli, it found 1,028 rifles and about 100,000 bullets in the hands of the Great Lama's officials and the Bar lineage and 850 rifles and 35,000 bullets in the three great gompa. In addition, the estate lords had significant armories, and nearly every peasant household kept its own arms.

PORTRAITS

In January 1924, Rock, a party from Nvlvk'ö, and two soldiers from Lijiang approached the great gompa of Muli, situated on a steep hillside and surrounded by some 230 houses and a wall. Some monks received the party and took it to lodgings in an empty house. Then a lama escorted Rock to an

Figure 35. Xiang Cicheng Zhaba in religious robes. Reproduced by permission of the National Geographic Society.

audience with the Great Lama: "The room was decorated with wall frescoes of very loud colors, all emblems of Buddha; the posts were red and decorated with a sort of applique work, and odd as it seems in all this lamanistic splendor, there was attached to the post cloth hooks with white knobs such as one would find in a cheap beer restaurant or in a German beer garden."

The Great Lama was seated, the light behind him, a row of lamas standing beside him. Like all the Great Lamas, Cicheng Zhaba had spent part of his youth studying in Lhasa, though Rock, knowing no Tibetan himself, would claim that his central Tibetan dialect was rudimentary. To converse, Rock spoke in Chinese to his cook, Yang Jiding, whose mother was from a Tibetan village on the distant Mekong. Yang spoke to the Great Lama's Tibetan secretary, Tungsen, in the upper Mekong version of the Khampa dialect of Tibetan, and Tungsen translated this into the Muli dialect for the Great Lama. The conversation was about photography and photographs: "He presently gave an order, and the lama brought a steriopticon glass and photos, and we went through the pictures one by one, and I had to explain to him what they represented . . . it was a motley collection, from Windsor Castle and the dining room of the White House, to a German beer garden in Berlin, the Panama Canal, Swiss views, and the land of the midnight sun." The Great Lama then had his own cameras brought out—three large cases full of fine equipment: "two large French cameras, 11 × 27, fine lenses, a splendid apparatus, and one post-card sized Eastman Kodak. But he had no idea how to use the outfit. I found that all the rolls of printing paper had been opened, as well as the boxes of plates which they tried to load in broad daylight. The Kodak had a film upside down, the paper to the lens. We all had a good laugh. He ordered one lama to come to my house and learn all about photography in an hour. I felt sorry for the lama." They dined on oranges and cakes, and then nine men escorted Rock out, bearing gifts of rice, flour, mutton, horse beans, eggs, salt, and cheese. On arriving at his quarters, Rock promptly discarded the wormy cheese: "it would have walked off by itself had I left it on the veranda." In return, Rock sent the Great Lama a collector gun, a hundred cartridges, and some photographs of the Thirteenth Dalai Lama whom, he was told later, the Great Lama despised.[26]

Rock's first visit to Muli lasted five days, and it seems he spent every waking moment photographing and developing photographs. To him, Muli seemed a world of surfaces. Everything was on display. Social relations were visible in the finery of lamas and nobles and the ragged hempen garb of peasants, as well as the exaggerated displays of deference he saw all about him. The public Tantric rituals he observed were "ridiculous in the extreme," outward displays of meaningless gestures, requiring no interpretation nor understanding. Even money held no mysteries. In most places in west China, competing coinages and fluctuating exchange rates made the price of the smallest necessity a matter of excruciating calculation. Here, however, everything had the price he chose for it. No money exchanged hands: instead, he returned an arbitrary donation of silver for a gift of grain from some peasants, an Austrian rifle for a gift of *thangka* and a silver-filigreed saddle from the Great Lama. Nothing hid anything else; everything posed frontally, open to vision and the camera. His photographs, published in *National Geographic,* were all in his preferred mode: figure against ground with no evident gaps between, no shadows, no reverse sides, each figure relating not to the others but directly to the camera. The most striking is of the royal bodyguard—ten men glowering shoulder to shoulder in robes trimmed with leopard skin, the two eldest, the guard's generals, in the center, wearing the robes of monks and an air of power. The guard stands before the palace temple, where the curtains have been pulled back and tied around columns to display the murals within.

Figure 36.   The Great Lama's bodyguard. Reproduced by permission of the National Geographic Society.

He photographed the gompa from a distance, perched on its hill, surrounded by its wall. He photographed its buildings, including the Great Lama's palace; he shot a fifty-foot-high bronze-gilded statue of the Buddha and a gigantic prayer wheel. "Finally [the Great Lama] appeared in his royal robes. Over his red lama garments and silver and gold brocaded jacket he wore a white and yellow silk cloak. The attendant lamas took off his Tibetan embroidered boots, and bare-footed he ascended the throne, sitting down in Buddha fashion. Over his knees in front they placed a long white silk scarf and on his head a yellow tamoshanter. . . . He sat perfectly motionless for nearly 20 minutes while I took many exposures of him." Rock gave Cicheng Zhaba his Colt automatic, receiving in return a *thangka* and prayer beads. Then he photographed the *tulku*, the young reincarnation of Muli's founder, Qiogyi Sangyi, seated on his horse, man and animal both dressed in finery. Two days later, he photographed the Great Lama again. For the first photos the monarch had dressed as befitted his role as religious leader of his kingdom. This time, he adorned himself in the robes of state, and he had the wall behind him decorated with a new set of tapestries, befitting that role.

They lunched on tea, meats, and clotted cream. Cicheng Zhaba watched

Figure 37.   Xiang Cicheng
Zhaba in robes of state.
Reproduced by per-
mission of the National
Geographic Society.

Rock develop his photographs in an improvised dark room, "with great de-
light," and he asked Rock to obtain field glasses for him. Rock asked if he
could photograph some of his subjects.

> He laughed hugely at the idea. He sent his prime minister down and
> ordered that three men and three women were to appear at my house
> tomorrow. The prime minister could not refrain from chuckling and
> laughing when he reported that he had given the order. The king joined
> in the laughter. I had to resort to this as all those I approached said they
> were afraid, and they could not have their pictures taken unless ordered
> by the lamas. The crawlings and bowings and speaking in whispers with
> folded hands, bowed in supplication, is getting on my nerves, but the
> King seems to enjoy it.[27]

That evening, Rock photographed a ritual held outside the gompa's main
gate, at which some forty monks burned *torma*, large pyramidal cakes, accompanied by trumpets, drums, cymbals, and firecrackers. Another day of
photography and gift exchange followed: a leopard skin, cloth, musk, silver,
and rings from the Great Lama, the much-desired field glasses from Rock.

The next morning the party was off, towards Yongning and Nvlvk'ö. Before long, they were overtaken by the Great Lama's minister, who presented them with two large brass ladles, rode with them for ten miles, and gave them a parting gift of walnuts and oranges. Rock fell into a sentimental mood, waxing nostalgic about the gompa they had just left.

> A peculiar loneliness crept into my heart as I rode through the somber, quiet forests . . . I longed to go back to quiet, peaceful Muli and its kind-hearted King with whom I established a real friendship. They are all buried in a sea of mountains . . . there they perform their rituals and prayers, cut off from the world beyond, in peace and quiet . . . These were memorable days. Nowhere have I ever been received as friendly as by the lamas of Muli. . . . Muli is now only like a dream . . . in which alas I did not tarry long enough. It was like living in a fairy land, gold and riches . . . contrasted with pine chips . . . and yak butter for oil lamps.[28]

When they reached their camping place, they found that the Great Lama had sent eight peasants ahead to prepare their camp, building a compound of fir branches and carpeting the ground with fir needles. For this visit at least, it had been a happy coincidence of social geometries of rule and ardently-sought geometries of subjectivity. And the medium of that coincidence had, of course, been the camera.

VISIBILITY

Muli might be called a "theater state," though not exactly in the sense that Clifford Geertz gave this term. In his study of nineteenth-century Balinese polities, Geertz declared that theories of the state in the West labored under the misconception that the semiotics of political power were "artifices, more or less cunning, more or less illusional, designed to facilitate the prosier aims of rule."[29] In Western theories of the state, political ritual and spectacle functioned merely to conceal, mystify, celebrate, or legitimize the true reality: the mechanics of political organization, competition for power, and social domination. In contrast, in nineteenth-century Bali, Geertz argued, the semiotics of state power *were* its reality. States in Bali devoted their resources to spectacles designed to dramatize the ruling obsessions of Balinese culture: "Court ceremonialism was the driving force of court politics; and mass ritual was not a device to shore up the state, but rather the state . . . was a device for the enactment of mass ritual. . . . Court and capital were a both microcosm of the cosmic order and the "material embodiment of the political order." The theatrical state politics of Bali operated upon the principle that by "the act of providing a model, a paragon, a faultless image of civilized existence, the court shapes the world around it into at least a rough approximation of its own excellence."[30] Political spectacle had as its ultimate aim the reflection and production of political reality.

Though in Muli, the theatrics of power were clearly the central activity of the state, there was rarely any lack of raw competition for social domination. The *Mu li chos 'byung* makes it clear that at stake in the polity's interminable wars was not merely who would dominate society but also, and crucially, what system of political/cosmological symbology would prevail. The semiotics of political power did not magically trump the "prosier aims of rule," as in Geertz's Bali. Instead, the ultimate goal of raw, violent social domination—be it through war or through rent extraction, tax collection, and corvée servitude—was to sustain the dramatic apparatus of the state, with its wealthy gompa, gold *chortens*, gold-clad *tulku*, pampered Great Lamas, and lavish ritual displays. The state's chief theatrical productions were Tantric rituals, public and private, which generated intricate cosmological models. In Geertz's Bali, the theater of the state served the purposes of rule by providing a model for reality to imitate. In Muli, rule served the purposes of the theater state by funding its dramatic displays of models of the cosmos.

Tibetan polities never found a single, stable design for power like Balinese states or like the Theravadin Buddhist polities in Burma and Thailand. Theravadin societies were organized into what Stanley Tambiah termed "galactic polities": an exemplary center surrounded by regional administrations that reduplicated the form of the center.[31] Even autonomous polities like the Shan states were formal replicas, on a smaller scale, of the political center. In Tibet, however, the central state could not provide a model for regional polities to replicate consistently. The center was surrounded by regional polities that took a bewildering variety of forms, some under its control, some autonomous, some only loosely incorporated into any political system. In Theravadin societies, the state worked diligently to keep the monastic community under control, periodically purging it of nonorthodox forms of Buddhism. Charismatic Buddhist movements did develop in peripheral areas, gathering large followings, and sometimes fomenting millennial rebellions. Yet these movements, dependent upon the charisma of their leaders, rarely lasted long after their deaths, and even more rarely dislodged the central state. In contrast, Tibetan states were never able to centralize and control the monastic community, which consisted of several autonomous sects, based in large teaching centers. These sects had a strong "clerical" element, oriented towards reproducing institutions through rules and procedures. But, as Geoffrey Samuel argues, they also had robust "magical" elements.[32] Political authority in Tibet derived both from established institutions and from direct personal contact with the cosmic sources of power. The stability of any Tibetan polity depended upon its capacity to create an effective mix of these two sources of authority.

In Muli, the three great gompa and their eighteen subordinate monasteries provided an institutional foundation for political power. But the hereditary personal authority of the Great Lama and the *tulku* was also crucial. Even as rulers, Great Lamas and *tulku* were primarily lamas. Lamas were teachers, tamers, and civilizers in relation to their disciples and to local deities and malevolent spiritual forces. Their relationship to the lay population centered on rituals that called on deities to subdue the wild forces inhabiting landscapes, houses, and bodies. The best-documented such rituals are *cham* temple dances and mass empowerment rites (and Rock gave detailed descriptions of both in his diaries of his visits to Muli). The original prototype of the Tantric master as tamer and civilizer was the eighth-century Indian yogin Padmasambhava who tamed the local deities of Tibet and bound them to Buddhist teachings.[33] In Muli, Padmasambhava's role was taken on first by the Dalai Lama's envoy Qoigyi Sangyi, who subordinated the local chiefs and founded the first gompa, and then by his reincarnation Jiangyang Rongbu ('Jam dbyangs rong po), who defeated the Mu soldiers, drove out the Karmapa sect, and reestablished the state of Muli. The *tulku*'s authority was founded on his claim of descent from Jiangyang Rongbu through incarnation; the Great Lama's authority was founded on his claim of descent from Jiangyang Rongbu through patrilineal bloodlines. It was the Lhasa government's reaffirmation of these principles of descent that ended the wars of succession after Jiangyang Rongbu's death and established the Bar clan's monopoly of power.

To be sure, competition between these parallel but asymmetrical lines of descent was among the points of politics most fraught with danger. In the early twentieth century, the Bar lineage seems to have taken pains to limit the *tulku*'s power by insuring that the incarnation was discovered in a peasant household rather than a noble lineage, by restricting his opportunities for education in central Tibet, and by discovering within the Bar lineage an incarnation of a minor line of *tulku*, of the Sakyapa sect. Cicheng Zhaba was the first Great Lama since the eighteenth century to be a *tulku* himself, of this minor line. Moreover, he was thirty-six when he ascended the throne, and the *tulku* was only eighteen. In 1928, Rock had a conversation about the *tulku* with the abbot of the large gompa in Yongning:

He said that the present Muli Hofu [*houfo*, the Chinese term for *tulku*]
could never go to Lhasa as he knows too little, not having been brought
up properly as a child. Should he go to Tibet where the incarnation of
the Muli Hofu is known, the old and big Buddhas . . . would question
him . . . and not having studied much . . . he would be unable to answer
them, and he would be made a laughingstock. . . . The former Living
Buddha of Muli was a venerable man, and the Muli King had to worship
him and kowtow before him. In the early days . . . the Living Buddha
was . . . made overseer of the land by the Buddha. . . . Now the Muli King,
who is only a very small Buddha . . . and who is not known in Tibet,
is playing the big role in Muli, and the Muli Hofu is treated like one
of the peasants of the King. Even the lower lamas do not treat him
respectably.[34]

Monks were taught to make their lama into an object of devotional meditation and to feel a warm affection for him, modeled on the affection a disciple might feel for the Buddha. It was a lama's duty to teach his disciples
the basic principles of Buddhist practice. In the Tantric case, this would in-
volve visualizing the cosmos as a *mandala* centered on a Tantric deity, or
yidam, with his or her entourage of other deities. The deity's consort and
hundreds of attendant figures all had to be visualized precisely. In principal, these practices required monks to develop extraordinary capacities of
concentration and visualization. The goal was the ability to see the entire
mandala, down to the whites of the eyes of hundreds of deities, in a drop
the size of a mustard seed, and to maintain this vision for hours. Frequently,
a monk combined the image of his personal lama with the image of the
yidam. Only the most privileged lamas in Muli would have had the *tulku*
or the Great Lama as their personal lama. But as the central figures of the
religious hierarchy, they were objects of devotion. The three great gompa
and their eighteen subsidiary monasteries, with hundreds of monks, were
workshops for the production of cosmological imagination on a grand scale.
The *tulku* and Great Lama were personally, by proxy, or by implication central figures in this cosmos.

Rock noted the care with which the Great Lama managed his visibility
during his first visit, as he watched him display himself for the camera, first
as the polity's religious center and then as its political and military center—
each time in different gorgeous robes, under a different silk-embroidered
blanket, before a wall with different embroidered tapestries. During his second visit, in 1928, Rock photographed the Great Lama again at his summer
retreat in Kopati: "The King was dressed in a gold brocade vest and red broad-
cloth. On his feet he wore yellow satin shoes and on his head a yellow miter.
Carpets and brocades were spread over the seats and tables. He sat motion-

less like a statue while I took his photos. The sun was at first obscure, but later shone brilliantly, and I secured magnificent color plates. I photographed him standing in his red robes and later, after he had changed his garments, in his yellow robes and purple jacket."[35] These were ritual events, modeled on the many occasions in which the Great Lama displayed himself, magnificent and immobile, at public Tantric rites. The monarch was sometimes incapacitated by this requirement that he be made visible precisely the right way. He refused Rock's request to photograph him outdoors, saying he could not be seen among the common people. Ah Yunshan, the administrator of Yongning, whose posture toward his subjects was far more relaxed, ridiculed this sensitivity in a conversation with Rock: "He said the King thought of himself as such a superior being that no one was worthy to look upon him. No peasant had ever had an interview with him, and they dare not approach him. When he goes out any peasant near must either flee and hide or bend so low that his head touches almost the ground, for his eyes dare not rest upon the King."[36] An accumulation of wealth, enormous for this small, poor, mountainous polity, was lavished upon staging the monarch as visible in the right ways at the right times. He was displayed in full, immobile finery in numerous public rituals. He was made visible as a central figure in countless meditative rites, either as himself or through the proxy of the leading teachers at his monasteries. And outside of ritual occasions, he was kept invisible to his inferior subjects, in the recesses of his palace or behind the curtains of his enclosed traveling chair.

It is not possible to know very much about the ways this imperative shaped Cicheng Zhaba's enthusiasm for optical devices. It is clear, however, that he was an enthusiast. In 1928, visiting Wacing gompa, Rock discovered that the Great Lama had sent a subordinate to Kunming to buy field glasses, film, and photographic supplies: "He was supposed to learn about taking pictures and loading films and etc., but by the time he reached Muli he had forgotten everything. He [The Great Lama] showed us the opera glasses, cheap affairs, which made near objects strange, made of ordinary glass and packed in paper boxes, worth perhaps one dollar Mex a piece." A few months later, Rock made an intriguing note on field glasses. The Great Lama had made it possible for him to visit the stateless, bandit-infested mountains of the Konka Ling. After his return, the abbot of the Yongning monastery told him about the reactions his visit had provoked: "[He] told me that the Konka people said that we had powerful glasses with which we could see all the precious stones in the mountains, while they could not see them. We also with the aid of our glasses could see the mountain gods and their hearts, and that it was that we were after. And . . . there would then be no use for anyone to circumambulate the mountains, for the gods would have gone. And thus should we attempt to come again they would kill us." Later, Cicheng Zhaba asked Rock for field glasses "that would look through a mountain to see what was on the other side," refusing to believe that such a thing did not exist. In the context of the Great Lama's ridicule of Rock's desire to photograph common people, these notes suggest at a minimum that optical devices may have been commonly associated with special powers analogous to the powers of Tantra to make visible Tantric deities and buried treasures.

Tibetan societies had rich traditions of religious iconography. Paintings of deities and divine or semidivine persons were central to institutionalized religious practice. Painters of gods *(lhabri)* created images of incarnate lamas in the styles used to depict the Buddha, with few distinguishing marks, and with eyes open to exchange glances with worshipers.[37] Painters mastered elaborate iconometric systems to depict divine persons, and paintings that deviated from these rules were thought to be harmful. In temples and monasteries, such paintings of divine persons *(thangka)* were employed to gain merit, commemorate death, aid in good rebirth, assist in meditation, and illustrate tales of bodhisattvas.[38] Like stupas, tombs, and printed volumes made in memory of incarnate lamas, painted portraits were venerated as visible manifestations of incarnate souls.

How might the photographic collaboration between Rock and Cicheng Zhaba be situated in relation to these traditions? Quoting Bernard Faure's observation that "the icons of the Buddha are sometimes compared with the 'original' shadow that he is said to have left at a cave at Nagahara," Christopher Pinney suggests that we might expand our histories of photography to consider the points at which it blends with other, sometimes ancient, ontologies of representation.[39] As Clare Harris notes, Tibetan Buddhists engaged in much semimechanical reproduction of images, from making stamped clay pilgrimage votives to printing prayer flags with wood blocks.[40] The Tibetan verb for "to photograph," *dpar rgyag* or *dpar rgyab,* is identical with the verb "to print." Printing involves carving a wooden block in relief and backwards, inking the block, placing a piece of paper over the block, then running a roller over the paper: the parallels to darkroom work are obvious.[41] In 1926, during his expedition to Gansu, Rock observed one such practice, which he found curiously upsetting, perhaps because it uncannily mirrored his own labors in the tent that served as his darkroom: "We met a lama on the river bank moving a board up and down the water, the board being fastened to a string; when asked what he was doing he said he was printing images of Buddhas into the Yellow River ... If that is not the height of absurdity for a full-grown man to occupy himself with such nonsense, then there is really no such thing as absurdity."[42] The incident imprinted itself in his memory: many years later he made it the topic of a manuscript titled, "Inane Lamaist Practices at Ragya Monastery," noting that on the bottom of the board were fastened "deep brass molds embossed with sacred images of various gods such as are used in making mud brick offerings to shrines."[43]

These resonances might help explain why Tibetans so quickly absorbed the photo-iconography of divine persons into devotional practices. In 1910, while in temporary exile in Darjeeling, the Thirteenth Dalai Lama asked Britain's representative to Tibet, Sir Charles Bell, to photograph him. Harris credits this photograph with overturning the Tibetan prohibition against creating portraits of divine persons before their deaths. Bell gave the Dalai Lama several copies of the photograph; the Dalai Lama made further copies to distribute to "monasteries and deserving people"; and soon thousands were in circulation in Tibetan communities. The Thirteenth Dalai Lama presented Bell with a copy, signed and imprinted with his seal, with his hat, robes, and the backdrop of paintings of the Buddha tinted by an artist. Harris notes that the artist did not touch the Dalai Lama's exposed head, face, arm, and hands: "The Tibetan painter, still wary of proscriptions against portraiture, allowed the camera to produce the simulacra of a religious body, just as according to Tibetan accounts, the living Buddha could only be depicted from a reflection in water or an imprint in cloth."[44]

Cicheng Zhaba had certainly seen the photograph of the Thirteenth Dalai Lama when he studied in Lhasa in the early 1910s, and he probably owned a copy. In any case, one of Rock's first gifts to him was this very photograph. When he asked Rock to photograph him, first as an incarnate lama surrounded by symbols of his divinity, then as the head of state surrounded by symbols of power, he likely had this famous photograph in mind. On his second visit, in 1928, Rock presented him with the results: two glossy full-page photographs published in *National Geographic.* He was delighted, and he made many further requests to be photographed. The portraits were in Rock's characteristic style: full-frontal, centered, and flat, with nothing dividing the monarch from the detailed ground but the outer contours of his clothing. The embroidered designs on the tapestries and blankets appeared to radiate from his seated figure.

In the 1920s, more photographs of the Thirteenth Dalai Lama and the Ninth Panchen Lama were reworked by painters, who surrounded their figures with mountains, lakes, and deities in the style of *thangka.* Rock's skill, his style of portraiture, and Cicheng Zhaba's deliberate arrangement of the room about him produced a very similar effect without the intervention of a painter. Yet, unlike a *thangka,* the portraits capture the 'aura' of this powerfully charismatic man. They create, as did the photographs of the Thirteenth Dalai Lama, " 'certificates of presence' attesting that what is seen has existed and that this very particular body had housed the incarnate godliness of a bodhisattva."[45] Cicheng Zhaba seems to have arranged his photographic sessions to produce precisely this intersection of forms: an individualized "certificate of presence" superimposed upon the enormously authoritative form of a religious painting. In the mid 1930s, German zoologist Ernest Schäfer photographed Cicheng Zhaba's successor, Zhaba Songdian. Schäfer's portrait shows the Great Lama close up, wearing his crown of state, with an unfocused outdoor scene in the background.[46] It draws on contemporary Euro-American portrait photography, focusing on the Great Lama's internal state as revealed by his facial features. It is a revealing contrast to Rock's portraits of Cicheng Zhaba, in which the focus is evenly distributed across a brilliant, flattened center and radiant periphery.

As I argued in the previous chapter, Rock used photography to stage himself as a social subject and to compose a world about him that he could inhabit. He used his camera to order the gazes of others, placing himself on a stage before them, his social presence thinned down to a depthless virtual body. The Victor came in handy here, silencing his subjects, suspending their relationships to each other, and aligning their senses. The presence of the Great Lama did exactly this: with bowed heads, deferential gestures, and captured or averted eyes, his entire world deferred to his presence upon the stage of state. Photographing Cicheng Zhaba, Rock borrowed this power. It was easy enough, given his experience in placing himself in his photos of others, to make the Great Lama's stage figuratively his own. Sometimes he did this literally. The most widely distributed photograph of Rock is the fabulously absurd self-portrait, taken (by Li Shichen) between the party's first two visits to Muli: the explorer stands haughtily in the robes of the prince of the Tibetan state of Chone (Zhuoni) in Gansu. When the director of the National Geographic Society mistook the robes for those of the "Muli King," Rock was uncharacteristically pleased and amused rather than irritated. In any case, he noted, the boots of the costume had belonged to the monarch of Muli.[47]

ABJECTION

Rock's love affair with Muli did not survive his second visit. In the spring of 1928, he and his party of twelve young men from Nvlvk'ö had completed their difficult expedition to Gansu. Not two weeks after finally arriving home, they set off for Muli, working for the National Geographic Society. The Great Lama sent the abbot of Muli gompa to escort them from Yongning and received them at his summer resort at Kopati, exploding three shells in their honor. They presented him with a Colt automatic, an automatic Winchester rifle, ammunition, a sword-cane, snow glasses, photographs, a torch light, and the *National Geographic* issue with his portraits. The *tulku* received some yellow satin. The Great Lama brought out photographs of European royalty before the war, which they viewed together. Two weeks later, they set off for their goal, the Konka Ling in stateless Daocheng and Xiangcheng Counties, accompanied by the Great Lama's Tibetan secretary, Tungsen, and bearing letters for the Tibetan chiefs who dominated that region. They circumambulated the four great peaks of the Konka Ling and returned to Wacing gompa safely in the autumn. After a final exchange of guns, silver, gold, rugs, blankets, *thangka,* and statues of the Buddha, they prepared to depart. Rock was grateful and sentimental. "I hate to leave this peaceful quiet place with its friendly King so generous and so helpful."[48] The Great Lama's bodyguards lined up to present arms and send them off with flags, bugles, and drums.[49] The party headed west for a final tour of rough, stateless Xiangcheng County, under the Great Lama's protection. On their way home, they passed through Muli again, via Muli gompa, where Rock's relationship to the polity and its monarch was abruptly transformed.

He had always assumed that as an intimate of the Great Lama he could see anything he wished. At Muli gompa he asked a young monk to give him a final tour of the Great Lama's palace. As they left, the monk pointed to a small triangular vent in the thick wall. Four people were imprisoned there, he said. Rock immediately insisted upon visiting them. He headed back into the palace; the monk followed. They descended into the basement kitchens and then into a guard room. The guards removed a post that rested on a trap door. They fetched a pine torch and led Rock down a notched log into a hole.

> The room was low and small, only about seven feet square. In one corner sat cross-legged on boards three prisoners, Lolo [Nuosu], their bodies bent forward. They were chained, and around their necks they wore a *kang,* or heavy, thick board, three and a half feet square, and through the center of which a head protruded. And what a head! The hair was matted and long, the eyes sunken, the faces pale and hairy. The board hid the entire body, one end of it resting on the floor, which forced the body and head forward. The neck was raw; flies, which hovered over their faces, had full sway to tease them, for their hands could just . . . reach to the margin of the square board. Thus sat these criminals tortured in this black hole. For months they had sat day and night with these heavy boards around their necks, impossible to move,

for the room was even too small for the three to turn around. They were surrounded by human (their own) excrement, for they were chained in addition and had to perform nature's call on the spot. . . . I was horror stricken; it was like a terrible nightmare; I could not for the moment realize that I was actually facing human beings.[50]

The three men had killed and robbed a Han trader, he was told. Their sentence was five years. With them, to his greater horror, was a Naxi man from Lijiang, who had harbored a thief. He was chained by the neck to a post and forced to stand day and night.

Memories of the sight tortured him. When he went to bed, he thought of the prisoners spending the night in their square neck boards, "one resting his [neck] on a corner, the other on the broad side bent forward, their hands stretched in order to leave room for their knees." He decided to write the Great Lama, asking him to release the Naxi man who, someone told him, had a starving, homeless daughter in Lijiang. In Yongning, he discussed the Great Lama with the abbot of the monastery, who had nothing good to say. The monarch was avaricious; he collected heavy taxes; his subjects were prisoners in his domain; he threw people in prison capriciously on suspicion of having stolen from him. And he was very rich, owning, personally, some three mule-loads of gold.[51]

After three months in Nvlvk'ö and Kunming, Rock and his party returned to Muli, in December 1928, hoping to travel from there to the great sacred mountain Minya Konka.[52] This time they went to Kulu gompa, where they were surprised to find the Great Lama resident. Rock presented him with a 32-caliber Belgian Browning, ammunition, batteries for his torch light, and casing for the rifles he had already given him. But the guest quarters in the old palace were prisonlike: dark, rough, thick walls, windows covered with newspaper to block the dust and wind. He walked through the great hall of the gompa, darkened, with a few butter lamps making faint light. Behind the hall was a dark room, with a huge figure of a Tantric deity, blue, many-armed, copulating with his consort, "hideous in the extreme." "These monasteries are ugly, filthy holes, physically and morally speaking," he lamented. In the palace courtyard, some thieves were being beaten.

They were stripped and placed flat on the stone floor in the courtyard. One man held their feet, another their head, and another gave them each thirty strokes with a broad wooden paddle. If the wooden paddle didn't break at the second stroke, the man who administered the beatings was beaten. The man howled pitifully until after the tenth stroke he could only groan. They broke thirty wooden paddles. The man howled pitifully, until after the tenth stroke he could only groan. . . . The men who were beaten were released, while those who were not beaten were chained and put in the dungeon.

He made more discoveries. Two of the lamas, having been caught stealing, were sitting in the dungeon with boards around their necks. A massive golden *chorten* in the Great Lama's reception room contained the remains of his father's paternal uncle. "The Muli King's palace is at once a residence, a tomb, and a prison. He lives with the dead, and below him are the living dead . . . in the black dungeon."[53] The entire kingdom, he reflected later, was an unfenced penitentiary, no one allowed to leave, and "peasants set on peasants to watch each other."[54]

Weeks later, after a brief tour of the Yalong River, Rock obtained the Great Lama's permission to photograph the prisoners at Muli gompa. The party made the three-day journey there with this as their sole purpose. The three prisoners were escorted out the palace gate into the courtyard. As it turned out, only two were Nuosu, a father and his twenty-year-old son. The third was a Tibetan from Muli, once a monk. It seemed that the Great Lama must have freed the man from Lijiang.

> They were unable to walk straight. Their legs shook, and the poor wretches staggered along, wearing around their necks the huge, heavy, thick boards, several feet square, several inches thick, through the center of which their heads protruded. . . . Their hair has not been cut since their imprisonment, and long, black strands hide most of their faces. . . . They are unable to move their hair out of their eyes, clean their nose, or wipe their mouths. They are fed like dogs by the hand of their dirty jailor, who kneads their barley flour with water or tea and shoves it in their mouths. . . . It was a terrible and ghastly sight never to be forgotten. I took several color plates and a number of panchromatic pictures . . . Their faces were a ghastly yellow like the pallor of death; their bodies thin and their fingers and nails like claws of beasts.[55]

After he photographed them, they were marched back to their dungeon. He sent the photographs to the National Geographic Society, accompanied by a letter. "Notice the boards are fastened permanently with iron clamps so they cannot be removed . . . When you receive this it will be over a year that they will have been wearing those wooden collars *day and night.*" He described the thieves being beaten at Kulu and remarked, "I think the State Department would be interested in this information, for . . . when extraterritoriality shall have been abolished, the Muli King might take it into his head to see if a foreigner can take such punishments," imagining himself, not for the first time, in the prisoners' place.[56]

Still, he had never liked Nuosu. Their reputation for banditry and slave raiding did not make them sympathetic victims. A few days later, the abbot of the Muli gompa told him that the Nuosu he had seen negotiating with Muli officials over stolen gold had raided the neighboring polity of Qiansuo, killing a lama and stealing livestock: it had been the *tusi* of Qiansuo who had informed the Great Lama of the theft of his gold, and the Nuosu, it was said, were intent on making his subjects reimburse them for their losses.[57] Rock immediately thought of the prisoners. "These Lolos are the most awful robbers and the most lawless wretches. They are quick to murder, which they perpetrate in the most cruel manner. A great many more should be wearing the *kang* or wooden collar, which the murderers in the Muli jail are carrying so meekly. When one learns of the depredations of these wild wretches, one loses all sympathy with the collar-wearers, and would feel inclined to increase the punishment, rather than mitigate it."[58]

That spring and summer, Rock and his party finally made their journey to the beautiful Minya Konka, which he trumpeted to the American press as the highest peak in the world. Returning, they passed through Muli once more. They stayed in the Kulu gompa for two weeks, where Rock saw the Great Lama nearly every day. Cicheng Zhaba presented Rock with a saddle filigreed in silver and his lovely traveling tent, which Rock had admired and photographed. The botanist gave him more rifles, including a 22 Longs, a gift from Kermit Roosevelt. His opinion of the monarch changed again: "He is a gentleman and a real friendly soul, although he holds his peasants in absolute slavery." But then the party went to Muli gompa, where they visited the prisoners again.

Rock had begged the Great Lama to remove the boards from the three prisoners' necks, and when they were brought up from the dungeon for his inspection he found that this had been done. But four more prisoners had been added. Two wore the board and the others were laden with heavy chains. Rock asked one his offense. He had, with three others, murdered a man, who years before had killed a member of his family. He laughed as he told his story, shocking Rock: "Such a psychology is to me incomprehensible. I was about moved to tears when I saw those wretches in the most abject misery that any man can possibly be, and yet the man could still laugh and tell his story, smiling all the time. . . . These seven wretches live like pigs on cold stone floors in absolute darkness, behind walls five feet thick, without ventilation, in filth and vermin, amid their own excrement."[59] He promised to write the Great Lama asking for the boards and chains to be removed, giving the new prisoners, he believed, a ray of hope to brighten their black, interminable futures. Later, wandering about the palace, he could not forget the sight of

Figure 38. Prisoners in Muli. Courtesy of the Royal Botanic Garden, Edinburgh.

the prisoners. "What an oppressed, lama-ridden country of slaves this is," he railed. "They are not human beings but kept in a condition worse than the worse slavery . . . I would go insane if I had to live among these people."[60]

His feelings about Muli continued to lurch between sentimental nostalgia and outraged fury. As he prepared to leave for what he was sure would be the last time, he felt sad affection for the little polity. But at the final moment a perceived insult from the Great Lama's secretary Tungsen, to whom he had grown quite close, sent him into a rage: "All the lamas are a greedy rotten lot . . . their morals are as filthy as their bodies—both stink. Their priestly garb hides venereal diseases; even the highest, next to the King, is afflicted with it. We left Muli while the lazy monks were congregating in front of the palace square for prayer . . . We had hardly left when a cloud swooped down and obliterated all, monks and monastery."[61] As they rode away in the rain, through seas of flowers and forests of giant pines, he grew sentimental again. He fantasized about asking the Great Lama for a piece of land where he could build a house and settle down with his Naxi "boys," far away from nationalism, militarism, prohibition, and other symptoms of the world's madness. He placed a *Primula* between the diary's pages, a last "souvenir of this land of flowers and forest."[62]

·   ·   ·

Figure 39.   Prisoner in Muli. Courtesy of the Royal Botanic Garden, Edinburgh.

The photographs are hardly empathetic. The three men squat or stand in the bright sunshine, throwing sharp shadows against the wall behind. With their hands they hold up the huge boards stapled around their necks. A rope is carefully coiled in the central prisoner's left hand. When they squat, all that can be seen of them are their heads in the centers of the boards and their fingers and toes at the edges. All three look steadily and directly at the photographer. Two photos are of a single prisoner. He squats, straining to roll his eyes to look up at the camera. His gaze is not despondent at all: it is more like a look of curiosity.

Rock used his camera to root his subjects as deeply as possible in exotic locales. In his portraits, he never merely posed a subject against a background coded with exotic signs. Instead, his subjects bore evidence of indelible indigeneity on their persons: in their clothing, their hair styles, their weapons. When there was a contextual surround, it appeared as an extension of the subject's body, like the Great Lama's throne of state or his beautiful traveling tent. Now and then, he would photograph members of his expeditions in the clothing of another species: his *National Geographic* article about the Konka Ling contains two wonderful photos of Nvlvk'ö men dressed in the leopard-skin-trimmed silken robes of Bar-lineage nobles, a wall draped with leopard skins as a background. The men inside the clothing serve as markers or placeholders in the same way as the men who appear in his hundreds of photographs of trees or mountains. This too his portraits share with specimen photographs: the individual is a token of the species type, except when, as in the photos of the Great Lama, only one individual of the species exists. In all this, his instincts were in accord with the tradition of colonial photography, which sought stable, indigenous identities for its subjects. As Pinney puts it, colonial photography "positioned people deep within chronotopic certainties"—giving subject and viewer stable identities in colonial space and time.[63] Rock's photographs did this very successfully, widening the distance between viewer and subject.

His photographs of the prisoners in Muli take these formulae so far that their effect reverses. On the one hand, the marks of the prisoners' exotic situation are literally locked onto their bodies. On the other hand, the purpose of the photographs is clearly to ask the viewer to imagine his own body in the place of the prisoners', slicing through the "chronotopic certainties" that put subject and viewer each in his own place. The photographs use the indexical power of the medium to do this: pointing to that body right there, illuminated by that bright sunlight, suffering in that particular way. If these photographs have a *punctum* in Barthes' sense of a point that cuts or wounds, I find it in the fingers of the central man's left hand and the coiled rope they grasp. One imagines the man, picking up his rope on the orders of his jailors and carefully coiling it so as not to trip and fall as he climbs out of his cell, into the sunlight, to stand before the Great Lama's friend.

The world was, as he found it, uninhabitable. Where possible, he employed the resources available to him to rearrange that world into a geometry that suited him better. For a time, it seems, this geometry aligned with the institution of kingship in Muli. A monarch who ruled by being visible on a stage of his own, with the power to center the gazes and voices of others. A monarch who found a congenial way to amplify his visibility in the mode of portraiture that the explorer developed to center, capture, and mirror the gazes of others. A mode of social engagement based on the exchange of gifts, in which value was assigned by the powerful party rather than a chain of substitutions of capricious and elusive values in a world of multiple, fluctuating currencies. That is to say, a mode of exchange parallel to that of the captured-gaze-mirroring-gaze that Rock sought with his camera. It seems almost inevitable that he would discover an underside to this arrangement, as if he were looking down, waiting for it. So that when he spotted that little triangular grate in the palace wall, he was ready to insist immediately upon going underground. When he did discover those abject bodies, his first response was to encounter them merely as suffering parcels of meat, cut off from the sociality of voice and gaze by the thick dungeon walls, and swimming in their corporeality extruded as their own filth. His second response was to bring them up into the light, to photograph them, and in doing so to encounter their eyes.

What was he to them? He had their chains and boards struck off, to be sure. And he gave them an hour or two of light—perhaps including more torture, given that their eyes must have adjusted permanently to the dark. Did he give them anything else? The evidence does not tell us. Still, there are the central prisoner's eyes, straining up to meet the camera's lens. I have called it a curious look. It is, more precisely, the look of one human meeting another's eyes, recognizing him and being recognized.

7.   The Mountain

He spent his life developing a singular cartography. The earth, for him, was layered with names, bones, histories, and genealogies. The master concepts that guided perceptual intervention in the earth were contiguity, proximity, and singularity. To record the earth adequately was to measure distances by the pace of feet, human and animal, to trace out routes, rivers, and watersheds on earth and paper, and to search for places with a distinct and unclassifiable existence. During his long adventure in Gansu and the Tibetan region of Amdo in 1924–1927, he made many maps of what he imagined to be uncharted territory, all without the latitude and longitude marks that would allow them to be coordinated with existing maps. In 1927, a cartographer for the National Geographic Society named Albert Bumstead tried and failed to teach him some simple methods for finding latitude and longitude in the field, reporting on his reactions. "He 'did not know much about mathematics,' 'All those things would be too much bother and take too much time,' and 'perhaps he shouldn't have bothered to make a map anyway.' "[1] Latitude and longitude could do little for him. His effort was not to assimilate unique places into the extensive, regular space of the world map; it was to multiply difference, exaggerate the boundaries that divided places from one another, and find places that were fully contained within themselves. These efforts were roughly synchronized with the axis of filth and purity he had developed to coordinate affect and experience. Filth dominated along major routes and thoroughfares. Purity was characteristic of singular places, isolated from these networks, and deeply rooted in textual histories and genealogies.

In June 1926, he visited a Ragya monastery on the banks of the Yellow River, deep into the grasslands that incline gradually up to the Tibetan plateau in Amdo. He photographed a painting on the wall of an incarnate lama's residence: a map of the world:

The center figure is red but of sienna red, and it represents Maitreya; the . . . Buddha to the right is Tsongkhapa; to the left is Chowuchi. Below the central figure is again, Maitreya, preaching. At the lowest part of the picture are the four oceans, the blue representing the south, the red the west, yellow the north, and white the eastern ocean. In these four seas are the four continents, represented in circular areas. In the center are the seven golden mountain ranges and the seven seas surrounding the cosmic mountain Rigyalwu Rirab. . . . On the top of the (cosmic) mountain are the 33 paradises and . . . the realm of the gods.

Ordinarily he dismissed Buddhist cosmologies with offhand contempt. This painting, however, he associated with a place like no other, which he had been striving to reach for years, and which he imagined to be so completely isolated from the rest of the world that it could only be represented as the world's center, in the vernacular of this painting, or its highest mountain peak, in his own.

As he gazed at this painting with Rock, the intendant for the incarnate lama whose residence this was described the route around that peak, the great mountain Amnye Machen, the earthly manifestation of the deity Machen Pomra and the goal of Rock's expedition:

Beyond the sand dunes on the west side of the mountains, one comes to a place called Girchu Chemo [Gos-sku-chen-mo]—meaning the great painting; this is on the steep slopes of the Amnye Machen [where there] are rocks of many colors on the face of the mountain like a huge painting. Near there are two conical hills called Mowa the diviner and the other Hdowa [Gdor-ba]; he is the one who throws *ts'amba* images into the fires as offerings. . . . From the Gitung pass one comes to the Gitung valley and the mouth of another valley with a white rock on which is an Obo. This rock is called Nuwodrandel gungsuch [Nu-bo-dgra-'dlu-rlung-shog] and is the younger brother of Amnye Machen.[2]

He scoffed at the Tibetan habit of animating the landscape with deities and demons, yet he was always searching for places like this, where the past had layered the earth with texts, maps, stories, and histories. In the end it mattered little to him whether that past was historical or mythological: what mattered was the way that words intertwined with the earth in such places, bringing it to life.

Many years later he included this description of the mountain in his monograph on the Amnye Machen. He never approached the mountain closer than thirty miles. Yet he reported the intendant's words as geographical fact, rewriting them to give the impression that they were an extension of his own journey: "We are now on the west side of the mountain, and we come to a place called Go-sku-chhen-mo (Gö-ku chhen-mo) or the Great Painting . . . "[3] He was not being dishonest; he merely sensed little discontinuity between this text, which explored the landscape's mythological depths, and his own record of the earth unfolding beneath his feet. In his private cosmology of filth and purity, the great mountain was a place of depth, in which mythology, history, and geography were mutually entangled. Eventually, this distant place of depth became for him a powerful node of textual interconnection over the immense distance that separated his adventures in northern Amdo from his adopted home in the Lijiang valley.[4]

Rock's expedition to the Amnye Machen was formative. He had been in China for only two years, and though he had traveled a great deal it was always over ground that many of the Nvlvk'ö sojourners with him knew very well. In Amdo, his efforts to record what he believed were the unmapped territories of the Minshan range and the grasslands beside the Yellow River were the real beginning of his great experiment in text and landscape. Amdo too was where he nurtured the household of twelve men from Nvlvk'ö which would, with subtractions and additions, accompany him for all his remaining years in China. This chapter relates some of the events of this long and hazardous journey, and chapter 8 focuses on the relationships between the botanist and the young adventurers from Nvlvk'ö that developed during this expedition.

He first heard the name Amnye Machen from a retired British Brigadier General named George Pereira. In 1921, Pereira had set out to be the first European to travel from Peking to Lhasa since the Abbé Huc in 1845. After making his way to Xining, on the edge of the Amdo grasslands, he set off for central Tibet with forty mounted soldiers and sixty loaded mules. On a plain southwest of the great lake Kokenor he saw, some seventy miles distant, a range of four or five peaks emerging from the grasslands. "It towers above everything else in its snow-clad grandeur," he wrote in his diary, "and must be well over 25,000 feet in height." Several months later, he visited Rock in Lijiang during an expedition to Kham that ended in his death. "General Pereira . . . swears that it is higher than the Himalaya," Rock wrote. "He said to me it is the most beautiful mountain mass he had ever seen. He had of course only a little vest-pocket camera."[5]

He immediately began to plan a journey.

Rock's party set off towards Amdo in December 1924 and returned to Yunnanfu, or Kunming, in May and June 1927. The events of this chapter thus overlap with some of those of the previous two. The last part of chapter 5 takes up Rock's photographic practice in nomad encampments and grassland monasteries in Amdo; the first two visits of Rock's party to Muli, described in chapter 6, take place shortly before and shortly after the Gansu/ Amdo expedition. This was without question the most expensive, elaborate, and difficult botanical expedition made in early twentieth-century China. Rock traveled in a very different way from his competitors. His huge caravans were magnets for bandits, making large military escorts necessary. His habit of identifying with princes, warlords, and incarnate lamas sometimes placed him and his companions near the center of the large-scale violence unfolding in west China. And, through ignorance and obstinance, he chose to travel in places where clashes between local armies were endemic. As soon as his caravan headed northeast from Yunnanfu, it plunged into a sea of violent conflict. Sichuan and Gansu were being torn apart by competing warlord armies; huge bands of disaffected peasants and demobilized soldiers roamed the countryside. In northeast Yunnan, the party was attacked twice by large armies of bandits. In Sichuan, the travelers barely skirted battles between the warlord Yang Sen and his competitors for control of Chengdu and its environs. In Gansu, they found themselves stranded by a complex war between a Muslim warlord and the Tibetan protectors of Labrang monastery. Later, they were trapped in the monastery town of Chone for a winter by clashes between forces of the powerful warlord Feng Yuxiang, who had declared himself the province's governor, and local rebel warlords. They escaped Chone for Sichuan and Yunnan only months before huge armies of starving Muslim rebel troops sacked that town and plunged most of the province into cataclysmic violence.

All this requires a different treatment than I have applied so far. In this chapter I attempt to embed the party's travels in the historical context of events they observed or narrowly missed observing. Gansu was an extremely complex place, however, and my efforts to provide context will necessarily be limited to outlines of local histories, particularly histories of conflict and trade between Chinese Muslims and Tibetans in the parts of the province where the party traveled most: the Tao river valley in the southeast, where their base in the monastery town of Chone was located, and the grasslands of the north under the informal jurisdiction of the monastery of Labrang, where they traveled to glimpse the Amnye Machen range.

THE ROAD

In 1924, after his first visit to Muli, Rock spent several months in the United States identifying his plant specimens at the Museum of Natural History, writing a *National Geographic* article about Muli, and planning his journey to Gansu/Amdo. He convinced Charles Sargent, who had directed Harvard's Arnold Arboretum since its founding in 1872, to fund a three-year expedition. Sargent had hired Ernest Henry Wilson to collect for the Arboretum in Hubei and Sichuan in 1906 and 1910, but Wilson had retired from field work in 1919. The Arboretum had found that most species from Yunnan, Sichuan, and Hubei could not survive Boston's cold climate: the mountains of Gansu seemed to be a logical place to search for new alpine plants that could.[6] Sargent agreed fund the expedition generously: a lump sum of fourteen thousand dollars for the first year and twelve thousand dollars for each of the next two. Ornithologist Outram Bangs had the Harvard Museum of Comparative Zoology contribute two thousand dollars to be used to collect birds. Rock ordered custom-made tents from Abercrombie and Fitch: two small traveling tents for his personal use and a large tent for his equipment and the Nvlvk'ö men. He ordered his camera, film, and photographic supplies directly from Eastman. And of course he bought a gramophone and a collection of recordings. By November 1924, he was back in Yunnanfu, buying provisions and hiring mules, muleteers, and soldiers in preparation for the expedition.

On December 13, 1924, Rock's caravan finally poured out the West Gate of Yunnanfu, through a crowd that had gathered to watch it depart. It included eight young men from Nvlvk'ö (four, delayed by bandits on the road from Lijiang, would join later), six muleteers, and twenty-six mules carrying equipment, nearly a year's worth of provisions, and several thousand dollars of silver and gold. As it moved through the dry, mountainous, and terribly poor northeastern arm of Yunnan toward the Sichuan basin, the caravan made eddies in the river of people, animals, and goods that ran along this interprovincial thoroughfare. The most significant disturbance was in the movements of soldiers and bandits. Rock carried letters from Tang Qiyao, the warlord who currently controlled Yunnan, ordering local magistrates and officers to provide him a military escort. The regular military escort employed to guard merchants along this thoroughfare, the *baoshangdui,* protected the party as it moved out of Yunnanfu. Two days later, the caravan picked up sixty soldiers in a county town; another forty joined it the next day. Rock paid about one hundred dollars a day for his military escort; he estimated that local governments collected some one hundred thousand dollars yearly for the services of the *baoshangdui.* "The soldiers themselves get very little of that money . . . [they] live off the people, who they also make carry their loads . . . thus exacting an additional tax."[7]

Long lines of porters walked the road, carrying cotton yarn from India. Many gathered behind the caravan, seeking the protection of the *baoshang-*

dui. Bandits began to follow a few days out of Yunnanfu. Four days out, they attacked its trailing end, robbing porters of baskets of cotton yarn before the *baoshangdui* drove them off. In the large town of Dongchuan, Rock asked the magistrate for more protection. Most of the forces stationed there were fighting another bandit army, some distance away, but the magistrate spared forty soldiers. When the caravan left the town after six days, hundreds of people departed with it: the porters who had followed it from Yunnanfu, a horse caravan carrying cotton yarn, many students returning home for their winter vacation, a traveling menagerie with animals in cages and a monkey sitting on top.

Two days out, bandits opened fire from the surrounding hills. The soldiers from Dongchuan fought while the caravan pushed ahead, passing a human head hanging from a pole. Arriving at a village called Banbian Jie, Rock and the eight youths from Nvlvk'ö hid in the temple, with ten soldiers at the gates. As many as two hundred bandits surrounded the village; the villagers buried their belongings and exploded two bombs to call in the local militia. Rock gave heavy underwear and a hundred dollars in silver to each of the Nvlvk'ö men. He hid on his person his passport, some gold, chocolate, biscuits, his toothbrush, and a tin of condensed milk, and then he lay down to wait. The bandits did not attack. The next day, the caravan proceeded, everyone nervously scanning the hills. More porters joined the hundreds already following it.[8]

The caravan was met by 250 well-armed soldiers who had traveled through the night from Zhaotong to escort it to that city. That night, Rock stayed in an inn with many of the hundreds who had come with him. He spent the night treating skin rashes, old wounds, sores, ringworm, hookworm, fever, colds. "The place was one squirming mass of humanity . . . soldiers everywhere, spitting, wiping their noses and their hands on the door posts, filthy dogs . . . water standing in filthy pools." In Zhaotong, the rest of the Nvlvk'ö men joined the caravan. Rock pitched his tent in a temple for eleven days, waiting out snowstorms. He paid ten soldiers fifty Yunnan cents a day to guard the temple. Though it was midwinter the soldiers wore only "dingy thin greyish-white cotton trousers, no underwear, one thin coat." They spent their days huddled over a fire out of doors, shivering with their bare feet in straw sandals. Their food was brought to them in buckets, one with rice, the other with some vegetables. Their regular pay was five Yunnan dollars a month.[9]

The mountainous road from Zhaotong into the Sichuan basin was too rough for mules. Rock engaged thirty-seven porters, loading each with sixty *jin* (just over eighty pounds), and paying each ninety Yunnan cents a day.

Carrying huge loads over the horrible roads of this thoroughfare was a way of life for thousands. Cotton yarn, produced in India, was the main product transported from Yunnan into Sichuan, where it was used in household cloth production. Porters walking the other way carried salt, noodles, kerosene lanterns, and slaughtered hogs. Their most common load, however, was tobacco. It was packed into long cylindrical barrels, each weighing fifty *jin* (about seventy pounds). A mule could handle two; porters carried three or four. A journey of eight days over the rough, snowy road earned a porter 3.60 Yunnan dollars. These were people who had no other possible way to make a living. "The condition of the coolies we meet is pitiable," Rock wrote. "Their feet are bare or . . . wrapped in palm fiber with toes protruding; their bodies are often much exposed through their torn and worn cotton garments. Every time they stop they sigh deeply. Their eyes are red and inflamed; their features haggard and drawn. Old and young are loaded down to the limit. The soldiers push them off the trail into the deep snow to let us pass."[10] The countryside between Zhaotong and Sichuan was famine-stricken; many people were living off the roots of bracken fern *(juecai)*.[11] When snowfall stranded the porters, they stayed in villages along the way, and as their money gave out they stopped eating. Throughout his journey, Rock had great difficulty hiring porters. Though traveling with his caravan was relatively safe and well paid, the journey back was hazardous; the porters were afraid of being forcefully employed by soldiers to walk many miles from their homes and then left without pay to starve on the road.[12] Even though porters were paid so little, the very high volume of traffic was an economic boon for villages along the road. Inns were everywhere, and many parts of the road were lined with food stalls, sometimes stretching for miles.

Most places he traveled, Rock rode a horse or walked. But on this rough thoroughfare, he soon gave up his horse and did as other wealthy travelers did, hiring four "chair coolies" to carry him in a sedan chair. It was miserable work. Three days out of Yunnanfu: "The trail leads through flooded fields to a rotten pile of rocks over which lean pigs are being driven by Lolos. Through this squealing mass, my chair coolies had to step lively."[13] Five days out: "The trail was ice-covered and . . . slippery . . . even my chair coolies with their straw sandals fell twice."[14] A month later: "my chair coolies slide about, their feet clad with remnants of socks over which they wear straw sandals. Most of the money they earn is spent on food, and what is left on opium, for they must smoke; it is the pain-killer of the poor; it is the only thing that keeps them going." The cantankerous botanist didn't make their lives any easier. He liked an early start and often ordered them roused out of makeshift beds in the streets or inn courtyards before dawn. On the morning they left Zhaotong, two bearers kept him waiting at the city gate. "I could not help myself, so I gave them a hiding with my rattan cane on the street. They took it without a murmur, opium-sodden wretches that they are. I made them carry me for a long distance as additional punishment."[15] Later, four bearers carried him from Chengdu all the way to Old Taozhou in Gansu. Deep into Gansu, they rebelled, asking to go home. Rock lost his temper, ordered the chair thrown into the river, and told the bearers to go home, vowing he would not pay them a cent more than he had already paid day by day. They gave in and carried him for another two weeks. When he finally dismissed them, he paid their salaries, and they made the long trek back to Sichuan.[16]

Within the Sichuan basin, travel was by boat: one day down the Daguan and Yangtze rivers to the city of Suifu, six days up the Min to Jiating and seven more to Chengdu. War had been nearly continuous in Sichuan since the 1911 revolution, when commanders of Qing military brigades had separately seized control of the province's regions. They had rapidly militarized the province: fifty thousand troops in 1911 grew to three hundred thousand in 1919, in dozens of ragtag armies. Armies from Yunnan and Guizhou had invaded four times, in favor of the revolution in 1911, against it in 1913, during the revolt against Yuan Shikai in 1916, and during the Protect the Constitution *(hu fa)* movement in 1917.[17] By 1925, a second generation of militarists was vying for control. Five eventually prevailed, carving the province into garrison areas and maintaining relative stability until the civil war and war with Japan.[18] In late 1924 and early 1925, Yang Sen, enjoying a brief reign in Chengdu, was ambitiously widening streets, building roads, and planning to make his city the center of an expansive territory.[19] While Rock's caravan was on the road south of Chengdu, Yang Sen's army was north of the city, invading Mianzhu and driving out the militarist Liu Lin, whose army took to the hills as bandits. Shortly after the party arrived in Chengdu, Yang moved south to Suifu, hoping to control a huge and lucrative salt producing complex at Ziliujing.[20] By 1927, he would be driven from Chengdu, to hold a small garrison area around his home town of Guang'an.[21]

On the rivers there was little evidence of all this military movement. Eleven of the Nvlvk'ö men, four soldiers, two boat owners, two steersmen, and the entire load of equipment and provisions fit with Rock into two houseboats. Rock's "horse boy" from Nvlvk'ö, He Ji, walked his horse along the bank with two more soldiers. Four trackers pulled each boat, each wearing a sling around his shoulders and another around his waist. Each sling was tied to a short rope, which could be attached, at any point, to the long bamboo cable that pulled the boat. The trackers received about twenty-five cents of a silver dollar (eight hundred cash) for each day's work, three meals of rice and vegetables a day, and meat every three days. Trackers waited along the river for boats; each boat added as many as ten extra trackers at rapids, paying each one hundred cash for a certain distance. "Every time a tracker is paid, there is the usual quarrel [with] the boat owner. . . . The result can be guessed; the boat goes on and the tracker is left behind; pull he must or he goes hungry, and he must accept what the boat man gives him."[22] "Over bad rapids the men . . . are practically flat on the ground with their stomachs, pulling for all they are worth, holding on with hand and feet; it is a hard life indeed."[23] Nevertheless, the trackers were remarkably cheerful, and Rock admired rather than pitied them, as he did the porters.

The party stayed in Chengdu for a month as Rock recovered from illness and planned his route. He called on Yang Sen, fresh from his victories in Mianzhu and Suifu: "a young man with little personality, a receding chin, but a pleasant face."[24] The countryside north of Chengdu was rife with disbanded soldiers turned to banditry. From Chengdu to the limit of Yang Sen's new territory near the Gansu border, 140 to 190 of Yang's troops, changed daily, accompanied the caravan.[25] The life of most Sichuan soldiers was pure misery. Victims of the recent battle at Mianzhu lay dead along the roads with no one to bury them. The wounded and sick dragged themselves along until they died of starvation. Deserters were beaten and their thumbs amputated. Rock noticed many soldiers who were only eight or ten years old. "They carry the bugles, spending their lives this way; when later discharged, they have no profession and become loafers or beggars."[26] In Gansu, the escort was changed to troops controlled by Lu Hongtao, who had been appointed by the Peking government to be the non-Muslim military governor of the province *(dujun)*, based in Lanzhou.[27] Gansu was poised on the brink of five years of cataclysmic bloodshed. But for the moment it was relatively peaceful, and these soldiers were better armed, better clothed, and far healthier than their Sichuan brethren.[28] They too had their miseries, however. "There is not a soldier that has not the itch or some awful skin disease neglected and ulcerated from scratching with black filthy fingernails . . . They walk and sleep in their only garment for week and months, and it is no wonder that they are diseased and filthy."[29] The inns were always packed, and soldiers sometimes unhinged the doors of houses and used them as bed pallets in the streets.

So many made these roads their home. Soldiers, bandits, drovers, porters, bearers, trackers: some spent weeks on the road, others months or years. Some no doubt a lifetime, probably short. Though many, particularly the porters, pursued largely solitary occupations, their lives were nevertheless social through and through, as they negotiated wages, shared food and beds, chatted with each other, and missed their families. The soldiers were all men, the bandits nearly all, the porters mostly men with a few women, but new relations of kinship and friendship must have flowered for many in the towns and villages along the road. Still, their daily experience was mostly a face-to-face encounter with the road. Heads inclined downward, wading through mud, stumbling over rocks, senses intertwined with the road. The earth lived its social life through these relationships: intense, repetitive, often unbearable. Nevertheless, many of the road's inhabitants seem to have found more reason to be cheerful than sour, tyrannical Rock, borne along in his bedding, grumbling endlessly to his diary. He suffered mightily—from fear, awful roads, horrid inns. Perhaps he found some kind of peace attending to the wounds and diseases of his traveling companions, though he never said so. As for the others, we know of their experience only through the eyes of wealthier and more educated travelers, and most of those eyes were looking the other way.

MUSLIMS AND TIBETANS IN GANSU

Though Gansu was much closer to China's central regions than Yunnan, it was seen from those centers to be an even more remote, impoverished, and violent periphery. Four ecological and cultural worlds met there: the desert of Central Asia, the steppe of Mongolia, the highlands of Tibet, and the arable loess of north China. The routes from China's core to Central Asia ran east to west through the Gansu corridor, a narrow gap between the mountains of northern Tibet and the Mongolian desert. The main routes of exchange between Tibet and Mongolia crossed the corridor south to north. This was a vast region, including until 1928 the present-day provinces of Ningxia and Qinghai, which includes most of Tibet's Amdo. It was fragmented into diverse cores and peripheries, each with several zones of cultural contact. These zones brought together Muslim Chinese of various ethnicities, non-Muslim Chinese, and Tibetans, some of whom were Muslim.

As Lipman points out in his elegant history of Muslims in Gansu, these groups were fluid and complex and contained many populations of mixed peoples: "Turkic-speaking Muslims, Mongolic-speaking Muslims, Chinese-speaking Tibetans, Tibetan-speaking Muslims, Monguor-speaking Muslims and non-Muslims, and more."[30]

Muslims had formed a substantial minority in Gansu since the eighth century. The towns of Hezhou, Old Taozhou, and Xunhua had histories of

Muslim political hegemony; Xining, Ningxia, and Guyuan had large Muslim minorities; and most other towns contained villages, streets, or suburbs inhabited by Muslims. A few thousand Muslim Turkic speakers known as Salar made their homes near Xunhua, where they developed reputations for banditry and mayhem. Before the seventeenth century, Muslim communities were centered on mosques with councils of elders who selected religious professionals—*imam* and *ahong.* Many mosques supported schools, which taught a formalized Islamic curriculum. This independent mosque structure was known as Gedimu. During the seventeenth century, Sufi orders entered Gansu, quickly gaining large followings and reorganizing many Muslim communities along very different lines. These reformist orders, the Khafiya and Jahriya (Hufeiye and Zhehelinye), did away with the Gedimu's parallel leadership of lay elders and religious professionals. They created lineage-based communities, known as *menhuan,* in which religious and political authority were inherited along patrilineal lines. "As a unique blend of Sufi and Chinese forms," Lipman writes, "the *menhuan* combined the appeal of prophetic descent with notions of Chinese family structure and socioeconomic competition."[31]

Local conflicts between Khafiya and Jahriya grew into the first of the great "Muslim rebellions" after Qing officials, then Qing troops, got involved. In 1781, the Qing provincial governor intervened in petty street violence between the two orders, rounding up Jahriyas in Xunhua and imprisoning the Jahriya founder, Ma Mingxin, in Lanzhou. A Salar leader known as Su Forty-One gathered some two thousand followers to rescue Ma Mingxin; Lanzhou officials beheaded Ma; Qing troops, joined by local Chinese garrisons and Tibetan and Mongolian militia, eventually wiped out Su and all his followers. In 1784, another army of Jahriya adherents, seeking revenge, attacked a Qing garrison and were brutally suppressed. After the 1781 "rebellion," the Jahriya order was stigmatized as the insurgent "new teaching" *(xinjiao),* as opposed to the "old teaching" of the Gedimu; the Khafiya sometimes shared both the label and the stigma.

This cycle—local conflicts exacerbated by official intervention to require military suppression which produced a reaction that grew into a large-scale rebellion—repeated in the 1860s and 1870s. Local society throughout China grew increasingly militarized in the mid-nineteenth century, as numerous rebellions threatened the weakened Qing state. In Gansu and Shaanxi, Muslim and non-Muslim armed bands began to feud in the early 1860s. Muslims built hill forts to defend against non-Muslim Chinese; non-Muslims built militias to defend against Muslims. The violence soon became epidemic. Their history of conflict with the Qing state caused Muslim communities to fear the label "rebel" above nearly anything else. In 1862, organized Muslim militias from across the two provinces attacked Xi'an, hoping to force the Qing to pardon them for the preceding violence and lift the "rebel" label. It took Qing forces over a year to relieve the city, five more years and many massacres to quell armed resistance in Shaanxi, and until 1873 to destroy the Muslim armies in Gansu. During these campaigns, Gansu's Qing governor did his best to wipe out the Jahriya, executing every leader he could find and thousands of followers.

A similar cycle began in 1895. A power struggle within a Khafiya *menhuan* in Xunhua escalated into lawsuits and armed feuding. The provincial governor sent in a small military force which executed the leaders of one faction and fired on the protesting crowds. Thousands of Khafiya surrounded the city; a Qing general marched from Xining declaring that he would kill all Muslims; Muslims throughout the province reacted, and armed violence spread to Hezhou, Haiheng, Didao, and Xining. After more than a year, Qing forces prevailed, with the help of militias of non-Muslim Chinese, Tibetans, and Monguors. Though all these wars are usually called "Muslim rebellions," none simply involved Muslims fighting Qing forces. Muslim communities were often pitted against other Muslims. Muslim military leaders often fought on the Qing side. And in both the rebellions of 1781 and 1895, militia commanded by Tibetan and Monguor *tusi* also fought alongside the Qing.

. . .

Tibetans were a far less significant minority in Gansu than Muslims, but they were central in the events that surrounded the visit of Rock's party. The province's population centers formed a rough arc, with Minzhou and the twin towns of Old Taozhou and New Taozhou in the south, Didao, Hezhou, and Lanzhou in the center, and Xining in the north. West of this arc, the topography tilted up through a series of long river valleys that led up to the Tibetan plateau. This higher country was the Tibetan province of Amdo. The valley land west of the population centers was farmed by agriculturalist Tibetans mixed with Muslim and non-Muslim Chinese and, in the north, Tibetanized Monguors.[32] The higher lands, approaching the largely uninhabitable plateau, were occupied solely by Tibetan pastoralists, or *drokwa*. Political affairs in Amdo centered on three large Gelugpa gompa. Kumbum (Da'er) was a few kilometers southwest of the city of Xining. It was the birthplace of Tsongkhapa, founder of the Gelugpa order, and the seat of the reincarnation of Tsongkhapa's father, the Achia *tulku*. In the Republican period, it housed some two thousand lamas.[33] Labrang (Labuleng) was south of Hezhou. It was the seat of another important incarnation, the center of a coalition of Tibetan chiefs, and the home of five thousand monks. Finally, Chone (Zhuoni) was situated on the Tao river, south of Taozhou, Old City. Since the beginning of the fifteenth century, it had been the seat of the Yang lineage of *tusi*, with authority over communities of agriculturalist Tibetans in the Tao river valley. Many other Gelugpa gompa, large and small, were scattered through Amdo, both in the agricultural and pastoral lands, some subordinate to the three centers, some independent. There were also several Nyingmapa gompa, including two large centers, Tar'tang and Dodrub Ch'en, in the grasslands north of the Yellow River.[34]

Amdo's collection of important gompa and its position on the route from central Tibet to Mongolia ensured many contacts with Lhasa.[35] But neither the province as a whole nor any of its individual centers were ever directly administered from central Tibet. The largest gompa were Gansu's wealthiest institutions. Many had extensive lands farmed by rent-paying tenants, large herds tended by tenant herders, and oil and grain mills. Gompa received offerings for services performed by monks; many received substantial contributions from associated *tusi;* some collected corvée and taxes from subjects. They also supplemented their income with expeditions to collect alms: one Westerner living in Kumbum in 1912 witnessed such an expedition returning from the Kokenor region with 150 horses, 200 cows, and 6,000 sheep.[36]

Gompa also focused political power. Kumbum, Labrang, and Chone each had very different political relationships with the surrounding populations. These differences proved very important to Rock's party, and I shall outline them briefly. Kumbum was the most important of a number of large gompa in the Huangzhong region, which stretched from the large city of Xining to the area around the great lake of the Kokenor. During the Ming dynasty, the imperial state had consolidated control over this frontier by moving thousands of non-Muslim Chinese settlers there, assigning many territories occupied by non-Chinese peoples to *tusi* answerable to the emperor, and encouraging the construction of gompa, which were given large territories to administer directly. The gompa of the Huangzhong region collected taxes from, served justice to, and provided religious services for large numbers of subjects. During the seventeenth century, many subjects of *tusi* switched allegiance to the gompa, attracted by their religious charisma and, possibly, their relatively lenient style of administration.

In 1723, Qing armies invaded the Kokenor region with the aim of crushing a "rebellion" that began when Lobzang Danjin, a grandson of Güshri

Khan, attempted to unite the Mongol princes of the Kokenor to declare him khan. The gompa of the Huangzhong region, including Kumbum and its flourishing sister institution Gunlong (Kuolong in Chinese), along with their subsidiary gompa, more than forty each, joined Lobzang. His forces battled Qing armies near Kumbum, besieged Xining, and attacked garrisons in Ganzhou and Liangzhou. Troops from Sichuan crushed the revolt, burned 150 villages, and destroyed Gunlong, killing its 6,000 monks and thousands of their subjects. In Kumbum, 500 monks, including two prominent incarnate lamas, were beheaded.[37]

After the rebellion, the Mongol princes of the Kokenor were made banner commanders under the supervision of the Qing military, and their subjects were absorbed into the banner system. The monasteries' territories were confiscated and their subjects placed under the direct authority of regular Chinese administrators. The gompa retained some private lands and the right to collect rent but not taxes. An official known as an a*mban* (formally, *Xining banshi dachen*) was appointed to administer all the pastoral lands in Gansu and Shaanxi not under the structure of regular administration. Of Gansu's three large gompa, Kumbum, situated very close to Xining, was under the most direct official supervision by the Qing *amban* and, during the Republic, by Ma Qi, the warlord who controlled that city.

Situated well outside of the influence of the Khoshot Mongol princes of the Kokenor, Labrang did not participate in the 1723 rebellion, and it continued to administer a large territory even after the reorganization of the Huangzhong region. The monastery was founded in 1709 by Jamyang Shepa, recognized as an incarnate lama by the Fifth Dalai Lama. Its supreme religious and political leader remained Jamyang Shepa's reincarnation. Many other incarnate lamas also made Labrang their home, in about forty well-appointed residences. The monastery was situated on the edge of a vast pastoral region, occupied by nomadic and semi-nomadic Tibetans. Its subjects were grouped into eight large coalitions of patron *dewa (sde ba)*—an administrative unit roughly corresponding to a village or encampment— known as the Eight Great Lhade Tribes. While some *dewa,* especially among the Golok, appear to have been organized patrilineally, most were loose organizations of related and unrelated descent groups under a chief or council of elders.

Labrang's relationship to its patron *dewa* took two forms. In some *dewa,* the gompa directly appointed a headman *(gowa)* from among the Jamyang Shepa's lay guard for a period of three years. The headman's duties included settling disputes, appointing leaders for smaller segments of the *dewa,* and arranging for monks to perform ritual services. These headmen did not collect taxes, though the *dewa* made frequent, substantial contributions to the monastery. The nineteen or more *dewa* administered this way were grouped into the Four Lhade Tribes *(lha sde shog bzhi)*. Many held indefinite leases on the gompa's farmland in return for a portion of the harvest and annual ritual duties and military service. Those in close proximity to Labrang took turns supporting the expensive annual assembly of monks.[38]

In the second arrangement, Labrang indirectly administered *dewa* through their chiefs, who shared power with councils of elders. Most such chiefs were hereditary; some were elected by elders.[39] Each of these *dewa* had a close relationship with a subsidiary gompa, making contributions and receiving religious services. Labrang did not tax these *dewa*, but it did expect contributions from them, and the chiefs were expected to heed its advice. Labrang had this relationship with a very large number of *dewa*. Even in the early twentieth century, after the ambitious Fourth Jamyang Shepa (1856–1919) expanded Labrang's influence over its patron *dewa* and subsidiary gompa, the monastery's political control over its patrons appears to have been very loose.[40] It could, however, call the chiefs together for military defense, as it did in 1927 during Rock's party's time in Gansu.

Chone, where the party resided for two years, was situated on the Tao river, twenty kilometers from Old Taozhou, surrounded by villages of agriculturalist Tibetans and non-Muslim Chinese. After 1723, Chone's Yang lineage of *tusi* became the most powerful in Gansu. Local gazeteers trace the lineage to an "aboriginal" *(fan)* founder named Cide to whom, in 1404, other *fan* tribes offered land and allegiance. The Yang ancestors governed a very large territory, with more than eleven thousand households as subjects. Their five hundred cavalry and fifteen hundred foot soldiers "guarded five secret entrances and twenty mountain passes."[41] The fourth *tusi* made a pilgrimage to Beijing, where he was granted the surname Yang by the Ming Zhengde emperor. The Yang *tusi* expanded their domain considerably in the seventeenth and eighteenth centuries, organizing it into forty-eight banners *(qi)*. The Yang lineage boasted a distinguished record of service for the Qing against "Muslim rebels." Yang Zhonghe led militias against Su Forty-One and his Salar troops in 1781; Yang Yuandao recaptured Old and New Taozhou for the Qing during the violence of the 1860s, and Yang Zuolin fought with the Qing against Muslim armies in 1895.[42]

The gompa, established on a site acquired by the Yuan dynasty ancestors of the lineage, was the *tusi's* seat of power. In a principle borrowed from the Mongols, the eldest son in each generation was supposed to become the *tusi* while a younger son became the intendant of the monastery—its chief executive. In practice, the genealogical records of the gompa's intendants show fourteen out of twenty also to have been *tusi*.[43] In the 1920s, the gompa housed about five hundred monks, including several incarnate lamas. It had seventeen subordinate gompa in the Tao River valley and pastoralist areas to the west, which supported the *tusi*'s power. The intendant appointed an abbot to manage religious affairs in Chone and the subordinate gompa, usually for a term of three years. And he appointed a disciplinarian with a very wide range of duties, including managing finances, collecting rents, organizing legal assemblies, enforcing religious discipline, and regulating the monks' relations with the outside world.[44] The monastery complex gradually grew over the centuries, each *tusi* adding or renovating. By 1928, when it was burned down, the gompa had some sixty buildings, including the *tusi*'s palace, dormitories, assembly halls, residences for incarnate lamas, and two large printing presses.[45]

From the sixteenth century forward, Chone was Amdo's center for the printing and dissemination of Buddhist texts. It was particularly famous for printing the two great compilations of Tibetan Buddhist scriptures: the Kangyur and Tengyur. The presses owed their prominence to many generations of patronage from the Yang lineage. They began as a small workshop, established at the end of the Yuan or beginning of the Ming. In 1538, the sixth *tusi*, Yang Zhen, sponsored the printing of six Buddhist classics. In 1658, the ninth *tusi*, Yang Chaoling, built smelting and paper-making facilities. Under his patronage, the monks copied and compiled a complete set of the 209-volume Tengyur, consulting a copy held in Beijing. His heir, Yang Wei, sponsored another Tengyur and the 109-volume Tengyur. The twelfth *tusi*, Yang Yousong, went to Beijing to receive a commendation from the Kangxi emperor, and on his return he built a press to print the Kangyur. The press was a two-story building that housed eighty thousand hand-carved printing blocks in order of volume and page number. Some fifty specialists, including printers, mounters, compilers, proofreaders, examiners, and supervisors worked in the press, taking three months of painstaking labor to print one copy of the classic. A second press, for the Tengyur, was built in 1753. To print this work was an immense task, requiring 120,000 hand-carved blocks and seventy specialists working steadily for five months.[46] The presses printed many other texts as well. When Rock's party resided in Chone, the price for one Kangyur was 1,000 silver dollars while a Tengyur cost 1,700. In 1928, the presses, with their approximately 300,000 hand-carved printing blocks, some of them four centuries old, were burned to the ground.

In the 1920s, the Yang *tusi* claimed 12,750 households as subjects, residing in some 520 villages, grouped into forty-eight banners. He could field two thousand militia. The *tusi* appointed two high officials, *toumu*, and three lower officials, *zongguan,* to handle civil and military affairs. The *toumu* appointed four judges, or *zhuanhao.* The *tusi's yamen* had a civil office of nine persons, including a personal secretary, a Confucian tutor, and various archivists and copyists.[47] The *yamen* appointed a chief for each banner, responsible for collecting taxes settling disputes, and raising militia. The seventeen subordinate gompa handled the banners' religious affairs. The *tusi* made it clear to Rock that he controlled not land but people: some villages had populations split between families governed by the *tusi* and others under regular Chinese administration, and who owed allegiance to whom seems to have depended mainly on long-established tradition.[48] The *tusi's* subjects paid taxes to him at significantly lighter rates than their neighbors, who were taxed through the regular administrative system.[49]

Though relations between Buddhist Tibetans and Muslim Chinese were often strained by enmity engendered by the role of Tibetan militias fighting for the Qing during the "Muslim rebellions," they were just as often cooperative. In particular, Muslim traders developed tight relationships with Tibetan pastoralists in the trade in wool, Gansu's only export to foreign countries. Nomadic and semi-nomadic Tibetans raised sheep on Amdo's immense grasslands. They sold wool to buy grain, tea, weapons, and household implements. In the nineteenth century, Muslim Chinese traders began to organize caravans from Xining, Hezhou, and Taozhou to buy wool. The traders, known as *xiejia,* "guests," often resided in Tibetan communities for long periods; many learned Tibetan languages; some married Tibetan women. In the early twentieth century, the largest trading firm was located in Old Taozhou, less than a day's walk from Chone. It was a unique community, the Xidaotang, or "Western Hospice." Founded around 1910 by a Muslim scholar named Ma Qixi, it was both an economic cooperative and a religious order. Four hundred "inner" members lived together in a huge compound and donated all their property to the cooperative, which invested it in profitable enterprises, particularly the wool trade. "They go in twice a year from Towchow with over 200 head of yak," Rock wrote after meeting with them for the first time. "They take in flour and tsamba and cloth, thread, needles, etc., and in return receive fur, musk, etc."[50]

WAR FOR LABRANG

For most of the Republican period, three lineages of Muslim Chinese warlords dominated China's northwest. During the 1895 rebellion, Muslim cavalry commander Ma Anliang led the Qing army to retake his native city,

Hezhou. When that army's commander, Dong Fuxiang died in 1908, Ma An-liang took control of its eastern division and became the provincial commander of Shaanxi, which he controlled until his death in 1918. Ma Fuxi-ang, another fighter under Dong Fuxiang in the 1895 rebellion, became a vice commander in Ningxia after the revolution. He expanded his power until he controlled western inner Mongolia and much of eastern Gansu. To curb the power of these warlords, Yuan Shikai sent his subordinates to serve as governors of Gansu and Shaanxi. His choice for Gansu, Zhang Guangjian, protected himself from Ma Anliang and Ma Fuxiang with his own army from Anhui, but he was unable to assert much military influence over the province. In 1920 he was succeeded by his assistant, Lu Hongtao. After Ma Anliang's death, military power in Gansu splintered. Some eight petty warlords vied with Lu Hongtao for resources in a context of earthquakes, famine, and epidemic banditry. The third significant Muslim warlord was Ma Qi. Once a *xiejia* in the grasslands of Amdo, Ma Qi was made a garrison commander in Xining after the 1911 revolution. With a large, disciplined army known as the Ninghaijun, Ma Qi and his brother Ma Lin and son Ma Bufang dominated Xining and the surrounding region until 1949. "I must say they are a strong, virile lot," Rock commented after meeting Ma Qi's black-clad soldiers carrying loot from Labrang, where they had just massacred Tibetan forces. "They are the Japanese of China."[51]

The big gompa, with their great concentrations of wealth and influence over large territories were enticing targets for Gansu's warlords. As he built his power base in Xining, Ma Qi quickly took control of Kumbum, not far from that city's walls. During Rock's visit, he maneuvered to control Labrang, precipitating a war that Rock's party observed first hand. Ma Qi had moved a garrison to Labrang in 1918, taking some lamas prisoner and demanding heavy indemnities. In early 1924, the regent and father of the eleven-year-old Jamyang Shepa reincarnation removed his son from Labrang in protest. The Jamyang Shepa traveled among remote gompa with his father, brothers, and tutor, under a guard of forty soldiers from Lu Hong-tao's provincial army. This was more than a mere gesture of defiance. Labrang was one of the centers through which wool from the uplands flowed into the hands of Muslim traders and eventually to foreign firms. Thousands of pastoralists made the pilgrimage to Labrang to be in the Jamyang Shepa's presence, making the gompa a natural place for Muslim *xiejia* to gather and the wool trade to develop. After he withdrew, this traffic ceased almost entirely, and the trade that depended upon it was devastated.[52]

In April 1925, the Jamyang Shepa's chief military officer, his brother Apa Alo, vowed that he would not return to Labrang until Ma Qi withdrew his forces and his demand for 200,000 *liang* (about 16,500 pounds) of gold.[53] That month, Rock's party met many mounted Tibetan soldiers on the roads, "many of them boys, dressed in red or green with silver ornaments studded with coral and fur caps."[54] Many were nomads who had additional reasons to oppose Ma Qi. The warlord had demanded that pastoralists in the grasslands south and west of Labrang pay him a "grass tax" of one yuan per year per head of yak. When the chiefs of the nomad *dewa* in the Golok region refused to pay, he sent five thousand troops to drive away their herds, leaving the nomads destitute and angry.[55]

By late May, the chiefs of all the *dewa* subject to Labrang had gathered in the grasslands between Old Taozhou and the monastery town of Heicuo.[56] Ma Qi instructed the Muslims living in Heicuo and Labrang to leave quickly.[57] The Jamyang Shepa's father told Rock that five hundred chiefs with seventy thousand warriors were determined to drive Ma Qi's garrison out of Labrang, though they had only three thousand modern rifles.[58] The provincial governor Lu Hongtao had promised the backing of the provincial army.[59] On June 8, 1925, a mounted Tibetan army of about twenty thousand attacked the garrison at Labrang, driving out Ma Qi's troops. Ma Qi's brother Ma Lin led three thousand Ninghaijun from Xining, meeting the Tibetans on a plain between Labrang and Xining, and defeating them with very heavy losses on both sides. A young Pentacostal missionary named William Simpson, informed by *drokwa* fighters, told Rock that the Tibetans had fought with spears, knives, swords, and axes against the Ninghaijun's machine guns.[60] Lu Hongtao's provincial troops never arrived: the governor had been playing to weaken Ma Qi without risking his own strength.[61]

Ma Qi's army reoccupied Labrang, slaughtering many of its inhabitants and looting the monastery. On August 14, the patron chiefs of Labrang led another army of nomad fighters to retake the gompa. The Ninghaijun defeated them again, executing their wounded and captured. Willam Simpson and Apa Alo attested to many gruesome atrocities, including Ma Lin's order to execute all Tibetans within sixty *li* of Labrang.[62] Passing through Labrang in November, Rock found the market deserted but for some Muslims slaughtering animals. Ma Qi had, however, ordered his troops not to enter the monastery, and it had not been looted again.[63]

The war with Ma Qi ended the system, in place since the early eighteenth century, in which the Eight Great Lhade Tribes took direction from Labrang while acting as its sponsors and military guardians. For the next few years, the Jamyang Shepa wandered among remote monasteries while his father and brother attempted to negotiate Ma Qi's removal from Labrang. In 1926, Apa Alo accepted the protection of Lu Hongtao's sucessor, Liu Yufen, allowing, in return, a regional government office and a force of provincial security officers in Labrang. The office, subordinate to the governor, worked to bring the patron *dewa* into the sphere of regular government, creating township- and village-level units from the *dewa* and selecting chiefs from among their councils of elders.[64] A couplet that decorated the gate of the government headquarters in 1936 spoke of that office's mission: "As the Xia [Chinese] civilize the Yi [barbarians], mountains and forests become a land of peace. Wide territories open up, bounded by the river, like the serene Peach Garden."[65]

Though the war between Labrang and Ma Qi captured the attention of Rock's party for many months, it was soon eclipsed by far worse violence. In the autumn of 1925, Feng Yuxiang, at war with the Manchurian warlord Zhang Zuolin, had himself declared governor of Gansu. He sent a general, Liu Yufen, to be acting governor, supported by fifteen thousand troops from his army, the Guominjun.[66] Liu Yufen drove out the petty militarists who had divided much of the province, leaving only the Muslim warlords. As drought pushed most of the province into famine and a huge earthquake struck, Liu Yufen imposed exorbitant taxes to raise funds for Feng's efforts elsewhere.[67] In the spring of 1928, Ma Tingxing, a Muslim militarist in Liangzhou, rebelled against Liu Yufen, attacking Hezhou. Around the same time, a young relative of Ma Qi named Ma Zhongying raised an army of ten thousand Salars and other Muslims, which also besieged Hezhou. The war soon engulfed most of the province. In November, some twenty-five thousand starving Muslim troops under Ma Zhongying swept south into the Tao river valley. They routed some three thousand Tibetan militia under Yang Jiqing, sacked Chone, burned the *tusi*'s palace, massacred the monks, and moved south, laying waste to most of southern Gansu.[68] Ma Tingxian also attacked the Taozhou area. After taking heavy losses from Tibetan militia, he appealed to Ma Zhongying, and the combined Muslim armies burned and destroyed the whole of the Taozhou-Chone region. They attacked Chone again, burning the temple and the famous printing presses. In January 1929, the Muslim armies, fleeing Liu Yufen's troops, swept back up into Labrang, where they massacred the monks and thoroughly looted the gompa.[69] Feng Yuxiang sent more troops as Ma Zhongying's armies moved through most of eastern and southern Gansu, destroying towns and villages, and leaving as many as two million dead.[70]

Feng Yuxiang's army eventually managed to quiet the rebellions, and Gansu achieved a fragile stability until the arrival of the Long March and its enemies in 1935. Yang Jiqing survived the chaos of 1928–1929 to rebuild a palace and temple in Chone. Alone among Gansu's prominent *tusi,* he retained a fighting force and continued to govern his territory.[71] Yet Chone remained a tempting morsel for those vying for power in southern Gansu. In 1937, with the encouragement of the provincial governor, the *tusi*'s troops mutinied. They attacked Yang Jiqing's summer palace with machine guns, murdering the *tusi*, his heir, Yang Kunfu, and seven of their wives. The provincial government declared that Yang Jiqing's surviving second son, Yang Fuxing, would become *tusi*, with a surviving wife of Yang Jiqing as his regent. This effectively ended the *tusi*'s rule in Chone. The provincial government took control of the militia, and set up a government office which created a system of townships and villages to replace Chone's banners, absorbing them into a regular system of administration.[72]

PICTURING HOME

On arriving in southern Gansu, the expedition's first task was to discover how to get to their goal, the great mountain Amnye Machen in the grasslands administered by Labrang. The party stopped for a few days in Chone, where Rock met Yang Jiqing—"a tall young man of 36 summers, well dressed in Chinese silk and satin gowns."[73] In Old Taozhou, a member of the Xidaotang approached Rock. Learning that caravans from the cooperative regularly traveled to Ragya, the gompa on the Yellow River that was the gateway to the Amnye Machen region, Rock called on the chief of the Xidaotang. Eventually, the cooperative agreed to escort the party to Ragya, and Rock gave them 337 *liang* (about twenty-five pounds) of silver to buy yaks. The party then visited Ankur gompa, less than a day from Old Taozhou, to call on the Jamyang Shepa. The twelve-year-old incarnation, "a cute little boy," received Rock, and his father and brother agreed to provide letters of introduction to Ragya. The incarnation's family gave Rock an earful about Ma Qi.

The party spent a month in Chone preparing for their journey to the Amnye Machen, while the patron *dewa* of Labrang gathered between Heicuo and Taozhou. Rock called upon the Jamyang Shepa to inquire about his intentions: was he going to war or not? It seemed that he was. Rock photographed the incarnation: "They arranged him on a chair . . . on a yellow carpet; he was carefully dressed in gold brocade jacket, red mantle, and large Tibetan boots. He sat very quietly." At dinner, the Jamyang Shepa's escort, jolly with wine, suggested Rock have no dealings with the Xidaotang. Instead, the incarnation's father would send him two officials and five men who knew the road to Ragya. But first the Tibetans had to deal with Ma Qi.

In Old Taozhou, the Xidaotang was building a new hall. Its members told Rock that they had bought yaks and saddles, but they were afraid that the Tibetan forces massed along the roads would attack him. They wanted him to wire Ma Qi for permission; he was afraid this would destroy his plans. Unsure whom to trust, he decided to use the Xidaotang as an escort while also accepting the offer from the Jamyang Shepa's father to send two Labrang officials with him. Reluctantly, the Xidaotang agreed to lend him six members. They told him that they had heard rumors that he was a Chinese agent sent to furnish the Tibetans with arms. Furious, he wrote to the governor, Lu Hongtao, to demand that this "malicious gossip" be quashed. Lu Hongtao instructed the magistrate of the Taozhou region to post a notice on the gates of both Taozhou cities proclaiming that any who spread such rumors would be arrested and punished.

The expedition would be expensive. He bought twenty-two yaks along with saddles, ropes, felt covers for the loads, three horses for himself, and shoes, clothing, and sheepskin coats for the Nvlvk'ö men, who were to walk. And he prepared presents for the Golok nomads who lived around the mountain: cloth, thread, needles, boots, brass locks, snow glasses, and pictures of the Dalai and Panchen Lamas. He needed silver. Since the fall of the Qing, silver, rather than paper money, had been the foundational currency in most of west China. Rock's money came from the Arnold Arboretum to Shanghai via American Express in the form of checks issued on a Shanghai bank and cashed at the post office in New Taozhou. He cashed a check for five thousand dollars. The currency came from Lanzhou escorted by four mounted soldiers—about 2,000 silver dollars of different forms and 2,081 *liang* of silver ingots, weighing a total of about three hundred pounds. It seemed to be short, but he could not tell.

.   .   .

Mostly his daily experience was a confusing muddle of one mildly or severely unpleasant sensation after another, accompanied by degrees of worry, loneliness, disaffection, or rage. But there were times when, simply by virtue of the way it was laid out, the landscape clarified experience for him, allowing him to coordinate it with affect along simple and forceful lines: suffering and happiness, ugliness and beauty, filth and purity. Quite suddenly, among the great limestone peaks just to the southwest of Chone, he found his experience transformed in this way. In mid-June, tired of waiting for news from Labrang, the party set off on an excursion to the Minshan range. Across the high passes of these mountains was the largest and most rugged district of the Yang *tusi*'s realm. This part of the Minshan was called the Dieshan; the region was Diebu in Chinese, Thewu in Tibetan. Across the Dieshan, gorges ran down into the Bailong river, enclosed by more mountains on the other side. From Chone, the region was accessible only through a few high passes. The chiefs of the Diebu region had joined the Yang *tusi*'s ancestor as he had subjugated the local "tribes" in 1404. Soon after, however, they rebelled, and the *tusi* did not recover the region until the end of the seventeenth century. Scattered through these mountains were 208 villages *(zu)*, gathered into fourteen banners *(qi)*, and the region was laced with a network of monasteries, some quite large. Rock called its people "Tebbus," and romanticized them as fierce, lawless, and uncivilized.[74]

The party left their twenty-two loads, packed, weighed, and balanced, in Chone. Though a rinderpest plague was killing off the livestock, they managed to secure six donkeys and two mules. Yang Jiqing sent soldiers from his bodyguard to accompany them. At first, Rock was in a foul mood, threatening to beat his favorite, Li Shichen, for bringing the wrong tent poles. But the flora was extraordinary: hillsides covered with yellow and crimson poppies, carmine *Rhododendron*, deep blue *Corydalis*, yellow *Pedicularis*. He had always loved trees, and these mountains were full of conifers, willows, birches, and berberis. Great limestone crags capped the range. The party filed through a narrow gap between cliffs and climbed over a pass. Limestone pinnacles rose into the sky, "so closely arranged that they appear as the leaves in a book standing on end." He was suddenly ecstatic: "Never have I seen such wonderful scenery and never have I been happier than today. It turned out after all a glorious day, billowy light grey clouds hung above the crags; the trail was clean and gravelly, the roaring waters of the stream as pure as crystal. The air was the purest which a human being can desire, and the whole ensemble was purity and beauty itself. One's heart was glad beyond words that it had not ceased to beat ere this." They came to another defile, cut into the limestone wall by a stream. The trail was laid above the stream on timbers supported on both sides by the rock walls. The defile opened into a long amphitheater, the Yiwu gorge, surrounded by forests and enormous limestone peaks. Far below, a gompa called Lhasam was perched on a mountainside; below it were the villages of Tongwu and Drakana, set among terraced fields of wheat and barley. Rather than descending, they camped in a meadow at 9,500 feet. Rock ate supper outside his tent, marveling at the view.

As always, he made his camera a foundation for his experience. He took photograph after photograph of the amphitheater and his camp: a few white tents at the brink of the gorge, nestled among enormous peaks. He had spent the previous summer camping in the Yulong range, in the meadow called

Figure 40.   Rock's camp in Deibu. Courtesy of the Royal Botanic Garden, Edinburgh.

Gvssugko Düman in Naxi. There, in tents pitched amongst the flowers, he had established the extensive household of Naxi men that would carry him all over China. He had spent that summer and the summer before wandering the Yulong range in a frenzy of floral collection and description. Yet he had collected nothing new: the range had been scoured down to bedrock first by the French priests then by Zhao, Forrest, and the older generation of Nvlvk'ö botanists. Still, during those two summers, Rock had consolidated a geographical practice that would sustain him for decades. For him, names of places and plants were bits of the earth he could hold onto; they were stepping stones through that fluid outpouring of revulsion and disgust which was so often his experience of the world. In the meadows of the Yulong range, the stones paved the stream from bank to bank, rendering it entirely knowable. Each step had a flower that spoke its name with precision. The earth and the great book of the herbarium folded together tightly, leaf by leaf, each completing the other.

Here in the meadows above Lhasam gompa it was the same. For the first time since arriving in Gansu he began to collect, and his writing became almost solid with floral names and descriptions: "The meadow gradually passes into magnificent forest, which extends to a mighty grey limestone wall; the

Figure 41.   Rock's camp in the Yulong range. Courtesy of the Royal Botanic Garden, Edinburgh.

forest is carpeted with moss and only a few steps from my tent. The trees are 150–200 feet tall, of a deep green with short, drooping branches. Birds abound in the forest. The undergrowth is composed of Rubus, Salia, Berberis, Paeonia, Rides, Prunus, Betula with glistening copper-colored bark, Cardamine, Anemone, Fragaria . . . "[75]

His photograph of his camp here and of his camp on the Yulong range are identical in every important respect. And both are, in many ways, replicas of the specimen photographs he had made for his masterpiece on Hawai'ian trees more than a decade before. The range is presented with luminous clarity, filling the entire frame. Everything is flat; there is no foreground or background; the range is divided from the viewer as though by a perfectly transparent pane of glass. Though the photographer had to travel out of the frame and then turn around to aim his camera, his path is not evident. There is no route leading from the viewer's place back into the photograph. And there, at the bottom of the range, are the tiny tents and the circle of Naxi boys, the empty mark, like the knife or hat, inserting the photographer into the photograph as an absence. Throughout his journey in Gansu, he would go to great lengths to recreate this scene. An enormous world, emptied of everything social, presented with the utmost clarity and attention to detail. Flat, in the sense that it is totally there, open to view, entirely surface, hiding nothing. And at the center, a tiny closed society, circle of tents containing his twelve boys, his own empty tent among them. He had been dreaming of the Amnye Machen range with great intensity for a long time. But here in Diebu, merely a few days from his base at Chone, he found without expecting it what he was always looking for. He would keep dreaming of the Amnye Machen and keep failing to get there, even when he did finally make the journey. But he would also keep coming back here where everything he wanted to find was already present.

.   .   .

That first trip to Diebu took only nine days. Back in Chone, the party learned that the Tibetans had driven Ma Qi's troops out of Labrang, then been heavily defeated in Ma's counterattack. Immensely disappointed, Rock postponed his plans for the Amnye Machen. Instead, the party packed up to return to Diebu. This journey was to last for three months—a thorough exploration of the region's flora. With much bullying and more confusion, the magistrate of New Taozhou requisitioned eight mules for them. They were escorted by soldiers from Chone and several men from Diebu who knew the region well. Rock brought his gramophone. Several days into the journey they met a man who had been attacked and beaten by robbers. The soldiers lit out to search for the culprits: they were ambushed and one was shot in the arm. Carrying the wounded man, the party returned to Chone only a week after setting out.

One source of Chone's fame in Amdo was its long-established *cham* dances. These public ritual entertainments, held yearly in most major gompa of every school, were among the ritual services that monks performed for the laity. They staged representations of deities, both local and universal, and drew on their power to destroy demonic forces. Chone held two different dances: *cham nyingma,* or old cham, in the sixth lunar month, and *cham soma,* or new cham, in the first and tenth lunar months. Rock's party was present for *cham nyingma;* they would see *cham soma* that winter. Rock took enthusiastic notes. It was his first sustained experiment in the ethnography of ritual. Perhaps because the dances' intricate complexity took all his attention, he did not comment, as he was wont, on the nonsensical nature of it all. His approach was broadly taxonomical: narratives and choreography passed him by, but he took careful notes on the beings represented by the dancers. Yang Jiqing, sitting beside him, ordered the dancers to appear one at a time before Rock's camera: "the lamas didn't like it, but there it was." Later, with the help of the lama who directed the dances, he went through his notes and penciled in the names of the deities and demons in Tibetan script. He would publish his images in *National Geographic.*

The party made a third journey to Diebu in early August, despite fighting between local village militia and a punitive expedition of fifty soldiers the Yang *tusi* had sent out in the wake of their last visit.[76] A few days after their return, they departed again, headed for the Kokenor region, leaving two of the Nvlvk'ö men behind to gather seed in Diebu. As always when traveling on thoroughfares, Rock was miserable. He was depressed by men with queues, by naked children playing in the roads, by the bare loess hills. He made a habit of shooting the dogs that kept him awake at night; one morning he awoke to find villagers skinning a dead dog: it had, after all, been someone's possession. In Lanzhou, he was put up in magnificent quarters, most recently occupied by the Sixth Panchen Lama, and he dined with Lu Hongtao, soon to be replaced as governor.[77] In Xining, he called on Ma Qi, "a tall, powerfully-built man . . . far from refined." He had long feared that Ma Qi would prevent him from going to the Amnye Machen, but the warlord offered to protect him on the journey. Rock immediately forgot the grisly tales he had heard of Ma's atrocities in Labrang: "We parted the best of friends."[78]

From Xining they set off for the Kokenor, escorted by Ma Qi's cavalry, passing yak caravans loaded with wool from the grasslands. They camped on the edge of the great lake, "a deep glorious blue merging with the horizon in the north." Near the shore, the lake was black with ducks and cranes; the sky was an "ethereal pale blue." Rock had planned to circle the lake, but the dizzying sun reflecting from the lake's surface and the vast landscape of dunes and grasslands were too much for him. The party made an arduous trek over high plateaus, through a severe snowstorm, to the trading post of Wobo, buried in the grasslands. Rock had heard that there were forests in the Nanshan, two ranges north of the Kokenor. So off they went, wandering through those great mountains in search of trees. It was mid-September and the weather was extremely cold in the mountains. Rock had officials force unwilling muleteers to accompany them at the point of a bayonet. "They whine continuously and groan so that it gets on one's nerves."[79]

Crossing the Nanshan, they went west to Ganzhou on the edge of the Gobi desert, through grasslands full of nomad camps and tens of thousands of yak and sheep. Rock photographed a *drokwa* wedding—men and women in silks and furs ornamented with silver and coral. Then it was back to Wobo, another arduous journey, and very cold. By November they were back in Xining. They took the route through Labrang and Heicuo to Chone, Rock issuing a stream of verbiage about filth. A flag printed with the character

"Ma" flew over Labrang, occupied and humiliated. The marketplace seemed a ghastly microcosm of the recent carnage. "Frozen dead dogs lay in the middle of the street; dead birds, wool, stagnant coagulated blood, large pools of it frozen, slaughtered animals, entrails of yaks and sheep, legs of cows and sheep, dirt and rubbish in which dogs and chickens looked for something to eat. Halfway into the village they were slaughtering something, and buckets of blood were dumped into the street."[80]

They reached Chone in early December, with an escort of mounted Tibetan cavalry. It had been a "hard, hard trip," and Rock didn't want to go anywhere else for a long time. The Yang *tusi* gave them comfortable quarters, the residence of an absent incarnate lama. It had space enough for all the Nvlvk'ö men and a foreign stove in Rock's room. They spent the winter there, with the exception of one quick trip to Diebu. The Nvlvk'ö men made boxes, packed and labeled specimens and seeds, and stuffed and packed birds. Rock experimented with photography and took notes on *cham* dances, butter festivals, and other public rituals.[81] Gansu was experiencing a brief, tense peace, enforced by Liu Yufen, and Chone was quiet as war raged in eastern China.[82] Rock bought the press's only copy of the Tengyur classic and one of its three copies of the Kangyur for the Library of Congress. The Nvlvk'ö men packed the 317 volumes into 96 boxes and loaded them onto seven big mules. Rock photographed the caravan as it paraded through Chone's streets on the way to Lanzhou.[83] Years later, he may have recalled this scene as he sat with a dongba translating texts about Ddibba Shílo, who descended from the skies with a caravan of yaks loaded with books.

THE MOUNTAIN

In the spring, Rock negotiated with the Xidaotang again for guides to Ragya gompa and the Amnye Machen. The war for Labrang had unsettled trade between the Muslim *xiejia* and the nomads of the wool-producing grasslands. Though the Xidaotang had not taken sides, the patron *dewa* of Labrang were not inclined to welcome Muslims of any stripe into their territories. The collective had written to the Jamyang Shepa, now residing in Ngura *gompa* in the remote grasslands, begging him to intercede on their behalf, but the incarnation's father had refused. Rock mulled over alternate plans while waiting for the snow to melt on the grasslands. Perhaps the party could go to Xining, rely upon Ma Qi to escort it to Ragya, and find its own way through the grasslands beyond. Or perhaps he could go to Ngura gompa and appeal to the Jamyang Shepa for protection to the Amnye Machen.

In the meantime, the party prepared again for the journey. This time Rock bought cheap, unshod Tibetan horses used to living on grass for all the Nvlvk'ö men: he had discovered how difficult walking in the vast grasslands could be. He prepared provisions for a full seven months. His companions lived on noodles and bread: they would need 2,000 *jin* (2,666 pounds) of flour. Rock figured forty yaks would be required to carry food, tents, bedding, silver, photographic equipment, winter clothing, grass paper for drying specimens, cotton for stuffing and packing them, cloth to trade with the Golok, and silk, thread, needles, sateen, and satin brocade as gifts. He bought more rifles and ammunition from the Yang *tusi.* He discovered that he would need beans and peas for the horses, since the grass would not be up on the grasslands for the first month. After packing, it turned out that forty yaks would not be sufficient: the party needed sixty. He arranged for the yaks to be bought at Labrang; thirty-three mules would carry them there. He used up all his money, including his salary; he begged Sargent at the Arnold Arboretum to send more.

The party set off for Labrang at the end of April without seeking Ma Qi's aid. Rock hired sixteen Tibetan men, mounted and armed, as an escort. As an interpreter and guide, he hired William Simpson, a young missionary who had grown up in the region, had traveled widely in Amdo, and spoke the *drokwa* dialect. The caravan of thirty mounted men and thirty-three loaded mules paraded through the string of small monastery towns that lay between Taozhou and Labrang. At Labrang, where the occupying garrison's bugle calls punctuated the days, Ma Qi's uncle watched the party with curiosity but did not stop it. The loads were transferred to sixty yaks, belonging to a group of nomadic Tibetans of Mongolian descent, whom Rock called the Sokwo Arik. Twenty nomads, armed and mounted, accompanied them as an escort to Ragya.

At Ragya gompa, situated on the Yellow River, they were received by the ranking incarnate lama, Shingja, the reincarnation of Tsongkhapa's mother. He sat on an air mattress and showed them a room filled with more than fifty clocks and watches, all ticking, all set approximately to the correct time. Rock presented him with a twenty-dollar gold piece and a letter of introduction from the Jamyang Shepa. The monastery's intendant explained to Rock that Ma Qi had mounted several attacks on the Golok descent groups *(shogpa)* that lived in the grasslands on the other side of the river, between Ragya and the Amnye Machen. Ragya, with no armed protection, had no choice but to cooperate with Ma, earning the resolute hostility of the Golok *shogpa.* Ragya could not offer Rock a protective escort, and the Golok enmity towards Muslims likely included "Russians" (foreigners) as well. Rock and the incarnate lama agreed that he should send the Golok chiefs the letters he carried from the Jamyang Shepa and await their response.

The party settled into the comfortable home of an absent incarnate lama, Rock occupying three rooms on the second floor, the Naxi man taking the ground floor. Around them were the dormitories for the *gompa*'s eight hundred monks. The party roamed the Yellow River's banks, collecting specimens, shooting partridges, photographing blue sheep, and filtering the silty river water to develop photographs. A month after their arrival, they received a reply to one of the Jamyang Shepa's letters. The chief of the Khangsar *shogpa* sent three plain sheets of paper marked with his seal and an oral request to the Ragya lamas to fill them in with orders to protect Rock's party, to be sent to the chiefs of three tributary groups *(ts'opa)* who resided near the mountain. The incarnate lama suggested that Rock join an expedition of sixty monks who were going from Ragya to the Ongthak *shogpa*, of which his brother was chief. Rock demanded twenty yaks and an escort of thirty armed men; they did not appear, and Rock's party set off alone, with fifteen yaks and no escort, for the Gyunpar range, divided from Golok territory by the Yellow River. A climb to the highest peak, at 14,300 feet, and a break in the clouds gave them a view of the Amnye Machen, seventy miles away, still too distant for a photograph.

In July, in the incarnate lama's absence, Rock resorted to threats. He told the gompa's treasurer that as the monastery had no authority over its Golok subjects he would seek the assistance of Ma Qi and his army. Results were immediate: the treasurer agreed to deliver a dozen yaks and twenty mounted men. But the day the party was to leave, Rock was told that fifty Golok were lying in wait for him en route. Angry and grieving, he fantasized about the Golok being beaten into submission. "Airplanes, to my mind, would be the only weapons that could subdue the Gologs, and it would be the least expensive method." It was a prescient suggestion: the PLA bombed and strafed Golok camps during the "democratic reforms" of 1956. Defeated, Rock prepared to return to Labrang. The next day, he met Gombo, the chief of the Jaza *dewa*, camped just across the river. No one was waiting to ambush the party, Gombo told him. Rock decided to go after all: he would get an armed escort of thirty men; they would take horses only, no slow yaks, and only three of the Naxi men would go. Rock would "rough it," taking only his tent, some bedding, some *ts'ampa*, rice, tea, and butter, "and some cake the cook has been making today." They started on July 13, sans escort, crossing the Yellow River on rafts of inflated sheepskins. But when a horse jumped from the raft and drowned, Rock lost heart and the party returned to Ragya.

They set off again the next day: Rock and four of the Nvlvk'ö men—Li Shichen, He Guangyi, He Qi, and Yang Jiding. With some hesitation, Rock invited Simpson, whose gentle manner with their Tibetan companions he blamed for the previous day's disaster. The party's guide was Gombo, who took them first to his camp then to the camp of an allied *dewa* closer to the great mountain. From there they walked up the Tarang valley to a summit thirty miles away from the Amnye Machen. Rock took a photograph of the distant range through his Zeiss binoculars. In his diary, he made the most of the moment:

> Before us lay one of the grandest mountain ranges of Asia. I should judge the Am-nye Ma-chhen to be about 21,000 feet or more in height. It was difficult to tear myself away from this sublime view, especially as I shall never see it again. The glorious mountain rose thousands of feet into the sky in purest whiteness, a real emblem of purity; Chen-zerig, the huge pyramid of snow is especially beautiful. I could have remained for hours at the summit of that mountain, never tiring of this grand spectacle which no other foreigner had the privilege to enjoy.[84]

The party walked back down the mountain and sped back to Ragya in three days. With the others, they made their way back to Labrang over the late-summer grasslands, thronged with blue sheep, wild yaks, wild asses, and festive gatherings of *drokwa*. In a few weeks they were back in Chone.

Years of effort, enormous expense, great worry, and intricate and prolonged negotiations had led merely to this: a ride of a few days to a summit, a routine ecstatic moment, a photograph through an improvised telescopic lens and a perfunctory (and inaccurate) announcement that he was the first foreigner to see the sight. With much strained prevarication he attempted to manufacture a triumph. "We did reach the Amne Machin range under the *greatest* difficulties," he wrote Sargent, from Chone. "The Ngoloks were waiting for us in the Gurzhung valley west of the Yellow river from Radja. They tried to intercept us. We were warned by some friendly person and this *saved all our lives*" (the emphasis is Rock's).[85] After his first sight of the mountain, he had written to Sargent, "I had a wonderful view, *my first*, of the Amne Machin range. I counted nine peaks, one a huge pyramid at least 28,000 ft in height; it may prove higher than any Himalayan peak including Everest."[86] And though he knew this estimate to be outlandish, he stuck to it in public: his article in *National Geographic* trumpeted: "I shouted for joy as I beheld the majestic peaks of one of the grandest mountain ranges of all Asia. We stood at an elevation of 16,000 feet, yet in the distance rose still higher peaks, yet another 12,000 feet of snow and ice. . . . I came to the conclusion that the

Amnyi Machen towers more than 28,000 feet."[87] In his monograph on his journey, published nearly thirty years later, he admitted that the mountain could not be much more than 21,000 feet.[88]

.   .   .

Rock spent more than three decades working his glimpse of the Amnye Machen into his private mythology of the landscape of west China, which he would center on the Mu lineage in Lijiang. In his monograph on the mountain, he included photographs of several paintings of its mythological layers. One, a gift from the chief incarnate lama of Ragya gompa, portrayed a seated Buddha. In the lower right corner was Machen Pomra, the deity embodied by the Amnye Machen—a warrior on a white horse, an eagle in one hand and a vase in the other. Another, a painting he found in Lijiang, showed the deity again as a warrior on a white horse. He translated a dongba text in which appeared a god named Muanmi Bpalo, which he tried to show was the Naxi name for Machen Pomra. The ancestors of the Naxi, he theorized, had migrated to Lijiang from the grasslands near the great mountain, bringing the deity with them.

He would spend the last half of his life finding ways to animate the landscapes of his experience with words from Chinese, Tibetan, and Naxi texts. His journey to the Amnye Machen began this project. Searching for the mountain, he learned that the language animating the earth was not confined to names and descriptions of plants. In many places, generations of humans had layered the earth with names, histories, and genealogies. The places where he found this layering to be densest compelled him in a profound way. They became the foundation for a developing relationship with the earth as a social being, which became more important to him than his relationship with any human being. This is one reason I think Rock's experience is worth knowing about, this relationship, forged from movements of feet over the earth and eyes over texts—experience and archive. Difficult, contradictory, and idiosyncratic though it may be, it is still a contribution. How are human persons shaped by the earth? This is one way: in its capacity to hold together diverse names, stories, histories, and experiences, the earth makes possible unlikely collaborations and convergences, here a vision of a mountain god and a personal sense, without the words to express it, of the deeply sacred nature of a place. In such convergences, our divergent pasts, languages, and senses of life and sociality impinge upon each other.

8. Adventurers

They stand in a courtyard in Chone in May 1925. Two orderly rows of six, those in back on packing cases offset precisely to allow them to appear between the shoulders of those in front: the photographer was a meticulous autocrat. All in the back but one have their arms crossed: this too was probably at the photographer's bidding. They are all dressed alike: double-breasted jackets, sturdy trousers, new boots—a mass purchase with expedition funds, still creased from the shop shelves in Old Taozhou. They must have needed the clothing, especially the boots, very badly. They had just survived the difficult trek from Kunming through Sichuan to southern Gansu. During the worst bandit scare, in the village of Banbian Jie, the eight then with the party lined up bravely with Rock's rifles to defend the door to his room in the temple until he took the guns away, telling them to leave the shooting to the soldiers. That night, as the bandits seemed about to overrun the village, each concealed a hundred dollars in silver on his person, hoping to hide and survive. Now they are preparing for another journey, more arduous by far, to the Amnye Machen, inhabited, they are told, by ruthless nomadic brigands.

Rock rarely wrote their names in the captions of his photographs. Some, however, can be identified by matching other photos with diary entries. Surly, handsome Li Shichen stands in the back row, far left. He Ji is also in the back, second from right, wearing his habitual heavy expression. He Shuishan stands hatless at center front, looking vulnerable and very young. I think that the intelligent and capable Zhao Zhongdian is on He Shuishan's right, but the only labeled photos of him are from ten years later, and he had changed a great deal. Somewhere in the group is the lively, intellectual He Guangyi, who became one of the unacknowledged coauthors of Rock's magna opera, including his two-volume *Na-Khi Naga Cults* and the two-

Figure 42.   In Chone. Courtesy of the Arnold Arboretum.

volume *Na-Khi-English Encyclopedic Dictionary*. The others are Yang Jiding, Lu Wanyue, Lu Wanxing, Li Shiwen, and three out of these five: Li
Shixi, He Zhu, He Zixiu, Li Wenzhao, and He Wenli.[1]

They were all very young. Rock's habit of calling them his "boys" and
writing of them often as "my children" was pernicious, of course. But for
all their experience, they were really not much more than boys. Their ages
are not recorded anywhere, but not many could not have been much older
than twenty. At least eight had worked for Rock before this expedition. A
few began working for him in July and August of 1922 in the meadows of
the Yulong range. More traveled with him that fall to the Tai Lü states on
the western border of Yunnan that fall. Eight, including Li Shichen, Lu Wanyue, and Yang Jiding, camped with him on the Yulong range in the summer
of 1923. Six others had traveled on their own to Yunnan's far northwest for
him, and on their return he had followed them back to the upper Mekong
and the Nu-Qiu divide. Finally, the core group of eight had taken him to
Muli in the winter of 1924. During this period, Zhao Chengzhang and his
own crew of eight tough, experienced botanical explorers were making their
final forays into the Nu-Qiu divide, while smaller groups of seasoned explorers from Nvlvk'ö were ranging widely through other regions. Some of

these men had been collecting in Yunnan since 1905. Many were old enough to be the fathers of the young men Rock hired. Some very likely were.

These youths did not have the experience of the older botanists, but they were remarkably skilled nonetheless. Rock seems to have spent little time deliberately training them. He simply expected them to be able, at the minimum, to discern minute differences between species and varieties, to collect what he asked for, and to press, arrange, label, and pack specimens. Lu Wanyue, the taxidermist, created bird specimens of museum quality; the two hired to hunt birds were fine marksmen; the carpenter quickly rendered rough timbers into light, sturdy packing cases (he was less successful in pulling Rock's teeth.) Their village had been the epicenter of botanical exploration in southwest China since before they were born. As children they had played in courtyards full of men and women sorting and drying specimens; their fathers and uncles passed botanical skills on to them in the same way that the carpenters, leather-workers, and basket weavers for which their village was locally famous had long apprenticed their own children. They were skillful, knowledgeable, and resourceful. But still, some were surprisingly vulnerable as explorers. A few spoke no Chinese, the crucial lingua franca for anyone who ventured out of the Lijiang basin.[2]

There were many differences between the two generations. None of the twelve "boys" were old enough to have been members of the councils of elders who decided local ritual and political affairs. None were among the heads of lineages who guided funeral rituals and the yearly *muanbpo* "sacrifice to heaven." In part 1, I argued that Zhao Chengzhang and his colleagues engaged in a rereading, through walking, of journeys they had heard dongba recite many times, sharpening the geographical knowledge they absorbed from the dongba archive into practical maps of the region. It is unlikely that the younger men who went to Gansu had much acquaintance with dongba texts, though some would later learn a great deal about them as scholars. On the other hand, they were beneficiaries of the extensive practical expertise in northern Yunnan's geography acquired by the elder generation in their two decades of exploration. They learned about this geography through stories of journeys, maps sketched on the ground, talk about routes and alternatives, and advice about places where new species were most likely to be found.

This inheritance of geographical knowledge made Rock's career possible. By 1924, he had already developed a reputation as an intrepid and knowledgeable botanical explorer of China. In only two years he had made two journeys to the upper Mekong (Lancang), one to the upper Salween (Nu), and one to Muli. These were regions it had taken Forrest many years to begin to explore. Rock had (by his own accounting) produced some 50,000 herbarium specimens, including 489 specimens of *Rhododendron*. He had collected seed of about 1,000 species, 1,700 bird specimens, and a few hundred mammal specimens. And he had taken 1,055 photographs.[3] None of this would have been conceivable without the knowledgeable guidance of the men from Nvlvk'ö, who were working over what was, for their fathers and uncles, intimately familiar territory. It was this reputation that convinced Sargent to stake forty thousand dollars on Rock's expedition to Gansu, from which flowed all his further opportunities for employment.

Why did they go? Less than a day after walking out the gates of Yunnanfu in late 1924, they had already reached territory into which no explorer from their village had ever ventured. All their geographical knowledge would avail them not at all for the next three years. Before he left for China to begin his expedition, Rock had telegraphed Li Shichen from Boston, asking him to gather seven others and meet him in Yunnanfu. He did not telegraph their destination, and when they left home, the young men had no idea they would be traveling in one of China's most dangerous regions for three years. Yet even after they did learn of the expedition's goals, they seemed eager to go. Long before Rock had gathered funding for the expedition, Li Shichen had written him volunteering to travel to Yunnanfu, Tengyue, or Bhamo with as many collectors as Rock wished whenever he might want them.[4] After the party arrived in Yunnanfu, in November 1924, Lu Wanyue, who had been working in Dali, telegraphed to ask if he might be wanted. When Rock assented, Lu set off with three others for Yunnanfu. There were bandits on the main road, and they took a long, mountainous footpath instead. While Rock's caravan was making its way along the harrowing road to Chengdu, stalked by hundreds of bandits, the four of them followed a few days behind, unarmed and inconspicuous. They too met bandits, and they took a mountain path to avoid them. They finally caught up with the caravan on Christmas day, as it waited out a snowstorm in Zhaotong.[5] Rock had not planned for twelve assistants: he ended up with this many because of the enthusiasm for this work that seemed to infect both generations of Nvlvk'ö villagers.

Certainly they were attracted by the money. Before Rock left Boston he sent a check of three hundred rupees to the China Inland Mission in Lijiang to be converted to silver and given to Li Shichen and the seven others. And he sent a hundred Shanghai dollars to the American consul in Yunnanfu to be converted to Yunnan currency to pay the men's expenses once they arrived. He promised to raise their monthly wages by a dollar every year of the expedition and pay them a bonus of a hundred Mexican silver dollars at the end.[6] He did not pay their wages directly to them, however. Every few months, he sent a check to the postal commissioner in Yunnan, who authorized the postmaster in Lijiang to pay the China Inland Mission, which handed the money over to their families.[7] He seems nowhere to have recorded the exact salary he paid them. But six decades later, people in Nvlvk'ö would recall that it was two and a half silver dollars per day—extremely high for the time and place.[8] Given that Forrest already had enough collectors that season, the only other paid employment they were likely to find was as muleteers, at a half a Yunnan dollar per day.[9]

Yet there may have been other reasons they were so quick to embark. In an analysis of gender narratives in the Lijiang basin, Sydney White notes that in the 1990s Naxi men spoke of proper masculinity as produced in fearless quests for fame as warriors, travelers, or scholars. Women emphasized suffering and self-sacrifice, bearing the burden of most agricultural production, while Naxi men specialized in consumption. In particular, men placed value on fame achieved during youthful exploits like epic journeys or participation in the underground Communist Party. In arenas of fame, Naxi men were extremely competitive: "Those who achieve fame are resented and envied, and those who either fall from fame or never attain it are ridiculed," White writes. "The long-term, long-distance trips and frequent disappearing acts [of] some basin Naxi men prior to 1949 were often inspired by a search for fame in the relatively 'wild' areas inhabited by Tibetans and other minority peoples." Such journeys, she points out, were often ways to avoid conscription—far more likely to end in ignominious suffering and death than in fame as a warrior.[10]

Rock's own fame must also have commanded their attention. They seem to have been attracted to him not only by the wealth that could organize caravans and buy soldiers but also by his air of learning. They called him *boshi*, "professor," a title he awarded himself as much for their benefit as for that of the officials he wanted to impress. Some, it is clear, looked on him as a teacher, with the respectful affection that, in the dongba and Buddhist traditions of scholarly apprenticeship, one owed to one's personal lama or dongba. For all this, he predictably, if inconsistently, worked to found his relations with them on the code of racial difference. In this unsettled context, in which waves of nationalist enthusiasm, critiques of Western imperialism, contempt for Western missionaries, and suspicion of foreign travelers were sweeping unevenly through China's western provinces, this code took enormous labor to reproduce. They all participated in this labor. Indeed, the public function of traveling with a personal escort of twelve young men was to insist in a loud voice and in the face of frequently open incredulity, upon the difficult fiction of racial superiority. Because of this difficulty, he was extremely sensitive to perceived slights by officials, military men, or high-ranking lamas. And he often worried that other Westerners were not doing their part. As his party neared Chengdu in February 1925, a young foreigner walked up to him and presented him with a letter from a missionary asking for employment. The man, named Gaskoff, told Rock he had been a lieutenant in the Russian army and had walked from Kashgar through Tibet to Dali and then to Chengdu. Suspecting that he was a Russian Jew instead, Rock gave him five dollars and left him in the road. "He had neither bedding nor baggage and he sleeps in the lousy Chinese Inns and the bedding they provide. My men could not understand how a white man can run around in such fashion . . . It is really embarrassing to meet such people, as it lowers the prestige of the white man considerably . . . it would not do to have him sleep with my men, and he probably had lice."[11]

He took it as given that they should be grateful for his protection, that they should obey his orders without hesitation, and that they should not question the salaries he chose to pay them. But still, in low moments, he wondered why they had come. In October 1925, the party was traveling from the Kokenor region to the city of Ganzhou to replenish provisions before setting off again to explore the great lake. They found a village of fortified farmhouses and a tiny temple, where they pitched their tent in the yard.

> The loads are spread out before the three solemn gods, and Yang is cooking on the terrace; the escort sleeps on straw on the other side, and my boys are cooking back of my tent against a pile of mud bricks. . . . I felt today like a lonely tramp when I looked at my men and said to myself, what fools to leave their lovely snow range near Likiang and their families for this, and I, well, I am alone and elected to be here. A new green moon looks on over the dusty dreary landscape, cold and full of misery, I can well understand why the Buddhist sees every living creature groaning for delivery.[12]

In fact, they were not always willing. In Zhaotong, after only twelve days of eventful travel, the cook Yang Jiding and another man to whom Rock had advanced sixty dollars, asked to go home. "I gave them a stern lecture and that was the end of the episode," Rock wrote in his diary. "When one has to deal with Orientals, one is never certain where one stands with them. A firm hand is necessary."[13] They worried about their distant families as news of warlord clashes and bandit armies in western Yunnan reached them. And they were often homesick. Nearly a year into the expedition, the party arrived in Wobo on the Amdo grasslands. They had walked seventy *li* (about twenty-five miles) that day; it was snowing, and the caravan of camels, carrying tents and provisions, seemed to have gotten lost. Zhao Zhongdian approached Rock and complained of the harsh country. A year was enough, he said: he and several others wanted to go home. This time, instead of lecturing, Rock grew despondent: "It made me very sad and I wandered up and down the dark street in the deep snow between the low mud huts to the dogs barking inside the courtyards. . . . I felt deeply hurt as I always hoped they would never say that they were tired and wanted to go back, and I—what shall I do, remain alone? In the bleak and dreary forlorn place I told them they could go if they wanted to, and I would do the best I could."[14]

The next day the party traveled through a deep blizzard at over 10,000 feet, Rock on his horse, the others wading through the snow in cheap shoes. The camel caravan got lost again, and two of the "boys" went out to search for it, fording a stream that caked their legs and feet in ice. They lodged in a small cave inn used by gold prospectors, where, "to keep my boys cheered up," Rock told them stories of meeting tigers and wild elephants in northern Burma and Siam.[15] Despite everything, none deserted him. Certainly, they displayed enormous loyalty. In addition, however, they knew that were they to leave, they would fare very badly on the road home. Without his mountains of provisions, his letters to local potentates calling up escorts, and his capacity to hire caravans they would be resourceless in this war-ravaged and famine-stricken landscape.

. . .

Though he had hired most as collectors, a few were specialists in other work. Foremost among these was Yang Jiding, the cook. Yang had worked for Rock since his first journey from Lijiang to the upper Mekong in 1923. He was the only one of the twelve who did not grow up in Nvlvk'ö: he was a native of nearby Wenhua village.[16] His mother was Tibetan, from the upper Mekong area: perhaps his father had met her during the long-distance trade that many Naxi carried out throughout the region. He spoke Naxi, Chinese, and Tibetan. Rock had expected him to serve as an interpreter, but he proved useless for this purpose as the Amdo dialects were very far from Yang's southern-Kham Tibetan. Rock considered him the most important member of his staff, and he fretted inconsolably when he was laid up for a few days in a hospital in Lanzhou. Except on the occasions when he was hosted by missionaries, magistrates, or hereditary chiefs, Rock ate alone, with Yang serving him, his only commensal companion. He was frequently peevish with his cook, complaining of neglect or bad planning. Just as frequent, however, were times like a March evening in 1926 in Diebu. It had been a very long day. Rock had not eaten since five that morning, and he was dead tired and angry. After a futile search for suitable lodging in monastery and village, he pitched his tent on the flat roof of a house. "I had only a piece of cake and some tea left. However at dark my cook appeared, good soul, with something to eat ... he came along through the snow with tea, bread, a pheasant, and some macaroni; that really saved the situation."[17]

Traveling with Rock in 1931, the journalist Edgar Snow ("a greenhorn of the first order and an impertinent sponger") wrote of the daily miracles of organization that Yang Jiding performed. "The tribal retainers divided into a vanguard and a rearguard. The advance party, led by a cook, an assistant cook, and a butler, would spot a sheltered place with a good view, unfold the table and chairs on a leopard skin rug, and lay out clean linen cloth, china, silver, and napkins. By the time we arrived, our meal would almost be ready. At night it was several courses ending with tea and liqueurs."[18] Yang was an exceptionally resourceful cook. In a miserable Sichuan inn, Rock wrote, "How the poor man can prepare meals in such places is a marvel. This morning he made fine cruller, and it did not take him long, only about 5 minutes. There is not a European or American cook who would cook a meal in such surroundings and after a hard day's travel. It is not an easy job, without a table or chair, and to cook everything on a charcoal brazier, baking cake or biscuits in a kerosene tin."[19]

After the Gansu expedition, Yang Jiding continued to work for Rock, participating in nearly every one of his journeys. When Rock settled down in Nvlvk'ö, Lijiang, or Yunnanfu, Yang lived in his house, his nearly continuous companion until his final departure from China. In 1935, in Yunnanfu, Rock fired him for having fallen under the influence of "an unscrupulous woman." A year later, Yang appeared on his doorstep, "looking like a tramp" and asking for forgiveness. Rock was relieved and delighted. When Rock left Lijiang for the final time in 1949, he declared to the American consul that he had been threatened by "4,000 bandits of a Communist nature." His cook had betrayed him, turning Communist, and leading a party of seventy to his house in Nvlvk'ö to search for hidden arms. Yang Jiding and his band left after the villagers persuaded them "not to act like bandits."[20]

Lu Wanyue, known to his friends as Lao Ru, was the expedition's taxidermist. He preserved over fifteen hundred birds for the Harvard Museum of Comparative Zoology. The museum's ornithologist, Outram Bangs, commented on his splendid skills.[21] For his taxidermical work, Rock paid him ten gold dollars a month; when he was not stuffing birds, he worked with the others, collecting.[22] He became a favorite of Rock's, who took him and Li Shichen on a brief tour of east China and America immediately after the expedition, in 1927. After the trip, Lu Wanyue married and worked his family's land in Nvlvk'ö. When Rock said goodbye to him for what he thought was the final time in 1936, he was still wearing the overcoat Rock had bought him in Shanghai. "He was young and strong, a boy of 19 summers," Rock wrote of him then, "and today he is 38 and looks as if he were 50. Illness and . . . hard work . . . have wrought many changes."[23]

In addition to cooking, stuffing birds, and collecting, the men did a variety of other work: overseeing the packing of the caravan, setting up and taking down Rock's camp, seeing to his personal needs. But perhaps their most important task was to provide him companionship—to take their places in the domestic circle he had arranged for himself. The burden of this labor was uneven, however, falling mainly upon only a few, most particularly Li Shichen and He Ji. And it had its hazards. He Ji saw to Rock's horse, feeding and watering it and leading it when Rock walked or rode in a boat or chair. He was often at Rock's side, and when Li Shichen was out of favor, he sometimes slept in Rock's tent to keep him company. In Gansu, this intimacy occasioned some vicious spats. In August 1925, traveling in the Diebu region, Rock woke one morning at five as usual and, having rousted the "lazy Yunnanese," he found his boxes of personal belongings and his tent bag out of doors, covered with ice. He scolded the men. Then "wild" He Ji tore his bedding while packing it. "I got angry and told him to look, giving him a push, whereupon he made out I had hit him. I then told him to leave the tent. He took a belligerent attitude and refused to budge. He tried to grab my arm, took hold of my thumb, and tried to twist it. I broke loose and quickly took my revolver (Colt 45), and if he had further attacked me I would have fired."[24] Rock turned the rebellious young man over to his military escort, soldiers of the Yang *tusi*. "I shall send him back to Yunnan under escort, reporting his murderous attitude towards me to the authorities," he declared. Two days later, the party camped by the Tao river, a day from Chone. Rock caught He Ji attempting escape, boarding a raft with his bedding. He threatened the raftsmen with jail, and they put He Ji off. He Ji marched to Chone under guard, his bedding on his back, then spent the night in the monastery town's extremely unpleasant jail. In the morning, Rock had him released, but when the party left for Lanzhou four days later, he was left behind "as punishment," with orders to return to Diebu to collect seed.[25]

Without doubt, the most significant relationship Rock formed with any of his companions was with Li Shichen. On his previous journeys Rock had employed as "head boy" an English-speaking Siamese he knew as Boomah, but before leaving for Gansu, he had discarded him as "unreliable."[26] Li Shichen who, like the rest, knew no English, took his position, becoming Rock's right-hand man on all his subsequent journeys. He frequently shared Rock's tent and, during the winter, his room in the Chone monastery. With Yang Jiding, He Guangyi, and Zhao Zhongdian, he was one of those who lived with Rock during his stays in Nvlvk'ö, Lijiang, and Yunnanfu. Under Rock's tutelage, he became an accomplished photographer. He is the author of many of Rock's "self portraits," including the fabulously absurd photo of the botanist on horseback, wearing the luxurious silks and furs given him by the Yang *tusi*. He clearly had some training in ritual healing traditions: he described to Rock in great detail an exorcism ritual performed by *lubbu*, Naxi spirit mediums. After Rock gave up botany and turned to the study of dongba texts, Li Shichen became, along with He Guangyi, his tutor and scholarly collaborator, working for years at his side, interrogating dongba and translating their texts.

ADVENTURE

At best, they provided uneven ground on which to found his sense of himself as a person in relation to intimate others. He often wanted to make them his children, and they acceded readily enough, accepting his patronage while occasionally rebelling against it as proper sons would. The obstacle was not so much in them as it was in himself. He always eventually returned to his foundational sense of himself as fundamentally alone, a wanderer at the edge of the social.

Upon returning to Chone after the journey to the Amnye Machen, the party learned that the non-Muslim warlords who occupied the province's southern and eastern regions had rebelled against the governor, Liu Yufen.[27] The warlords Zhang Zhaojia and Han Yulu had advanced towards Liu Yufen's base in Lanzhou from the east; Kong Fajin was moving in from the south. Two thousand of the Yang *tusi*'s Tibetan militia had gone to defend Lanzhou. Feng Yuxiang's Guominjun troops were under siege in Xi'an, to the north, and in the south Sichuan was engaged in another of its interminable wars. The people of Gansu were suffering: to support the wars, the governor and warlords taxed their populations mercilessly and requisitioned many thousands of draft animals. Food prices were spiking, and a violent hailstorm destroyed the crops around Chone. Rock was out of money. Sargent had promised him funds, but he had attempted to send money directly to Gansu rather than to the Shanghai bank on which Rock was issuing checks, and it stalled in a bank in Peking. The Yang *tusi* reluctantly lent Rock eight hundred dollars until a sympathetic Italian postal commissioner at Lanzhou sent six thousand Mexican dollars in a special camouflaged caravan, on a check that had no funds to back it in Shanghai. Rock was frightened. Anti-foreign nationalism had finally taken root in Gansu, and he heard stories of missionaries being killed in Sichuan. There was no safe route out.[28]

The thirteen of them made another trip to Diebu. "If the writer of Genesis had seen the Tebbu country, he would have made it the birthplace of Adam and Eve," Rock soliloquized. They explored forests of huge conifers and collected maples, cherries, apples, picea, juniper. Back in Chone, Rock fretted about getting his collections out. A mule caravan to Chengdu was possible, but the mules were likely to be requisitioned by some military commander and the loads dumped on the road. And parcel post no longer existed. He made a plan to repack the collections in small boxes and send them by registered letter post from Lanzhou all the way to Boston. The Nvlvk'ö men fit the thousands of plant specimens into eighty-six boxes, the 1,044 birds into twenty. They took the packages, twenty or thirty at a time, to Taozhou by yak cart; mule transport carried them to Lanzhou; the postal service took them from there by circuitous routes to the coast.[29]

· · ·

His long experience in west China had taught him the importance of lineage and descent to the rooted concentration of wealth and power. By now, he saw west China as a vast sea of war on which floated a few tiny islands of peace and stability. On these islands, a few princely lineages of *tusi* and incarnate lamas, having taken root centuries ago, were barely surviving the cataclysms that wracked China as it emerged into nationhood. Though much diminished, their wealth, power, and influence still made possible a kind of domesticity that he could find comfortable: luxurious surroundings, obsequious servants, an aura of privilege. This existence depended upon the capacity to display their lines of descent in records carefully crafted and guarded. Their cramped old palaces at Chone, Muli, and Yongning were hardly the most comfortable places he knew: his houses in Lijiang and Yunnanfu were better suited to the style of life he loved, and indeed he stayed in those cities for much longer periods. But he never fantasized about settling into them to live out the rest of his life as he dreamed of each of the old palaces, and even of the lovely meadow in the Lijiang range just above the site on which the Mu kings had built their summer villa. In this landscape, alternating for him between the purity of depth and the filth of extension, these were the deepest places. Even their physical form gave abundant evidence of this. The walls of Chone were centuries old, and they held the graves of innumerable monks. "The scenery has not changed for 4,000 years," he mused one lonely winter evening there: "all is as it was that long ago; only the buildings show their age . . . peace, quiet, and oblivion reign supreme."

In December, slowly emerging from a deep depression, he attended the *cham* dances, taking notes. That evening, he mused in his diary,

> Since the Lamas' dance, the Akus [young monk recruits] . . . have been blowing their trumpets during the night like the moaning of a lost soul . . . For centuries, the walls of this monastery have reechoed . . . the self-same sounds . . . long after this and other generations shall have gone forth, these trumpets shall resound through the medium of life . . . The word "who" . . . What potentialities it embraces . . . Life is not continuity . . . but periods . . . the inanimate alone is continuous. Transmigration of the soul is more logical than the chance existence of an individual for a brief span of years. [The latter] would be cheating innumerable beings out of realization and experience: the freak chance of self-restraining bachelors or old maids who lapse without leaving offspring. When one contemplates the fact that "one" "I" is the product of a continuity since the time when man came upon the scene it is truly startling . . . Does this mean that I am the tyrant who denies existence to what may account thousands of generations? The thought is indeed overwhelming. . . . If [trans]migration of the soul is considered, the thought of being the end of experience, realization, and life to thousands is less overwhelming, and as it soothes my guiltless conscience, I deign to believe in a wandering soul.[30]

The endless, inchoate current of writing generated by this lonely wanderer's movements through the world rarely explored the foundational machineries on which his effortful life turned. I have attempted to understand his writing as an effect of these machineries rather than a synthesis or analysis of them. I am tempted to treat these musings about lineage, descent, and reincarnation in the same way—to attend to their form, if possible, but to dismiss their content as the erratic ramblings of a lonely man, little more consequential in themselves than any of the tens of thousands of passages that describe turning up a valley, passing a conifer, crossing a stream, and proceeding twenty paces more. Yet this passage is unusual. The Rock I have come to know would have been more likely to make a bad-tempered comment about the trumpets keeping him awake, belittle the notion of "transmigration of souls," and move on. But this evening, after shivering a little at the moaning of ghosts trapped in the monastery's old walls, he made a choice to commit, for the moment, to belief in this idea because it gave him a wider framework in which to insert the "period" of his life. It was an alternative to the more familiar framework of lineages linked by "blood" to which he, because of his deliberately undefined sexual orientation, could not commit his life.

Fifteen years before, the sociologist Georg Simmel had published an essay titled, "The Adventurer." It is an opaque few pages. Yet in its compact expository brilliance, it has seemed to some to encapsulate the core of Simmel's nuanced philosophy of society.[31] For Simmel, an adventure was "a particular form in which the fundamental categories of life are synthesized." This form was a dreamlike island in the sea of a lifetime. It had coherence, a beginning and an end, even a perceived meaning, and it was felt to form a whole. But it was also interwoven with one's uninterrupted existence. "Indeed it is an attribute of this form to make us feel that in both the work of art and the adventure the whole of life is somehow comprehended and consummated—and this irrespective of the particular theme either of them might have." In the essay's most perplexing passage, which few of its many commentators have taken on, Simmel declared that in some cases a whole life might be an adventure: "Occasionally . . . this whole relationship is comprehended in a still more profound configuration. No matter how much the adventure seems to rest on a differentiation within life, life as a whole may be perceived as an adventure. . . . To have such a remarkable attitude toward life, one must sense above its totality a higher unity, a super-life, as it were, whose relation to life parallels the relation of the immediate life totality itself to those particular experiences we call adventures."[32] An analogy, then. Adventure is to life as life-as-adventure is to a higher unity. It is because it is embraced by the inchoate sea of a life that an adventure can sharply synthesize the elements of life—such as activity and passivity, certainty and uncertainty, chance and necessity—tightening them into a unity.

It is the adventure's connection to the center of the life in which it is embedded that makes it possible for adventurers to treat the incalculable as though it is calculable, to "risk all, burn our bridges, and step into the mist, as if the road will lead us on no matter what" and still find in that gamble a sense of unity.[33] For those who live life as an adventure, the same formal relation exists, but now between the whole of life and something else that surrounds it—a higher unity. "Perhaps we belong to a metaphysical order," wrote Simmel,

> perhaps our soul lives a transcendent existence such that our earthly, conscious life is only an isolated fragment as compared to the unnamable context of an existence running its course in it. The myth of the transmigration of souls may be a halting attempt to express such a segmental character of every individual life. Whoever senses through all

actual life a secret, timeless existence of the soul, which is connected
with the realities of life as from a distance, will perceive life in its given
and limited wholeness as an adventure when compared to that transcendent and self-consistent fate.[34]

"Certain religious moods," he continued, encourage such a perception. When
we see this earth not as a home but as a temporary asylum, then life, though
merely a fragment, can become, like a work of art, a whole. "Like a dream,
it gathers all passions into itself and yet, like a dream, is destined to be forgotten; like gaming, it contrasts with seriousness, yet, like the *va banque*
of the gambler, it involves the alternative between the highest gain and
destruction."[35]

Translated into Simmel's terms, Rock may have been finding a way to
understand his life as an "adventure," in relation to a "higher unity" while
at the same time confronting directly what was one of the constant, unstated
undercurrents of his unconscious life: his fate never to marry and his con-
sequent inability to insert himself into a lineage. Here in Chone, surrounded
by the solidly materialized effects of a lineage conscientiously cultivated over
eighteen generations, he took the idea of lineage to its extreme, pondering
the millions of forbears he was disappointing by depriving millions of future souls of life. This in itself was a way to frame his life within a higher
unity. The idea of the transmigration of souls was simply a variation on lineage that made it easier to fit his life into such a framework without disso-
nance. He had, during his recent encounter with depression, the deepest he
had ever recorded, been at the point of suicide. And he had never before in
his life confronted in any direct way the fact of his unconventional sexual-
ity and its consequences. He was, perhaps, arriving at an intuition similar
to Simmel's: that it was possible, even in the absence of a specifically religious attitude, to live life "as an adventure." That is, to be freed to embrace
its contingency and treat it as a gamble or a dream by the understanding
that it is merely a fragment of a much larger whole. His life had been an alternation between the timid rigidity of guiding formulas like that of filth
and purity, on the one hand, and the capacity to toss the dice with unusual
courage on the other. Here, perhaps, he was finding in himself permission
to venture further into realms of contingency.

.   .   .

In January, with enormous relief, Rock determined to leave. On what he
thought was his last day in Chone, he made a completely uncharacteristic
gesture. He made up some fifty tickets, distributed them to the people of

the monastery town, then received them back at two dollars each. "There were many real happy people today . . . the lame, blind, and lepers came, a pitiful sight. There were many opium sots which I drove out with a stick . . . Crowds of children came; children are the best crop of Chone, much quantity but no quality . . . it was a mad scramble, but all were happy and so was I."[36] But he could not leave. Bands of robbers roamed the roads; one of the rebel warlords had destroyed the bridges on the main route to Sichuan; his chosen alternative road was occupied by soldiers who were looting the countryside. The party stayed on until March.

Their route of escape was over the Minshan range through the Diebu region to Songpan, Sichuan, then south to Chengdu. Rock chose Diebu because it was free of Feng Yuxiang's Guominjun troops, whom he feared as "reds." He bought more rifles from the Yang *tusi*, secured a caravan of mules with Muslim muleteers, and hired an escort of twenty armed men from Diebu. A few days into Diebu, some twenty men attacked the caravan with rocks and sticks. Yang Jiding took a large boulder in the abdomen. One of the hunters shot an attacker in the stomach with a riot shell, killing him; the caravan's escort opened fire, killing one attacker and wounding two more; a muleteer wounded another bandit with an axe. The caravan took refuge in a gompa full of armed men carrying on a feud with a neighboring village. Rock was terrified and miserable.[37] The party crossed the Baishui river, into which the mountains of Diebu drained, and climbed the opposite range, cutting through snow banks with swords. In this unadministered part of Sichuan, they spotted a group of armed Tibetans hiding beside the road. Unable to decide whether they were robbers, Rock and the Nvlvk'ö men rushed them and threw their arms—a flintlock rifle, a sword, and some clubs—into the stream. Hoping to impress them with the quality of his own rifles, Rock shot at a distant cliff, splintering the rocks. It seemed to work: though other groups of armed men were watching from the hills above, the caravan was not attacked again. That night, sleeping in "a black windowless hole surrounded by pigs human and otherwise," Rock dreamed of being awakened in an elegant hotel by the room boy inquiring how hot he would like his bath water.

There was a magistrate at Songpan, and the caravan got an escort of thirty soldiers. Then it was down the Min River valley toward Chengdu: haggard porters overloaded with huge tea bales, captured Tibetan bandits roped together, ragged soldiers, half-starved militia. Northern Sichuan was chaotic— an uprising of militia against a warlord—and Rock felt hostility from officials, soldiers, and the people alike. In the town of Zongning, after being followed by a "jeering" crowd, he fantasized about someday being insulted by a "Chinaman" in America, where he could feel free to take revenge. Near Chengdu, he traded his horse for a bridal chair and hid behind its curtains the rest of the way.[38]

. . .

He could not bear to think of retracing the road they had taken up from Yunnanfu. So he left the Nvlvk'ö men in Chengdu with instructions to travel that road without him. When it came time to part, Li Shichen began to weep, then Rock, then the other "boys." "We had been together for six years and had been like a big family and now we were to separate, perhaps never to meet again. . . . Now there is before me only loneliness and sadness. . . . I should never have left them. . . . I feel that I have lost all my children, for such the boys were to me."[39] Perhaps they wept as much over the danger to which he was abandoning them as over their parting. He traveled thirteen days by chair to Chongqing. There he was evacuated with the Japanese consul general. He went by river to Shanghai, by sea to Haiphong, then up the French light-gauge railroad to Yunnanfu.[40] He was very ill on the way, losing some sixty pounds.[41] Having braved the overland route from Chengdu, Li Shichen and the rest arrived safe in early June, more than two months later, to find Rock waiting for them. He had intended to go to Nvlvk'ö with them, but his Arnold Arboretum sponsors insisted that he return to Boston. Sargent had died, and Ernest Henry Wilson, who succeeded him as director, had determined not to spend any more on the wayward botanist. Rock used the last of his funds to take Li Shichen and Lu Wanyue on a whirlwind tour of Haiphong, Hongkong, Shanghai, Honolulu, San Francisco, Chicago, Washington, New York, Boston, and Seattle, returning to Kunming in November.[42] Several of the others remained in his house in Yunnanfu awaiting their return.

In Washington, he persuaded the National Geographic Society to fund an exploration of the huge mountain range of Konka Ling in the Xiangcheng region, west of Muli, and the gigantic Minya Konka range, west of Kangding. In March 1928, he and the Nvlvk'ö men who were waiting for him left Yunnanfu for their home. Rock brought as a companion an eighteen-year-old American named W. T. Hagen, of whom he soon tired—"a swaggering youth from the grasslands of North Dakota where he has been brought up by cattle and wolves."[43] The roads of west Yunnan were very dangerous. Tang Qiyao had died, and Long Yun was vying for control of the province with three other warlords. Bands of bandits several hundred strong roamed the region, attacking even the largest towns. Long Yun had just taken Dali, and

Rock's party traveled there with an escort of nearly four hundred of his troops before turning north to Lijiang and Nvlvk'ö. For most of the Nvlvk'ö men it was their first homecoming in nearly three years. Their relatives paraded them triumphantly to a house they had prepared for Rock then sat down to watch the forty-five mules of the caravan make their way over the Lijiang plain. For several, it was a very sad occasion. He Guanyi had lost six members of his family, including both parents. Li Shichen's father and elder brother had died even before the party had set off to Gansu, but his mother had asked that he not be informed. He learned of the deaths when he entered the village, prepared to joyfully greet his family, with rolls of satin he had bought his father and brother.[44] A week after their arrival, the twelve men went to a temple some five *li* below the village to make offerings to Sadon, the god of the Yulong range. After another four days, the entire party left for Muli and the Konka Ling range.

LOVE AND LACK

Filed with Rock's photographs at the Royal Botanic Garden, Edinburgh is a series of prints without notes or captions. Rock had made most of his photographs, through various, conflicting arrangements, property of the U.S. Department of Agriculture, the Arnold Arboretum, the National Geographic Society, and the Harvard Yenching Institute. These, however, he had kept to himself until his death. Li Shichen stands in a silk kimono with a broad white sash around his waist. His hair is combed and oiled, and he cradles a Lady Amherst's pheasant in his arms. He stands before a background of trees and bushes, probably in Nvlvk'ö, where his kimono would be extremely unconventional wear. His expression might be described as a mild pout. The portraits superficially resemble specimen photographs. Rock made several portraits of one of the hunters on the Gansu expedition standing in a Nvlvk'ö street, also holding a Lady Amherst's pheasant. But the hunter's expression is dead serious, and he wears his rough daily clothing and a battered hat. In contrast, the photos of Li Shichen invoke a metaphorical relationship between the silky beauty of the bird and that of the young man. Around the same time that Rock made these portraits, he translated an improvised song, whispered into a bamboo mouth harp. It was a dialog between a boy and a girl, forbidden to marry, contemplating suicide together in a meadow on the Yulong range, so as to roam forever freely as lovers. The girl's part of the song mentions the Lady Amherst's pheasant:

Figure 43. Li Shichen
with pheasant. Courtesy
of the Royal Botanic
Garden, Edinburgh.

those people, who do not like us
    will search for us like a needle lost
they will search for you, my boy
as they will search for me . . .

nun muan nyi ma nnü
    gkò p'í gkò shù bä
zo shù nằ ssä dsàw
mí shù nằ ssä dsàw . . .

on nine mountains at the pine's feet they rise
    the stone pheasants call
    the Amherst pheasants call
I love you, my boy
    do not show remorse

ngv ngyù t'o k'ö dtü
    fu llü̃ dgyú lä dsì
    khữr llü̃ dgyú lä dsì
yu dsì zo gkyí mùan
    muan ggò dyu muan t'à[45]

Figure 44. Li Shichen with mouth harp. Courtesy of the Royal Botanic Garden, Edinburgh.

Rock accompanied the song's published version with a pair of portraits of Li Shichen with oiled hair and a brilliant white shirt, playing a bamboo mouth harp.[46]

Rock's relationship with Li Shichen was the most intimate of his adult life. Li spent twenty-seven years collaborating with the splenetic botanist, accompanying him on every expedition between 1922 and 1949. Except for the periods when Rock was traveling outside of China, the two men were rarely out of each others' sight. The relationship of affectionate, fatherly patronage, grounded, in the last instance, on racial difference, that Rock labored to craft with his circle of "boys" did not always sit well with this intelligent, mercurial man. Rock complained of his sharp tongue, sometimes imperiously insisting that "it will not do to accept incivility from one's servants."[47] Nearly everything that can be known of Rock's relationships with the Nvlvk'ö men must be gleaned from his own self-absorbed and self-justifying ramblings. But in the 1990s, the ethnographer He Jiangyu, whose mother was from Nvlvk'ö, recorded a few stories that had been passed down to the men's grandchildren. In one of those stories, the entire crew spent a month in Rock's house in Yunnanfu, preparing for the trip to America:

Once, because Li Shichen had not noted the plants collected clearly, Rock grew indignant; his old malady returned, and he kicked Li Shichen fiercely, twice. Li Shichen flared up and returned two ferocious kicks. And then the two had a fistfight. Rock, beaten until his nose was black and his face swollen, was ill with rage. He pulled out his pistol to do away with Li Shichen. Li Shichen's brother took the gun away. He brought the gun to the American consulate, asking the consul to advise Rock to give them their salary so they could depart. Li Shichen said to the consul: "Rock is a man; we are men. He has a big foreign degree, but he must learn from us Naxi how to conduct oneself as becomes a man. We have no choice but to educate that half-breed dog."

The next day, Rock apologized, and they made peace, Rock exclaiming, "We shall be friends forever!" Li Shichen answering, "We are friends; that is enough."[48]

Rock's feelings for Li Shichen grew deeper and more complex over time. At times, his diaries seem to be coded records of overtures of intimacy rejected, the cyphers signaled by his rare use of underscoring. During that winter trapped in Chone, he had endured a deep depression that seems to have been precipitated by such an incident. "Today," he wrote one January evening, "I have found out at last that I cannot depend on anyone. . . . Life is hard to bear among such people in whom one cannot trust, *not even in one whom I had thought to be above all most trustworthy*, but they all reveal their nature sooner or later and one realizes that one is *all all* alone."[49] Were this an ordinary breach of confidence, he would have gone on about it, rancorously, for pages. But he wrote not a word more, for weeks.

In September 1928, having made their exploration of the Konka Ling, the party was in Yongning, headed back to Nvlvk'ö. As usual, Li Shichen was sharing Rock's tent. After a day that opened with thoughts of suicide and gradually got better, Rock wrote in his diary: "I expect to have quite a peaceful night. Lissuchen and I have been talking about *various things* tonight."[50] And the next morning, "as miserable as I was yesterday, my situation has not changed one whit, so it is merely a matter of the mind, brooding on the pinings of the heart. . . . "[51] Over the last four years, Li Shichen had spent a total of two weeks in his village. His parents, anxious that he had not yet married, arranged a wedding less than a month after his return from the journey to the Konka Ling. Rock attended the wedding, then maintained a stony silence in his diary for weeks. Then,

The days of October 2nd to 8th were to me the most terrible I have ever experienced. I was deeply depressed and melancholy on the verge of ending it all, and I must confess that I am still not myself. . . . The 21st of the eighth Chinese moon witnessed the wedding of my head boy Li Suuchen. I saw it all and had a feast at his house. I am still sad and despondent and I doubt if I will live to see the end of this year, although it is not far off.[52]

His marriage did not prevent Li Shichen from continuing to collaborate with Rock. Not a month later, Rock and his party were off again, back to Yongning and Muli, fleeing a notorious bandit named Zhang Jieba, who was advancing north from Dali with a small army. Rock photographed Li Shichen frequently during that anxious winter of wandering, once in a Tibetan garment, leather coat, and cartridge belt: "He reminded me of pictures in Lederstrumps Erzahlungen aus meines Kindheit" (James Fennimore Cooper's *Leatherstocking Tales*, popular in Germany and Austria when Rock was a child).[53] And he continued to record moments of immense despondency connected to the younger man. Li Shichen simply could not be everything Rock wanted: brother and child, companion and servant, soulmate and racial inferior. It is possible that the two did have some sexual encounters, though it is far more likely, given the evidence, that Rock dared a few unwieldy advances that were firmly rebuffed.

It must have been a confusing situation for both of them—particularly, one guesses, for Li Shichen. Rock, after all, had his resources. He could always fall back, savage and forlorn, on the easy panacea of racism: "These people, like most Orientals, are an ungrateful lot, and the more one does for them the worse it is," he wrote one December evening in 1929, in the Nvlvk'ö house he was sharing with Li Shichen. "One cannot take them and treat them as white men: they are ready to take everything, but when it comes to have any consideration for oneself or giving a little thought to one's wishes, they are as cold as stone and as responsive . . . I am really sick at heart; it is a great disappointment; that is all I can say."[54]

·   ·   ·

In the long term, his relationship with them was defined simply, by a lack. They were all adventurers; they were all out for fame and glory, variously defined. But he could see clearly that the dream-islands of their adventures were interwoven with another existence, with its own unity. They married and had children; they became the bones of their fathers' houses; they climbed the mountain behind their village each year to celebrate their patrilineages with a sacrifice to heaven—a ceremony he studied intently. And he could see just as clearly that not only did he lack all of this: he negated it in his life. His constitution and life choices made him a broken link in the great chain of forbears who had given him life. Fantasies about the trans-migration of souls aside, he had no other way to relate the adventure of his life to a broader coherence. In the end, what he wanted of them was a place in a setting of human relations of the kind they so easily demonstrated with their families. Many were likely willing enough to comply, on their terms. Later, he would receive repeated, emphatic offers from one young Naxi man to insert him into family and lineage as an adopted father and grandfather and to care for him until his death. But these were not terms he could accept. Quite simply, the residual racism to which he clung, and on which he relied when other avenues of relation failed, imprisoned him, allowing no full expression of human warmth and commitment.

His was a wandering generation. He took his place among thousands of Europeans and Americans who left home after the Great War to search for other ways of living abroad.[55] And ill ease with conventional sexual life as lived at home was an extremely common reason for wandering. What made Rock's life particularly difficult was this unusual combination: rootlessness with virtually no family at home; long and intimate acquaintance with places where continuity of family in the form of patrilineage was both the ultimate goal of life and the foundation for wealth and power; and a conflicted form of racism that never allowed him to be fully at home even with his most intimate of others.

9.   The Book of the Earth

He had seen many of the old books, with their strange, vigorous script, and he even owned a few. Yet he had paid little attention to them. Six years of travel with the young men from Nvlvk'ö and long periods of living in their village had not changed his mind about dongba rituals: they were nonsensical drivel and a great waste of time and energy. In late 1929, this changed abruptly. He and the twelve adventurers from Nvlvk'ö had completed a two-year-long expedition funded by the National Geographic Society. They had gone to Muli in April 1928, where Cicheng Zhaba had given them permission to explore the Konka Ling. They had made two tours of the great range, then returned to Nvlvk'ö for the winter. In the spring of 1929, they had gone to Kangding and explored the gigantic Minya Konka range, returning through Muli in the fall. A few days after their return, Rock was sitting in his house in Nvlvk'ö working on his expense account ("a nightmare"), when he heard a man calling. He went out, and someone told him that a woman had drowned in the little pond behind the village:

> I went out to see and lo and behold two strong husky girls tied
> together at the waist had jumped in to commit suicide and were
> dead. They looked like fairies through the greenish blue crystal
> waters; one lay somewhat on her back and the other . . . held her . . .
> about her shoulders, her face bent down over her dead friend; thus
> they were balanced semi-erect . . . in the water. Their goat jackets lay
> on the banks of the pond, their shoes in the aquatic plants; their hair
> was loose and some raw silk thread from tassels floated on the water. . . .
> People stood around but no one made a move to pull them out. The
> whole reminded me of a tableau of wax in a panopticon . . . a mirror
> fell out of the lower girl's bosom and reflected the face of the one which
> lay on top of her.[1]

It is no wonder that this scene excited Rock's imagination: death always did. But in addition his description makes the scene remarkably like the kind of photograph he was always trying to achieve: clear to the minutest detail, divided from the viewer by a transparent surface giving the effect of an extensive flatness while allowing the figures behind a full, fleshy particularity. In the final detail, the *punctum* of the mirror falling from the girl's bosom, his description even manages to make both girls face the viewer, despite their face-to-face embrace.

He asked Zhao Zhongdian to twist a long stick in the girls' hair and pull them to the bank. No one seemed to know the reason for the suicide. The next day, distraught relatives arrived from nearby Baisha. The girls' bracelets were missing, they complained, and what business was it of the foreigner's to pull them out of the water? Perhaps they had not killed themselves but been murdered by the people of Nvlvk'ö. Rock, furious, threatened arrest and jail; they fell on their knees to beg pardon. He learned that both girls were to be married in a few days to men they did not like. Two others had planned to join them, but they had not been able to slip away from their homes.[2]

Once he began investigating he did not stop. He learned about the funeral rituals the girls' parents were planning. Instead of going to see them, he decided to stage one himself. Two days later, he hired twelve dongba to perform a rite called *szichungbpö,* intended to prolong the lives of the elderly after the death of a loved one. Nearly everyone in Nvlvk'ö crowded into his courtyard and watched, silent and absorbed, for three days. The dongba created a map of the cosmos, with altars for the gods of earth, sky, and underworld. They journeyed into the sunless regions, captured the demon king Muanllüssùndzi, dragged him out of his prison-like hell, and executed him. The third day, a woman who had drowned herself forty years before in the same pond spoke to the rapt audience through the mouth of a a *lubbu,* spirit medium.[3]

Afterwards, he huddled with two of the dongba for two days, taking notes on the ritual. It was his first real encounter with dongba script. It was hopelessly complex, he concluded. No one could learn it.[4] Still, his attention arrested, he began to collect dongba texts: in a few days he had 261. He hired a dongba to explain the writing system to him and make a list of all the rituals. The ritualist told him that at a funeral for a suicide, called *harlallü,* a spirit medium describes a world among the peaks of the Yulong range where suicides spend their afterlives, never growing old, never being reborn, never descending into hell, "a blissful, joyful time, a couple of lovers rejoicing forever, young and happy."[5] He walked the foothills with Li Shichen, dreaming of living out his days in Nvlvk'ö studying dongba writing.

The two suicides changed the course of his life. After late 1929, he undertook no more expeditions, plunging instead into the study of dongba texts. The old books opened up a new world for him. In this world, the landscape was replete with texts to be discovered, collected, translated, described, and organized. Aided by Li Shichen, He Guangyi, Zhao Zhongdian, and some exceptionally erudite dongba, he undertook ambitious projects of translation. He published many works on ritual, including two strange masterpieces, the two-volume *Na-Khi Naga Cults*, and the posthumous two-volume *Na-Khi–English Encyclopedic Dictionary*. The young men from Nvlvk'ö led him around the Lijiang valley to the sites associated with the Mu lineage, and they collected inscriptions, gazetteers, and rubbings for him, which he gathered into his idiosyncratic geographical history of the Mu chiefs, the unreadable *Ancient Na-Khi Kingdom of Southwest China*. After leaving China for good in 1949, he moved from a world structured through travel over a text-filled landscape to a world centrally structured by textual movements alone—in journeys, translations, exchanges, and transformations. He had found the landscape of the Lijiang valley to be an archive of texts; now he rebuilt his own living archive of this landscape, rediscovering the lineages of kings, demons, and deities that were rooted there. This chapter traces the gradual emergence of this creation—this great, mobile, intertwining of earth and world that he brought into being.

MAPPING

Only a month after his first frustrating encounter with dongba writing, Rock left Nvlvk'ö for Washington, D.C. to report to the National Geographic Society on his expeditions to the Konka Ling and the Minya Konka. In preparation, he sent a telegraph: "Minyakonka highest peak on globe 30250 feet Rock." He had made wild claims about the Amnye Machen too, however, and his chief editor and the Society's president were not inclined to believe him. The Society had spent $68,000 on his two expeditions, netting half an article, 1,811 black-and-white negatives, and 890 autochromes, useless without words to accompany them. The stock market had crashed, and his editors declined to fund the expedition he proposed, to the headwaters of the Mekong, Salween, and Yangtze rivers. Instead, they suggested that he write the articles he owed them: for this he would not need his twelve Naxi men, and his salary could be reduced. Furious, he returned to Yunnan on his own savings. He would write the articles, he decided, but most of his efforts he would devote to exploring dongba texts.[6]

By March 1931, he was back in Nvlvk'ö working on dongba manuscripts. He did not attempt full translations. Instead, he worked on taxonomy, translating titles and making brief synopses. He conceived of each manuscript as a contribution to an archive, numbering and labeling it like an herbarium specimen. He tried to list the books for each ceremony, sometimes as many as seventy. But the taxonomy rapidly grew confused. He discovered new rituals daily; he found that different dongba read different books at the same ritual. It would take him years merely to accomplish this simple task, he despaired. Many years later he would decide that it had been quixotic. "It is impossible to give a synopsis of the content of a Nakhi manuscript because the texts do not always deal with one subject but contain much irrelevant matter only indirectly connected with the main theme," he wrote in a scholarly publication the year of his death. "From a perusal of the first page . . . it is impossible to determine to what ceremony it belongs, still less to define its title. Only after a study of the text . . . in its entirety is it possible to make sure of its title and then not always successfully."[7]

He fired the dongba who had been helping him—"a lazy ignorant fellow; he knew most of the chants by heart but could not explain what the meaning was of what he read."[8] For dongba, texts were practical tools. Their value lay in their efficacy: their meaning was secondary. Most texts could not be read off the page in any simple way. In many cases a combination of pictographs served merely to remind a dongba of an entire verb phrase, previously memorized. Recitations often contained many intricate formal features not visible in the writing, features as crucial to the text's efficacy as the stories it told or ritual procedures it outlined. Rock was unaware of such features: he wished only to extract semantic meaning from the text, a procedure most dongba must have found absurd. Still, the dongba cult had been in decline for nearly a century; many dongba were not very skilled or knowledgeable, and there is no reason to think that Rock's first dongba teacher was particularly accomplished.

Rock and his household of Nvlvk'ö youths walked into the Yulong range to spend the summer in the alpine meadows. A week into their stay, He Guangyi appeared, accompanied by a dongba named He Huating, whom he had been sent to fetch from his village Gkvnawua (Longpan in Chinese), on the Yangtze. The two men led a donkey loaded with manuscripts. He Huating was forty-nine years old. He was reputed to be among the most skilled dongba in the Yangtze loop, and he could recite many texts that had not been performed for centuries. He quickly seduced Rock, impressing him with his intelligence—no easy feat—and astonishing him with a display of shamanistic aplomb, washing his hands in boiling water after reading a protective text. He would work with Rock until his own death in 1943. Trying to read with him those first few weeks, however, drove the botanist into a rage.

> They never give a direct answer but always . . . talk and talk . . . They are very bad listeners . . . Everything is done haphazardly and care-lessly; their writing is so sloppy and slovenly that one has the greatest difficulty to come to any conclusion what the character really repre-sents, although the character is a real picture. There is no system in their mode of writing; this minute they write a character this way and in the next again entirely differently . . . Their writing is so concise, in fact so much is unwritten and left for the memory, certain charac-ters acting only as prompters, that a book of twenty pages would mean perhaps 20 long typewritten pages, while one of their pages has no more than 50 pictures.[9]

He decided that dongba were cunning tricksters, masters at eliciting offer-ings from an unwitting populace. For instance, at the end of the five-day-long *harlallü* rite for suicides, a final chicken, the last of many, was sacrificed to feed the elderly lame demons who had not arrived on time and the old toothless demons who had not eaten fast enough. It is an example of the earthy wittiness that glimmers from within the texts' arcane phrasing. But it astounded Rock into a lengthy diatribe on dongba wickedness.

He escaped the frustration of these sessions by wandering the range with his companions. As he walked he mapped, sketching the mountain skyline and taking bearings on each peak. In his diaries of that summer, drawings of dongba characters indexed to manuscript numbers alternate with line drawings of mountain peaks indexed to bearings.[10] Reading and mapping oriented experience in a similar way, dividing it into two moments. On the one hand, both activities required an immanent and absorptive orientation. Hand and eye followed the jagged contours of the mountain skyline and traced the intricate figures of dongba script: immanent as experience arose moment by moment, absorptive in attention to exact details of ridges and characters. On the other hand, both also required an abstract and mediat-ing orientation. The movements of hand and eye were indexed to numeric bearings and semantic meanings. Bearings mediated between these move-ments and the imagined archive of maps of the world, coordinating a rela-tionship between maps and earth. Meanings mediated between these move-ments and the imagined archive of the dictionary, coordinating a relationship between the semantic worlds of Naxi and English languages.

As texts and maps came together for him, he began to explore the land-scape in a new way. Before ascending the range for the summer, he had made a hasty trip to Bberdder (Baidi), the sacred center of the dongba cult, to avoid

Forrest, who he had heard was en route to Nvlvk'ö. He was shown the cave where Ami Shílo, a reincarnation of the founder Ddibba Shílo, had lived and where many dongba had been buried. After descending the range in the autumn, the Nvlvk'ö men took him to the family compound of the descendants of the Mu kings. It had been turned into a primary school, but the direct patrilineal descendant, Mu Qiong, who bore the official, powerless title of *tutongpan,* lived nearby with his wife and son. His paternal uncle and several cousins also lived in the neighborhood. They received Rock indifferently until he showed them a printed copy of the illustrated Mu chronicles made by the French Tibetologist Jaques Bacot.[11] They brought out the original. Rock photographed the entire text and hired the painter who had copied the illustrations for Bacot to make copies for him as well. Later, Mu Qiong's paternal uncle Mu Shu showed him an earlier chronicle, with a somewhat different genealogy of the Mu chiefs.[12] He would eventually discover that the first part of this text, which appeared incomprehensible, was actually a phonetic transliteration in Chinese characters of a Naxi text, telling of the twelve mythical generations that preceded the first historical chiefs.

He had always collected texts as he traveled, copying into his diary fragments of gazetteers and inscriptions on tombstones, monuments, and temple walls. He now began to discover an astonishing density of texts hidden in this landscape that he thought he knew so well. Both the recognized literary elite of Lijiang and the unrecognized literary masters, the dongba, had been filling it with writing for centuries. Rock sent his companions to make rubbings of all the old stone inscriptions on the Lijiang plain. One poem, cut into a cliff behind a village called Ak'ö, spoke of the first residence of the Mu ancestors, at the foot of the cliff. It was signed by Mu Gao, among the most literary of the Mu chiefs. Next to it was a poem in Tibetan by Mu Gao's father Mu Gong, describing the forests that surrounded the cliff like evening smoke settling on the hills.[13] One of the Nvlvk'ö men remembered a folk song that listed the twelve residences of the Mu chiefs: Rock set out to find them all.

After a year of assiduous labor he had, he thought, acquired a copy of every dongba book used in every ritual, more than five thousand in all. He wintered in Yongning, in the small palace of the polity's hereditary administrator, the *zongguan,* on the island of Nyarop'u in the beautiful Lugu lake. Then, after falling seriously ill, he decided to go to Yunnanfu, arriving in March 1932. After a brief hospital stay, he settled into a new house with Li Shichen, He Guangyi, Yang Jiding, and the dongba He Huating. Zhao Zhongdian led the rest of the crew on an expedition to the upper Mekong and Salween valleys for the University of California Botanic Garden. They returned to Yunnanfu in the winter with thirteen mule loads of specimens and seeds.[14]

For nearly two years, Rock, He Huating, and He Guangyi sat every morning in Rock's room or on the veranda deciphering texts. The dongba read and explained; He Guangyi translated from Naxi into Chinese; Rock worked his way gradually into the intricacies of the script. He learned how characters were built up from simple roots, some used as pictographs, some as phonetic symbols. In written Chinese, phonetic symbols were often paired with pictographic symbols to create characters. Dongba script, however, was far more complex. Each graph involved many elements; a single graph was often recited as an entire phrase. He compared the graphs to picture puzzles. "There doesn't seem to be any system or rule for forming such combinations," he complained. "Phonetics are used at random, and one can find several characters or combinations . . . expressing one and the same word or idea, while the pictographic or original value is entirely lost." Many elements in a recited phrase were not written; some graphs were not read; some, though written once, were read two or three times. He began to call the unit of several graphs inside a box a "rubric." He found that many rubrics did not contain a verb: he had to search for the verb in the next rubric. He began to notice elements of the texts' formal architecture, such as repetitions of phrases building into a climatic phrase that finally delivered the verb.[15] In order to make translations, he decided, he would first require a phrasebook; the team set about making a dictionary. They arranged the texts into a complete taxonomy, listing those read for each ritual in the order they were read. This was He Huating's taxonomy—other dongba would have arranged the texts differently—but Rock considered it definitive. Then they began reading and translating texts for the rituals they thought most important: the various funeral rituals, the *harlallü* ritual for suicides, the *dtona* ritual for combating the demons of slander.

In the autumn, Rock and Li Shichen went to Peking and Shanghai, returning to Yunnanfu by way of Hong Kong, Haiphong, and the French light railway. Rock made long bibliographies of books on west China in Chinese, English, German, and French and ordered them from bookstores in Shanghai and Peking. He searched Chinese texts for mentions of Lijiang or the Mosuo (Naxi), and assiduously copied everything he could find. In February, after shipping their specimens to Berkeley, Zhao Zhongdian and most of the Nvlvk'ö men left for home, leaving only Li Shichen, He Guangyi, Yang Jiding, and possibly one or two others. In the autumn of 1933, Rock decided he had enough of Yunnanfu: he would return to Europe, perhaps for good. He asked Li Shichen to go with him. Li refused: he had been married for five years, but he had rarely lived with his wife, and he had not fathered a child. He would return to Nvlvk'ö. Rock settled his affairs, sent eighteen crates of books and effects to Haiphong on the French railway, then followed. He left his belongings in Hong Kong and sailed to Shanghai and Europe. He went to Paris, London, New York, Chicago, and San Francisco, then sailed back to Peking. By the summer of 1934 he was back in Yunnan: Europe had not suited him, nor had America.

TRANSLATING

For him, every text had an antecedent—an "original" in the sense Walter Benjamin gave this term in his famous essay on translation.[16] For Benjamin, an original was a text, by nature more or less translatable, whose life was extended and amplified by the translations that issued from it. Early on, the original for all Rock's texts had been the "book of the earth." He collected leaves from that book in the form of plants, rubbings, and genealogies, to insert in his own books; he read that book with his feet, transcribing it into endless, rambling prose. Now and then, he did natter on for pages about his inner life—memories of childhood, thoughts about desire and suffering, musings about ancestry, progeny, and reincarnation. But once he caught his breath he dismissed these asides as nonsensical and chided himself for wasting paper. The real work of manufacturing text—the motor, if you will, of his textual production—had little to do with recording "thoughts." It was far more direct. It was about tracing lines inscribed in the earth onto his sheets of paper: transcribing the movements of feet with the movements of a pen, following the arc of flowing movement, inscribing a line of sight as it followed a jagged skyline. The energizing positive and negative poles of this motor were, directly or implicitly, that dialect of filth and purity: the animate sputum of crowds, the immobile perfection of snow peaks.

   Now, he was beginning to map texts. He wanted to create lines of English prose that would follow the contours of those enigmatic pictographs as accurately as a route survey and as thoroughly as a traveling diary entry. At their best, all his translations were "bad" in Benjamin's sense. They did not deliver an "essence" of the original; they did not rise into a higher and purer linguistic air. Yet dongba texts, for all their difficulty, were eminently translatable—again in Benjamin's sense. "Where a text is identical with truth or dogma, where it is supposed to be 'the true language' in all its literalness and without the mediation of meaning, this text is unconditionally translatable."[17] In the most translatable texts, meaning was no longer the "watershed for the flow of language." It is true that in most dongba texts a thread of semantic meaning could be pursued through laminate narratives, genealogies, and ritual instructions. But this thread was insubstantial and illusory. The stories were merely arenas for unseen beings to manifest themselves, and the texts were merely stages for these manifestations. The pictographic drawings were emanations of and substitutions for these beings in the same way as were the intricate effigies of twigs, leaves, and paper assembled for rituals. As stages for the acts of these beings, the texts did not so much mean as do: to read was to propel the beings manifested by the texts into acts upon other unseen beings—the dead, the demons at the root of disease and calamity, the effects of scurrilous words. Meaning was not at all the "watershed for the flow of language."[18] In practice, that flow had its effects even in the absence of any listener competent to understand the texts' narratives and instructions. Rock's chief complaint about dongba was that most had no capacity to step back from a text and give a gloss or interpretation of its meaning. To most dongba a text was merely itself, and it acted not through understanding or interpretation but in recitation. The great translator He Huating's signal accomplishment was to achieve a truly cross-cultural orientation towards his texts, using them as raw materials to manufacture the meaning that Rock desired.

For Benjamin, the most translatable of texts was Holy Writ, in which there was no essential meaning to be conveyed, and in which the plurality of languages alone called for translation. "Just as, in the original, language and revelation are one without any tension, so the translation must be one with the original in the form of the interlinear version, in which literalness and freedom are united."[19] This describes Rock's best (and least readable) translations quite precisely. Rock put much effort into extracting essential meanings and core narratives, but he gradually came to consider all this work preliminary, experimental, and inadequate. In the end, there was simply no good way to sum up the meaning of a dongba text. His truest translations might begin with such an attempt, but very soon they spiral inward, discovering more narratives embedded within the first, finding others worked into the structure of the graphs themselves, and finally abandoning narrative lines altogether, transmitting instead massive blocs of non-narrative information, digressive and disorganized.

Together, Rock and He Huating developed an elaborate mode of interlinear translation that showcased the great dongba's erudition: his ability to explain with precision the details of each graph and his capacity to create summary glosses of narrative meaning tailored to Rock's needs. First, He Huating copied each page of the pictographic text in his elegant hand.

Figure 45.   Rubric from Joseph Rock, *The Zhimä Funeral Ceremony of the Na-Khi of Southwest China.*

Then he recited the text for Rock to transcribe with the idiosyncratic orthography he invented for this purpose. Third, he dictated a summary translation of each line of the recited text in colloquial Naxi to Li Shichen or He Guangyi. Li or He translated this into Chinese, and Rock rendered their Chinese into inelegant English. Fourth, the translation doubled back, plunging into the pictographs with blocs of text that mapped out each rubric. Here is a relatively brief example, from *The Zhi Mä Funeral Ceremony of the Na-Khi:*

> Rubr. 3: The first two symbols or rather the phonetic ^{2}ggo-baw letter ^{2}yi and the symbol ^{1}bi = to twist (it shows a man twisting a rope with his hands) are both used phonetically for ^{2}yi -^{1}bi = the Yangtze which encircles the Li-chiang district at a height of from 6,000 to 5,000 feet to the west and east respectively, is rich in gold, hence it is called by the ^{1}Na^2 khi ^{2} La-^{2}ler-^{1}ha-^{2}yi-^{1}bi = Vast gold stream, and by the Chinese Chinsha Chiang = River of the golden sand. The next symbol is ^{1}ha = gold, ^{1}gu = load, ^{3}ssaw invite. The second half shows the following symbols: ^{3}khyü = a juniper tree to the top of which the symbol ^{1}gkyi = cloud is attached, below the juniper symbol is that for needle = ^{1}gko, here it stands for ^{3}gko = branch, the branches of the juniper, ^{1}gu = load, ^{3}ssaw = invite.[20]

Literalness and freedom are indeed united here. Paragraphs like these, joined to each rubric, are analagous to route surveys, recording as exactly as possible a track through the original, while also digressing frequently to take in prominent orienting features to the track's sides (here, the river with its many names). The fifth layer of the translation unfolded in abundant footnotes. The notes further decomposed the narrative gloss of the third layer, diverging in two directions. Many mapped out further details of the pictographs, describing the various tools, plants, materials, clothing, food, houses, peoples, and so on, included in the drawings. Thus, for example: "The ^{1}Mun-^{2}ss = *Rhododendron decorum* wood; ^{2}dsho-^{1}bpa is the name of the utensil. It is actually called ²djo-¹bpa, but as there is no symbol for ²djo, ²dsho which comes nearest to it is used. The hoe-like wooden utensil is of one piece and may not always be of the species of rhododendron mentioned above, it is however the nearest to villages as it always grows in the pines and oaks; other large-trunked species occur at much higher elevation as *Rhododendron adenogynum* also called ¹Muₙ or ²Muàn."[21] Others worked to pin text to landscape by opening up place names, again in two directions. On the one hand Rock worked to assign the names to places in his experience of the landscape; on the other hand, he worked to open up the names to the archive of texts in Naxi and Chinese layered into that landscape:

> ¹Ng'a-²k'ö is the name of a village meaning at the foot of ¹Ng'a, it is also called ¹Gka-²kö. There is also a ¹Ng'a-²gkv or ¹Gka-²gkv, that is at the head of ¹Ng'a or ¹Gka. Here there is a mountain pass called in Chinese Hsi-kuan or Western pass. A wall was built across the spur here in the time of Mu Sheng-pai with a soldier guard to watch the southern approach from ²Lä-bbu land. It was the border between Tung-yüan hsiang and Ch'i-ho hsiang and the ¹Na-²khi and the ²Lä-bbu. See *l.c.*, p. 157, note 198; also p. 247, note 493. See MBC, p. 118, also p. 64, note 157.[22]

The string of references at the end of this note are to other notes in this text and others. In his translations, Rock increasingly cross-referenced his texts, particularly in notes regarding place names. He had always searched for the textual qualities of places, first by unearthing names in multiple languages, then by collecting more extensive texts embedded in places—steles, maps, genealogies, histories. Now he was working place-texts into textual networks of his own making.

At their best, then, his translations decomposed "originals" into the most concrete possible components. They blew pictographs apart into small bits, each bit matched up to particular parts of the lived world: wood, leather, cloth, plants, bricks, tools, etc. Narratives, perfunctory and disorganized, began their disintegration as soon as they emerged: interlarded with names of places and beings, interrupted by digressive notes, overwhelmed by a mass of route surveys of pictographs. While many dongba texts did have a narrative structure, many others were organized instead by routes: beings moving through a landscape of place names. Rock's translations were always alert to routes, pinning the masses of concrete detail, wherever possible, to place names. For him any place name was a node in a textual network, a door that opened between one translation and many others. An operation that began as a way to anchor texts in experience became, as translations accumulated, a method to open texts up to one another.

. . .

Dongba texts were fundamentally genealogical. They described lineages of spirits, plants, lands, texts, and peoples as routes through the geocosmological landscape of the Mu kings' former domain. They created a spatial and temporal imaginary for a ritual by tracing the descent through time and space of the beings involved. They did this in a huge variety of ways. Here, as an example, is an excerpt from a manuscript called *dtv khi,* "lowering the *dtv* tree." The manuscript was read at funeral rituals to take beings over the spruce and fir trees set up in the courtyard as a pathway to the sky.

> One day, on the horizon of the gods, K'wuadtvmberddvgyi died;
> his dongba, Nabbussangu, led him over the turquoise plowshare,
>   over the *dtv* tree
> and defeated the nine *dtv* and *dsa* demons of the sky
> in the next generation, Lalerdundzi died;
> his dongba, Ssawbbussawla, led him over the golden plowshare,
>   over the *dtv* tree
> and defeated the nine *mun* and *ghugh* demons of the earth
> in the next generation, on the top of the great mountain of Shilo,
>   Lvndzhwuagyibbu died;
> the dongba Tutugkowua led him over the white clouds, over the
>   *dtv* tree . . .
> in the next generation, in the dense black forest, Llumunk'osi died
> his dongba K'omunmiuggu led him over the black *dtv* tree to the sky
> and defeated the demons of the black clouds and winds . . .
> in one generation, on the great mountain of Shilo, Ddibba Shilo
>   died . . . [23]

This text traces the descent of Ddibba Shílo by telling of the deaths of his ancestors along the route from the celestial realms to earth, by way of the great mountain, origin of all earthly things, to which he gave his name. At the same time, it traces the origin of the powerful words that guide the dead past the tree of spines, by telling how each of Ddibba Shílo's ancestors had his own dongba who guided him, on his death, back to the celestial realms. This genealogy is in continuous reciprocal motion, each line and each generation joining descent with ascent.

The texts, particularly those used in funerals, were full of this mode of genealogy, in which the descent of an ancestral line is literally a descent, first from the sky and then from the mountainous north. The task of textual recitation was to render this temporal and spatial descent reversible, guiding the deceased back along the route, to join his ancestors. The texts always joined ancestry with movement. It was not merely that the deceased was sent to join his distant ancestors in the north. Many texts also told of how parents and grandparents had traveled the path before the deceased, preparing it for him:

> They go north calling you
> from south to north, neigh like a horse
> your father is the one in armor
> your mother is the one in the pretty dress
> your grandfather is the one armed with a spear
> your grandmother is the one in beautiful clothing
> see them, laugh with them, speak to them.[24]

Ascendants became ancestors by walking north; they forged their connection with the living by preparing their road.

The mode of genealogical imagination embedded in the texts worked its way into Rock's preoccupation with the coincidence of text and landscape. He refused to consider the ancestral migrations outlined in the texts as merely mythological. Wherever he could, he stitched text to earth by locating place names in experience. He searched the texts for clues about the lands from which the original ancestors may have migrated. The migration must have begun in the grasslands of northeastern Tibet, he speculated, the site of his own epic journey to the Amnye Machen. He had seen groves of stately junipers in the grasslands of Amdo, preserved because the nomads used yak dung rather than wood for cooking. Perhaps it was in these groves that Naxi ancestors had created *muanbpo*, the "sacrifice to heaven," now performed in juniper groves behind villages in Lijiang. *Muanbpo* texts mentioned yaks and tents: Naxi must once have been nomads. He found further evidence in a group of texts once used at funerals for warriors, which mentioned cranes, yurts, and herds of yaks and sheep: "[The texts] tell us ... of the time when they lived in the far North, in the valleys and among the mountains of north-eastern Tibet, south of the Mongolian border in yurts made of felt of the wool of white sheep. There they suffered no hunger, they milked their yak and had an abundance of butter and cheese. They dressed in *p'u-lu* garments such as the Tibetans still wear, and their women were adorned with carnelian, red coral, turquoise, silver, and gold ornaments."[25] He argued that all the textual evidence placed the ancient Naxi ancestors in the Kokenor region, south of the Nanshan, bordering on the Amnye Machen range. To clinch the case, he told an anecdote that directly linked experience and text: when his companions from Nvlvk'ö had seen yurts for the first time in the northern Kokenor, they recognized the style of house from dongba books.

He had already found references in the texts to the sites of his other adventures: Muli, the Konka Ling, Minya Konka, the upper Mekong, even the ranges beyond the upper Salween. The texts read at funerals for warriors also seemed to refer to the southern and southwestern parts of the province, through which he had passed several times. They mentioned armor made from rhinoceros hide, prompting memories of his first trek into China from Siam in 1921: surely the rhinos for the ancient armor had come from these regions. Now, speculating that the texts also described the Kokenor and Am-nye Machen, he extended the archive of dongba writing along the furthest-flung routes of his wandering.[26] The map of his own experience could now be cut to nearly the exact shape of that archive. And considered in the light of Naxi migrations and wars, this map could be reoriented to center on what had become his only true home in China, the Lijiang valley.

ARCHIVES

In 1935 and 1936, as nationalist troops pursued the First and then the Second Front Armies of the People's Liberation Army through Yunnan and Sichuan towards Shaanxi, Zhao Zhongdian, He Zhihui, and Yang Jiding lived with Rock in his house in Yunnanfu. In May 1935, as the First Front Army approached, Rock fled on the French railway to Kaiyuan, a hundred miles south, while the Naxi men waited and packed in Yunnanfu. He hired an entire railway car for his library, now almost four thousand volumes in addition to the thousands of dongba texts: the car was taken to the border with Vietnam. After a few days, learning that the communists had bypassed the city, Rock returned and his library followed. A year later, the Second Front Army approached, and the drill was repeated. Rock repaired to the railway station with the other foreigners in the city and then returned home when news arrived that an invasion was unlikely.[27]

He spent most of his life savings on books—every Chinese gazetteer from Yunnan and Gansu he could find, every work on west China in European languages, everything he could find on Tibet and Tibetan Buddhism. And he worked on a new manuscript which he would eventually publish with financial backing from the Harvard Yenching Institute as the two-volume *Ancient Na-Khi Kingdom of Southwest China*. It was a strange text, of no established genre, awkward, haphazard, badly edited, brilliantly illustrated, and brutally unreadable. Yet in many ways it was a mature expression of the orientation toward the earth that he had been developing for many years.

"In this book, I describe the Na-khi region as it passed in review before my eyes," he wrote in the preface. The book begins in Kunming (once Yun-nanfu), for no better reason than that is where Rock started walking, and describes the caravan route, stage by stage, to Lijiang. The first part of the book resembles an historically informed travel guide, mixing evaluations of the relative sanitary standards of inns with notes on the dynastic history of places and monuments.[28] Rock inserted into the text translations or sum-maries of the historical sections of the gazetteers *(zhi)* of the towns through which he passed. Stage by stage, the text is anchored in this way: deep, ge-nealogically structured chronicles of place names alternating with fleeting descriptions of the contingent experiences of the road.[29]

Rock's editors despaired of his prose. He had no imagination, they com-plained. His writing followed no theme and gave no unified impression. He had no idea of how to build a narrative in logical steps toward a climax. His sentences, of six or eight clauses, ended without expressing a thought. "He can do more, with surplus words, to confuse readers, than any contributor whose work has ever come to my attention," wrote the editor of his *National Geographic* articles.[30] In fact, he was capable of writing lean and vibrant prose: his works about Hawai'i were sturdy and straightforward, and some of the writing was sharply evocative of the islands' lovely, lava-strewn land-scapes. In China, however, where writing became a daily habit, even an ad-diction, the more he wrote the worse his prose became. One editor, comment-ing on a manuscript about the Deibu region, gives a clue as to why. The style was "muddy and incoherent," he wrote, "laden on every page with mean-ingless proper names." As an example, he quoted the following. "South of the Dakhe La rises another river called the He ho or Black river, which some days journey further south joins the Pe ho which is another branch of the Pe shui kiang. The former joins the latter as the Wen hsien ho forty li west of the town of Pi Kou in south Kan-su. All present maps of this region are incorrect as they give the junction of these two rivers to the east of Pi Kou. Some distance north this river joins the Kia ling in Szu-chuan territory."[31] Few readers would care to navigate this for long, it is true. But it was not written to be read. It merely traces in words the course of a river, charting the changes of its name as it moves over the earth. In China, Rock struggled to bring his prose closer and closer to the earth. He aspired to make his words move over the landscape, following its contours first in space and then in time. He did not think in terms of unified impressions or logical narratives, for journeys were full of irrelevant minutia; they did not take the form of narratives; they did not build towards climaxes. Even the form of his sen-tences echoed the rhythms of caravan travel: clauses strung together endlessly with few full stops, little effort to make causes cohere into a unity. What mattered to him most was "meaningless proper names"—the foundational form of the earth's textuality.

After the first section, the remainder of *Ancient Na-Khi Kingdom* is not a travel account: it is an archive. At its center is the entire illustrated text of the chronicles of the Mu chiefs, reproduced in photographs. Many other texts surround the chronicles. Foremost is a partial translation of the eighteenth-century history of Lijiang from the Chinese dynastic point of view, the *Lijiang Fu zhilue,* by the region's second appointed magistrate, Guan Xue-xuan. Then there are photographs and translations of all the inscriptions on stelae, cliffs, and temples that the Nvlvk'ö men collected in the Lijiang valley. And there are more genealogies: of the chief incarnate lama of the Karmapa sect, which had a temple in the Lijiang valley; of the last remaining regional chief of the former Mu kingdom, the Wang *tusi* of Yezhi; of the ruling lineage of Yongning; and of the various *tusi* in the surrounding districts. A network of route surveys is thrown over the mountains separating these places. This web's threads are most concentrated at the text's center, the Yulong mountain range. Here, the lists of Chinese place names suturing the landscape to local gazetteers are replaced by Naxi place names opening up into dongba texts: "Between Dü -gkv and Nvlv-k'ö is a meadow called Mba-mä. Here a large spring called Bao-shi gko-gyi issues from the mountain-side under a grove of century-old maples *(Acer cappadocicum sinicum)* where Na-khi sorcerers perform *Zä-mä,* a ceremony for the propitiation of the Llü-mun (Serpent spirits)."[32] In this landscape, lines of relatively thin textuality, anchored with genealogies of place names, connect deep and extensive pools of textualilty in the centers of the former Mu kingdom. Like Rock's herbarium in Hawai'i, his book was an attempt to be "great illustrated volume" of this landscape, opening up in every phrase to the material earth.

For most of the rest of his life, he attempted to carry his huge library about with him, like the lost sinologist Professor Kien in Elias Canetti's *Auto da Fe.* Though he rarely managed this, his fate intertwined closely with his library's fate. In 1937 he went to Vietnam, Cambodia, Hong Kong, Shang-hai, and Peking with Zhao Zhongdian and his youthful secretary He Zhi-hui. He collected more books; most of his library remained in his house in Yunnanfu, guarded by other men from Nvlvk'ö. He thought about going to Peking or Hawai'i to live, but he ended up back in Kunming (formerly Yunnanfu), with his books and "faithful servants."[33] As the war with Japan began, he packed up his most valuable belongings, including his oldest books, and sent them to Hong Kong for storage. In January 1938, the Nvlvk'ö men crated the library, and Rock set off for Europe, leaving instructions to ship the books to Hanoi when the Japanese reached Kunming. He flew to Berlin and Paris, where, at the Conciergerie, he was "overcome with grief" in the dungeon where the prisoners of the Revolution had sat awaiting execution.[34]

Then he went back to Kunming. That autumn, he and the men from Nvlvk'ö sat in their house while the Japanese bombed the city. After that, he asked them to pack up the books, paid them enough to return to Nvlvk'ö, and took the train to Haiphong. The library followed on a sealed railway car. He went to Hawai'i, where he arranged a trade: the University of Hawaii appointed him research professor with a stipend; in return, he would donate his library before his death. Because he felt he could not move the library back to Yunnan while the Japanese were bombing he looked for a place to relocate, relatively close, where his books would be safe. He chose Dalat in southern Annam. He had the library shipped from Hanoi by rail—over 1,780 kilometers. Eventually Zhao Zhongdian, He Zhihui, and two others joined him. They worked together on translations of dongba texts for a happy year. When it became clear that the Japanese would invade French Indochina, the Naxi men took the train back to Yunnan, and Rock went to Hawai'i. The library went to Saigon by rail, to Manila by boat, then to Honolulu, where the University of Hawaii took charge of it. When Rock visited his books, he found them lying about haphazardly in a storage room. He complained to the university president; they argued; Rock resigned his position, withdrew the offer of his library, removed the books to commercial storage, flew to Hong Kong, Shanghai, and Kunming, and took the new motor road to Dali and a sedan chair to Lijiang.[35]

He rented a comfortable house in Lijiang, planted a vegetable garden, and gathered as many of his old companions from Nvlvk'ö as would come. The dongba He Huating joined them, and they set to work again on the dictionary.[36] Though the Communist Party and the Guomindang were fighting a two-front war against Japan and against each other, Lijiang was relatively peaceful. When the Japanese bombed Pearl Harbor, he panicked about his library in Honolulu, but it proved safe. He stayed in Lijiang; he felt he had nowhere else to go. But he was living on the last of his savings, and he knew he needed employment. In February 1944, he had the Naxi men pack up his books, manuscripts, and other belongings, went to Kunming, and caught a China National Aviation Corporation (CNAC) flight to Assam and Calcutta. There he ran into the CNAC's American president, who arranged for him to work for the Army Map Service, mapping the Chinese side of the "hump," the air route from India to China along which the Allies supplied the Guomindang in its war against Japan. A few months later, to his immense distress, he learned that the ship carrying his belongings, including eighteen volumes of notes, hundreds of dongba texts, and his manuscript dictionary, had been sunk.

The Army Map Service seems not to have known quite what to do with him. He had his own ideas for his contribution to the war effort: a pamphlet on west China for pilots illustrated with his photographs, and "giving an account of the topography, trails, mountains, names of native chiefs or rulers, names of tribes who are friendly and who are not. . . . " Though the Army paid him a salary and a travel stipend, he found himself low on funds. But he had his library. He offered to will it to the Harvard Yenching Institute in return for a salary of six thousand dollars a year until his retirement and funding to return to Lijiang after the war to redo the work he had lost to a Japanese torpedo.[37] When the Institute proved uninterested, an old friend, E. D. Merrill, now administrator of botanical collections at Harvard, offered shipping and storage for the library, details to be worked out later. After months of haggling, Rock negotiated a deal: the Army Map Service, also interested in the library, would bear the expense of shipping it from Honolulu to Washington, then eventually to Boston, where the Arnold Arboretum would provide storage space. The Harvard Yenching Institute would appoint Rock as a research associate, give him $4,500 in expenses for his return to China, and publish *Ancient Na-Khi Kingdom,* which had been stalled for years with a publisher in Peking.[38]

He had rarely corresponded directly with the Nvlvk'ö men, using missionaries and priests to relay funds and instructions to them from abroad. But his young secretary He Zhihui now became an enthusiastic correspondent. He Zhihui had worked for Rock for at least fifteen years, and as far as he was concerned, he continued in the botanist's employ through the war. He wrote Rock letter after letter in serviceable Chinese, the most personal records left by any member of the two generations of Nvlvlk'ö botanists and explorers. In 1944, He Zhihui and his wife were living in Wenhua village, not far from Nvlvlk'ö, with two children, a girl and a boy. He Zhihui rode a jeep on a new road from Wenhua to Lijiang every day. His son, Manhong, went to school in Lijiang; his wife, frequently ill, and his daughter remained in Wenhua.[39] He suffered from malaria; he worried anxiously about being conscripted. Rock wrote him back from time to time. He sent pictures of bears and monkeys for little Manhong. He Zhihui begged him to send pesticides and an American-made gun to shoot the porcupines that gobbled up the crops in his village. Rock sent money, which He Zhihui retrieved with difficulty from the American consulate.

When Rock's books and manuscripts were lost to the sea, He Zhihui consoled him: "I will continue the professor's work. Please don't lose heart.

Please don't weep. If only the professor will return to Lijiang soon, all will be well." He knew exactly what would comfort his patron. On his own initiative, he set out to replace one of the lost books, the rare eighteenth-century gazetteer, *Lijiang Fu zhilue*. He went to the public library in Lijiang and copied the work daily, page by page. He sent Rock notes from the gazetteer on routes over the Yulong range; he wrote to descendants of the Naxi *tusi* of the Mekong valley asking for genealogical information; he had acquaintances make rubbings of inscriptions on stele; he had kin search for dongba manuscripts and some much rarer books written by a related people, the Zhongjia. His letters were models of dutiiful, solicitous longing for his "adopted father." Often he repeated, "When a plane flies over Wenhua village, my son Manhong cries out, 'The professor's plane has come!'" In the autumn of 1946, Rock finally wrote that he was returning to Lijiang. "Please bring an American-made leather jacket like the ones American soldiers wear," He Zhihui replied. "And please buy a coat for Manhong to wear. And when you come to China on the airplane, please bring an American-made bicycle. And a cane, the kind that has a gun hidden inside it." Rock should sell all his belongings in America and plan to live in Lijiang forever. They would live and work together; Rock could teach Manhong English. "Adopted father; we will live together and die to together as a family in Lijaing. As you grow old in Lijiang, my family will take care of you."[40]

With great difficulty, Rock did get himself, more than a thousand dongba manuscripts, and half a ton of household supplies back to Lijiang on the last day of 1946.[41] He rented a large house and gathered his entourage. He Huating had passed away. Rock hired his elder cousin, decided he knew little, and fired him. Then he hired and fired a younger dongba. Eventually he found another relative of He Huanting's who satisfied him. The worst battles of the civil war were far away, but Lijiang was chaotic nonetheless. The stupendous inflation of the period made daily life extremely hard. Ragtag bands of soldiers from elsewhere in the province flooded the town. Medicines and medical care were nonexistent: Rock spent much time in his courtyard treating minor ailments and handing out medicine. He suffered from excruciating facial neuralgia—for several months, he could eat nothing but liquids. Still, Rock, He Zhihui, and the dongba worked daily on translations.

· · ·

In 1947, as the social chaos brought to west China by the civil war worsened, hundreds of dongba gathered in Lijiang to perform the *dtónà* ritual to drive away demons of slander. It seems to have been an unprecedented occasion. People in Lijiang knew of only two other times when the grand *dtónà* had been performed (there was also a small version). The first was during the reign of Mu Zeng (1587–1646), when a live slave was used as a substitute *(dtó)* for the demons of slander and driven into the forest on an old packhorse. The second was in 1929, when twelve dongba had performed the ritual for Rock in his Nvlvk'ö courtyard, using effigies of willow and rhododendron as substitutes for the demons of slander.[42] The dongba who gathered in 1947 must have sensed that this time of calamity was also a time of transition in which relationships among humans and the unseen forces of the earth were at a point of fundamental reordering. As it happened, this was the final ritual performance for most. All dongba ceased to perform in public after Lijiang was liberated, and all faced severe criticism for having employed superstition to cheat the people.

The grand *dtónà* rite of 1947 did not focus on a specific household: it took place in several locations around the Lijiang basin. For seven days, the army of dongba battled a great variety of maleficent beings. They built effigies to embody the forces of violence and calamity; they sacrificed cattle, sheep, and goats; they offered up pine, oak, bamboo, juniper, and rhododendron; and they read texts. The texts described the origins of the demons of slander in ancient battles between gods and demons, Naxi heroes and their enemies, the tiger and the leopard, the stag and the antelope, the wild boar and the bear. One was a story of four brothers who lived in the land's four corners. The brother in the west, having dreamed of being carried off by the wind, attacked his Tibetan enemy, smashed his nine houses, destroyed him, and threw his soul into a dog's skull. The brother in the north, having dreamed of drowning, attacked his Golok enemy, smashed his nine houses, destroyed him, and threw his soul into a yak's hoof—and so forth in the east and south. At the center, the four brothers smashed the nine houses of the P'u, Naxi, Boa, and O, smashed their spears, arrows, and armor, and destroyed their souls. In another text, the nine sons of Ser commanded ten thousand soldiers to make ten thousand arrows, spears, knives, shields, and suits of rhododendron armor. The soldiers were as dense as stars in the sky; their arrows were a swarm of bees; their spears were a forest of hemp stalks. They attacked the villages of Haw, destroyed their lands, killed their people, and bound their souls.[43] Several times during the seven days, the dongba dealt with the *ngawbä*, weapons against and substitutes for the demons of slander. They built the effigies of fir, willow, and rhododendron, danced with them, slew them with swords, cursed them, burned their heads, chopped them into pieces, carried them out on stretchers to drums and gongs, and threw them away on the roads.

. . .

Conditions in Lijiang quickly grew desperate for nearly everyone. A new paper currency caused monetary chaos. A bandit army four to five thousand strong moved on the town from the south. In June 1949, anticipating the arrival of the People's Liberation Army, the underground Communist Party, which had been organizing armed demonstrations against the Nationalist military, seized power in the town. There was, immediately, a great deal to do: military order to establish, the elite of the old regime to deal with, a world of social hierarchies to upend. Rock had departed in February 1948 on a CNAC plane, leaving everything behind. He had flown to Europe and New York where he got surgery for his facial neuralgia. Then, deciding he wanted to die in China, he had returned to Lijiang four months after leaving. As the bandit army approached, he fled to Kunming, then returned in July, shortly after the Communist Party takeover. He Zhihui and several other men from Nvlvk'ö, perhaps unaware of the risk they were taking, met his CNAC plane on an airfield not far from their village. Rock worked on his dictionary with He Zhihui, afraid to leave his house. In Nvlvk'ö, a band of men, led by Yang Jiding, Rock's cook and companion for more than twenty years, now a member of the Communist Party, broke into Rock's house to search for arms. Three weeks after arriving, confused and frightened by the new order, Rock flew out for the last time, with his trunks of books and manuscripts.

He went to Rome to meet with the Tibetologist Giuseppe Tucci, whose Instituto Italiano del Medio ed Estremo Oriente agreed to publish his *Na-Khi Naga Cult and Related Ceremonies* (1952) and his monograph on the Amnye Machen (1956). He made his way to Kalimpong, India, in the foothills of the Himalayas: perhaps Yunnan, now under Lu Han, would shake off the Communist Party and declare independence, allowing him to return. He found many Naxi there, some of whom he knew from Lijiang.[44] Upper-class Tibetan refugees, including many incarnate lamas, flooded the town. Rock bought books from this abruptly destitute elite and kept working on his dictionary. In 1951, he gave up on the idea of returning to China. He began wandering again: Europe, Honolulu, the Pacific Northwest. In Seattle, in 1955, the University of Washington made him an honorary research scientist in its Far East and Russian Institute and agreed to buy his library, still occupying a room at the Arnold Arboretum. He was seventy-two, and he had no income; the twenty-five thousand dollars the library brought was a godsend.[45]

By 1956 he had settled in Hawai'i where he became honorary associate in botany at the Bernice P. Bishop museum. He arranged for another, older archive—his Hawai'ian specimens, notebooks, photographs, and glass-plate negatives—to be deposited at the museum, receiving in return a copy of his out-of-print masterpiece, *The Indigenous Trees of the Hawaiian Islands*.[46] In another trade of archive for printed text, he donated part of his very large collection of dongba texts to Giuseppe Tucci's institute to help defray the costs of printing his dictionary.[47] For eight years, he stayed in Honolulu, accumulating friends and working on plants and texts, with trips to Japan, Hong Kong, Burma, India, and Europe. On December 5, 1962, at the house of his friend Lester Marks in Honolulu, he died of coronary heart failure. Ten years later, the Instituto Italiano del Medio ed Estremo Oriente completed printing the final volume of the remarkable *Na-Khi–English Encyclopedic Dictionary*, with elegant pictographs by his still unacknowledged collaborator, the great dongba He Huating.[48]

THE BOOK OF THE EARTH

In the last two years of his life, knowing he had little time left, he had envisioned a final project. He would have his diaries typed, edited, and mounted in beautifully bound large albums. Opposite each page of text, he would display the photographs that corresponded to each diary entry—which he had already indexed to the diary with the numbers of his negatives. The work would fill sixty large volumes. A separate volume would contain the maps he had made of the entire region covered in his journeys, as well as a gazetteer giving place names in Chinese and Tibetan characters.[49] For the maps, he would work with a cartographer, John A. Sherman, who taught geography at the University of Washington and who had prepared the five maps in Rock's monograph on the Amnye Machen from sketches Rock gave him.[50] Sherman set his graduate cartography class to work, eventually producing ten beautiful hand-drawn charts of Gansu/Amdo.[51] Rock died before making much progress on the rest of the project.

It was not a new thought. He had been trying to do just this all along, from the day he first stepped over the border from Siam, lunched in a temple with the people crowding around him, "filthy beyond all description," and took solace in recording his passage through the beautiful Muang Hun valley.[52] He was making an equation, simple enough: experience becomes archive; archive orders experience. Perception makes the earth into paper; paper makes the earth social. He wanted his pen, his camera, his compass, to be conduits for this simple reciprocal exchange, nothing more. As he walked further into Yunnan, the chaotic sociality of flowing excreta, crossed gazes, and heterogeneous voices exploded about him. He found a way to use this exchange to remake the earth as a stage, a geometric reordering of social experience. He joined his pen and camera to his gramophone, creating lines of force to redirect gazes and voices, compelling them into a one-to-one correspondence with his own gaze—into the same reciprocal exchange toward which he had been working all along. I argued that, faced with the sense that his body was not his alone, that it was given him by the gazes and voices of others in which his own gaze and voice had its origin, Rock found a way to make a virtual body, composed of camera and gramophone, raised above the slime of normal social exchange on tripods and packing cases. In this dream of a last project, the nib of his pen, the lens of his camera, the point of his compass were to be the flesh of exactly this kind of virtual body, interposed like a completely transparent membrane, without substance or character, between earth and paper. A membrane around which this creation organized itself.

Not a new thought at all: each step, as it moved his body over the earth, should generate a paper road, of words, images, and charts. In Amdo, aspiring to create route surveys, he had begun to invest this idea with more intensity. Among his effects willed to the Royal Botanic Garden, Edinburgh are stacks of small, worn pocket notebooks, three inches wide by five long. Red lines divide the pages into columns: they were made for keeping logs of daily expenditures. Rock filled them with an endless penciled scrawl: " . . . dist 3 li, elev 10,900 ft then turn north 1/4 li & cont. W 20, 1/4 li to right Lolo huts here to left deep depression the Lolo settlement. W 27, place called Gutzer ½ li, then on again W 20—W 25 reget. Pines Rhum cot. Querc Del. F. Arm. (zig zag) W 20 to pass called Tokalo W 21 25.50—28.50 dist to pass 5 li elev 11,600 ft. Hlidgi comes for E 3 from pass go W 7 1 1/8 li then up and down to depression . . ."[53] On the page opposite this one, and many others, the horizon is represented with a continuous penciled line, the highest peaks marked with compass declinations (E 29.80, E 1.40). A *li* was a measure made for walking or riding and for calculating by feel—about a third of a mile. On foot, at a pace sustainable on a rough trail, a *li* would take about eight minutes to cover; it would be faster on a walking horse. The notes were made at the average rate of about one quarter *li* per phrase, or one phrase every two minutes.

These are gross approximations, but they make it clear that he could not have written these notes anywhere but his moving saddle. Writing, he needed one hand to hold the notebook, the other to hold the pencil. His prismatic compass took two hands to operate, so he had to pocket the notebook and pencil. As the trail rounded a corner, he estimated the distance since the last turn, recorded it in his notebook, put pencil and notebook back in his pocket, retrieved the prismatic compass, took a bearing, placed the compass back in his pocket, retrieved the notebook and pencil, recorded the figure, noted the shape of some limestone rocks, recorded that, took some notes on plants, then began the whole procedure again at the next turn. He kept this up hour after hour, day after day, first in Diebu, later in the Konka Ling and Minya Konka regions. He was still creating for himself a virtual body, now a machine for rendering the earth into text. At night, he rendered the day's journey into successive legs, each with a bearing and a distance, the line of script that retraced his path often meandering off into a map-like graphic of the mountain skyline. Pencil, notebook, and compass were always juxtaposed between his eyes and the moving ground; he carried the mediating coordinates of archive everywhere he went. Later, he hoped to convert the text into route maps, the foundation for maps like those Sherman prepared. With a few exceptions, however, these route maps did not take shape: he hadn't the skills.

It was a more compelling structure for experience than botany had been. The problem with botany was that the text that sprang out of the earth's floral carpet was too thin. There were only a few places—the meadows of the Yulong range, the alpine valleys of Deibu—where he could find a new name at each step, assuring the tightest possible link between his experience and the herbarium archive. In most places, gaps of a few steps to many days intervened between new specimens, during which the miserable, confusing business of social life with other humans intruded again. As he rode along with pen, notebook, and prismatic compass, however, no narratives arose, no catalog of disgust, no polar structure of filth and purity. All that dropped away to make space for this record, moment by moment, of experience rising up from the earth. It was a record of immanence, and it demanded that the attitude of his eyes and hands towards the landscape be one of immanence—of attending to what was arising at the moment. His anticipated task was to transform this open, motile perspective on the earth into a map—compiling all these moments into a single panoramic abstraction which would describe with precision the relationship of each moment and each day to each other—closing the circuit between experience and archive tighter than botany ever could.

Still later, as he began to collect and translate dongba texts, he discovered dimensions of the earth's textuality which he had only glimpsed so far. It turned out that the earth was littered with texts. Until now, the earth's sociality had been an insubstantial thing for him, brought newly into being with his feet, eyes, pen, and camera. But as he collected books, read them, copied them, translated them, absorbed them into his own texts, he found that everywhere he went the earth was already an archive. He had read the earth by walking; now he read other layers in books—dongba, Chinese, and Tibetan, writing as he went. Throughout, he paid careful attention to those places where the earth opened into text and text into the earth, places marked on the earth by written inscriptions, marked in texts by place names. As he added to all this text, now with translations and descriptions more than with specimens, diary entries, or maps, he attended to openings from text to text. Again these openings were marked by place names, and the names of the beings, human and nonhuman, whose lineages were anchored to the earth in those places. He had worked the seam between earth and archive first in letting the earth generate text as he moved over it, then in moving from place to place to relate texts he had found with the places they named. Finally, in exile, he let the place names move him from text to text, generating more texts as he went—culminating in the completion of that final great exercise in naming and connecting, the *Na-Khi–English Encyclopedic Dictionary*.

My account of Rock's explorations has been biographical in structure: this was the only way I found to write it. But the story I have really wanted to tell is of this creation—this accumulative interleaving of earth and word, for which Rock, more than anyone else I know, was a vehicle. I have tried to show that he was not the only vehicle, even for those parts of this creation most closely associated with his name. There were many other participants. The Ming dynasty Mu chiefs who, in decades of warfare against their neighbors, unified Yunnan's northwest under the name of their lineage, making it possible to conceive as a single, coherent region. The eighteenth- and nineteenth-century dongba who invented a script to write that region and created a populous world of nonhuman beings to wander its paths and ranges. The French who began shipping the flora of the lovely Yulong and Cangshan mountain ranges to Paris. The dongba of the early twentieth century who, as their cult waned, kept the old Mu domain alive through their archival practice, collecting and copying texts, and reciting those that the people in mountain villages like Nvlvk'ö still wanted to hear. Zhao Chengzhang, He Nüli, Zhao Tangguang, Lu Wanyu, and the tens of others who set out from Nvlvk'ö to seek plants along the routes inscribed in the dongba archive. Ganton, his son, and the other Tibetan Catholics who participated in the furthest flung of these journeys in Yunnan's remote northwest. George Forrest, Isaac Bailey Balfour, William Wright Smith, and the other scientists and herbarium workers at Edinburgh who remade the region again in their own great archive. And of course He Guangyi, He Zhihui, Li Shichen, Yang Jiding, He Ji, Lu Wanyue, Zhao Zhongdian, He Shuishan, Lu Wanxing, Li Shiwen, Li Shixi, He Zhu, He Zixiu, Li Wenzhao, He Wenli, and the dongba He Huating. All were vehicles for this open, moving, living process of using bodies and their technological extensions to generate words, images, lines, and texts from the earth; of folding them back into the earth; of making them part of the shape of the earth that is experienced by other bodies, other eyes and cameras, ears and pens. In the end, this is what I wish to name when I say earth. For this living, social, creative process is what the earth has always been, ever since humans first began to walk over the land and speak to it.

Notes

INTRODUCTION

1. This description of how Zhao packed specimens is drawn from George Forrest, "Notes on Collecting," in "Field Notes, 1904, 1906, 1911," Forrest Collection, Royal Botanic Garden, Edinburgh (hereafter, RBGE).

2. Guan Xuexuan and Wan Xianyan, *Lijiang Fu zhi lue* (1743).

3. The map and Davies's account of his travels comprise H. R. Davies, *Yunnan: The Link Between India and the Yangtze* (Cambridge: Cambridge University Press, 1909).

4. Heinrich Handel-Mazzetti, *A Botanical Pioneer In South West China: Experiences and Impressions of an Austrian Botanist during the First World War.* trans. David Winstanley (Chippenham: Antony Rowe, 1996).

5. For descriptions of public and private elementary schools in the Lijiang valley, see Li Ruming, ed., *Lijiang Naxizu Zizhixian zhi* (Kunming: Yunnan renmin chubanshe, 2001), 676–85.

6. The three biographies are John Macqueen Cowan, *The Journeys And Plant Introductions of George Forrest, V.M.H.* (London: Oxford University Press, 1952); Scottish Rock Garden Club, *George Forrest, V.M.H., Explorer and Botanist* (Edinburgh: Stoddart & Malcolm, 1935); and Brenda McLean, *George Forrest, Plant Hunter* (Edinburgh: Antique Collectors' Club, 2004).

7. The ledger pages are in the Forrest Collection at the Royal Botanic Garden, Edinburgh. They are not dated with a year, but they give several dates with month and day in both the Gregorian and the Chinese lunar calendars, which, when compared, reveal the year to be 1925. Edward M. Reingold and Nacham Dershowitz, *Calendrical Tabulations, 1900–2200* (Cambridge: Cambridge University Press, 2002), 52.

8. "Bamboo rats" is what Forrest calls them in his marginal note, but to the collectors they were simply rats *(laoshu).*

9. Forrest to Williams, February 26, 1925; Forrest to Williams, April 9, 1925; Forrest to Williams, July 19, 1925, all at RBGE.

10. Paul Carter, *The Road to Botany Bay: An Exploration of Landscape and History* (New York, Knopf, 1988), xxii.

11. E. Bretschneider, *History of European Botanical Discoveries in China* (London: Marston, 1898).

12. Fa-ti Fan, *British Naturalists in Qing China: Science, Empire, and Cultural Encounter* (Cambridge: Harvard University Press, 2003), 18; John Gascoigne, *Science in the Service of Empire: Joseph Banks, the British State, and the Uses of Science in the Age of Revolution* (New York: Cambridge University Press, 1998), 140–42.

13. "Letter of appointment to Robert Fortune, 1842," Horticultural Society of London. Quoted in Tyler Whittle, *The Plant Hunters* (Philadelphia: Chilton Book Company, 1970), 187.

14. Fan, *British Naturalists*, 74. P. D. Coates, *The China Consuls: British Consular Officers, 1843–1943* (Hong Kong: Oxford University Press, 1988) lists all of the consulates; Stanley Fowler Wright, *Hart and the Chinese Customs* (Belfast: W. Mullan, 1950) gives a detailed account of the workings of the Chinese Maritime Customs Service, established in 1854 to oversee maritime trade at the treaty ports.

15. The consular records of the Kunming, Tengyue, and Simao consulates contain voluminous correspondence on Augustus Margary and the diplomatic crisis his murder provoked: Foreign Office, Public Records Office, Kew (hereafter PRO). On Hosie see E. Bretschneider, *History*, 767–71 as well as Hosie's own books, *Three Years in Western China* (London: G. Philip & Son, 1890), and *On the Trail of the Opium Poppy* (London: G. Philip & Son, 1914). On Henry, see Bretschneider, *History*, 774–94; Ernest Newlmers and William Cuthbertson, *Curtis's Botanical Magazine Dedications, 1827–1927* (London: Bernard Quaritch, 1931), 298–300; and Henry's book, *Notes on Economic Botany of China* (Shanghai: Presbyterian Mission Press, 1893). Brief biographies of all three are also given in E. H. M. Cox, *Plant Hunting in China* (Hong Kong: Oxford University Press, 1986).

16. Thomas Richards, *The Commodity Culture of Victorian England: Advertising and Spectacle, 1851–1914* (Stanford: Stanford University Press, 1990), 27–28.

17. Thant Myint-U, *The Making of Modern Burma* (Cambridge: Cambridge University Press, 2001), 198–218; a detailed account of these military, cartographic, and ethnographic adventures is given in Dorothy Woodman, *The Making of Burma* (London: Cresset Press, 1962).

18. See Chen Zonghai, *Tengyue Ting zhi* (1887).

19. These are the articles mentioned in the Tengyue consul's trade report to the Foreign Office in 1908, "Report on the Trade of Tengyueh for the year 1908, by Mr. Archibald Rose, H.M. Acting Consul," Foreign Office 228/1733, PRO. On the species of musk deer and their distribution in China, see Colin P. Groves, Yingxiang Wang, and Peter Grubb, "Taxonomy of Musk Deer, Genus Moschus (Moschidae, Mammalia)," *Acta Theriologica Sinica* 15, no. 3 (1995): 181–97.

20. Farrer's accounts of his two-year sojourn in Gansu are *The Rainbow*

Bridge (London: E. Arnold, 1926) and *On the Eaves of the World* (London: E. Arnold, 1917). Farrer died during his journey to the Yunnan-Burma border; the account of his expedition by his companion E. M. H. Cox, with Helen T. Maxwell, is *Farrer's Last Journey* (London: Dulau, 1926). For biographies, see Nicola Shulman, *A Rage for Rock Gardening: The Story of Reginald Farrer* (London: Short Books, 2002) and E. M. H. Cox and William T. Stearn, *The Plant Introductions of Reginald Farrer* (London: New Flora and Silva, 1930).

21. On Ward, see Erik Mueggler, "The Lapponicum Sea: Matter, Sense and Affect in the Botanical Exploration of Southwest China and Tibet," *Comparative Studies in Society and History* 47 no. 3 (2005): 442–79 and Erik Mueggler, "Reading, Glaciers, and Love in the Botanical Exploration of Southwest China and Tibet," *Michigan Quarterly Review* 44, no. 4 (2005): 722–53. For a light biography, see Charles Lyte, *Frank Kingdon-Ward: The Last of the Great Plant Hunters* (London: J. Murray, 1989). Ward's other biographer was his sister Winifred Kingdon-Ward, whose lovely unpublished manuscript "The Flower Chief" resides in the archives of the Royal Botanic Garden, Kew. Ward wrote two dozen books and nearly six hundred articles about his journeys: they are all cataloged in Ulrich Schweinfurth and Heidrun Marby Schweinfurth, *Exploration In the Eastern Himalayas And the River Gorge Country of Southeastern Tibet: Francis (Frank) Kingdon-Ward (1885–1958)* (Wiesbaden: Steiner, 1975).

22. Harold Roy Fletcher, *A Quest of Flowers: The Plant Explorations of Frank Ludlow And George Sherriff Told From Their Diaries And Other Occasional Writings* (Edinburgh: Edinburgh University Press, 1975).

23. This information about Forrest's forbears and early life is from McLean, *George Forrest*, 16–20.

24. William Adamson, *The Life of the Rev. James Morison* (London: Hodder & Stoughton, 1898).

25. Catherine Hall, *Civilizing Subjects: Metropole and Colony in the English Imagination, 1830–1867* (Cambridge: Polity, 2002).

26. John Abercromby, "Excavation of Three Long Cists at Gladhouse Reservoir, Midlothian," *Proceedings of the Society of Antiquaries of Scotland* 38, no. 4 (1903): 96–98.

27. Balfour to Forrest, September 1, 1903, RBGE.

28. Forrest to Balfour, February 19, 1904, RBGE. The biographical information in this paragraph is from Cowan, *George Forrest;* Scottish Rock Garden Club, *George Forrest;* and McLean, *George Forrest.*

29. James H. Veitch, *Hortus Veitchii* (London: James Veitch and Sons, 1906).

30. Veitch, *Hortus Veitchii,* 79–84, 96.

31. On Wilson, see Ernest H. Wilson, *A Naturalist in Western China* (New York: Doubleday, Page, 1913).

32. Benoit Dayrat, *Les Botanistes et la flore de France* (Paris: Muséum national d'histoire naturelle, 2003).

33. On Sir Joseph Dalton Hooker's explorations of Sikkim and Nepal, see W. B. Turrill, *Pioneer Plant Geography: The Phytogeographical Researches of*

Sir Joseph Dalton Hooker. (The Hague: M. Nijhoff, 1953). W. B. Turrill, *Joseph Dalton Hooker* (London: Thomas Nelson and Sons, 1963); Ray Desmond, *European Discovery of the Indian Flora* (Oxford: Clarendon Press, 1992); as well as Hooker's own *Rhododendrons of Sikkim-Himalaya* (London: Reeve, 1849) and *Himalayan Journals* (London: J. Murray, 1855).

34. On Bulley, see Brenda McLean, *A Pioneering Plantsman: A. K. Bulley and the Great Plant Hunters* (London: Stationary Office, 1997).

35. Bulley to Balfour, April 28, 1904, RBGE. McLean quotes this letter in *George Forrest*, 30.

36. Information about the Catholic community in Cigu is from the author's interviews with its members in 2007.

37. Forrest gave his account of the attack on Cigu in his letters to Clementina Trail and Isaac Bailey Balfour (Forrest to Balfour, October 1, 1905, RBGE), and a published article, George Forrest, "The Perils of Plant Collecting," *Gardener's Chronicle*, May 21, 1924, 325–26; May 28, 1924, 344. The magistrate of Lijiang made an official report of incident in a letter to Shi Hongshao, trade commissioner at Tengyue (Foreign Office 228/1604, PRO). A more comprehensive narrative of the Tibetan uprising against the foreign missions at Adunzi and Cigu, including an account of its suppression by Qing troops and the reparations paid to the French, is given in Yang Xuezheng, *Yunnan zongjiao shi* (Kunming: Yunnan renmin chubanshe, 1999), 374–75. See also William M. Coleman, "The Uprising at Batang: Khams and its Significance in Chinese and Tibetan History," in *Kamps pa Histories: Visions of People, Place and Authority*, ed. Lawrence Epstein, 31–56 (Boston: Brill, 2002).

38. H. H. Davidian, *The Rhododendron Species*, vol. 1, *Lepidotes* (Portland, OR: Timber Press, 1982).

39. J. B. Stevenson, J. Hutchinson, H. V. Tagg, and A. Rehder, *The Species of Rhododendron* (Edinburgh: The Rhododendron Society, 1930).

40. These early biographical details are from Alvin Chock, "J. F. Rock, 1884–1962, *Newsletter of the Hawaiian Botanical Society* 2, no. 1 (1963): 1–13; and S. B. Sutton, *In China's Border Provinces: The Turbulent Career of Joseph Rock, Botanist-Explorer* (New York: Hastings House, 1974).

41. Joseph Rock, "A New Hawaiian *Scaevola* (*S. Swezayana*)," *Bulletin of the Torrey Botanical Club* 36 (1909): 645–50.

42. Chock, "J. F Rock," gives a thorough bibliography of Rock's botanical publications.

43. Joseph Rock, *The Indigenous Trees of the Hawaiian Islands* (Honolulu, 1913).

44. Joseph Rock, "List of Hawaiian Names of Plants," *Hawaii Board of Agriculture and Forestry Botanical Bulletin*, 2 (1913): 2.

45. Joseph Rock, Diary, September 14, 1913, RBGE.

46. Ibid., October 10, 1913.

47. Ibid., October 24, 1913.

48. Ibid., December 10, 1913.

49. Joseph Rock, *The Ornamental Trees of Hawaii* (Honolulu, 1917); Joseph

Rock, "The Ohia Lehua Trees of Hawaii," *Hawaii Board of Agriculture and Forestry Botanical Bulletin* 4 (1917): 1–76.); Joseph Rock, *The Arborescent Indigenous Legumes of Hawaii* (Honolulu, 1919); Joseph Rock, "A Monographic Study of the Hawaiian Species of the Tribe Lobelioideae, Family Campanulaceae," *Bernice P. Bishop Museum Memorial* 7 no. 2 (1919): i–xvi, 1–395. Joseph Rock, *The Leguminous Plants of Hawaii: Being an Account of the Native, Introduced and Naturalized Trees, Shrubs, Vines and Herbs, Belonging to the Family Leguminosae.* (Honolulu: Hawaiian Sugar Planter's Association Experiment Station, 1920).

50. Chock, "J. F. Rock."

51. Sutton, *In China's Border Provinces*, 44.

52. E. H. Bryan, "An Anecdote Concerning Joseph F. Rock," *Newsletter of the Hawaiian Botanical Society* 2, no. 1 (1963): 16.

53. See John Tayman, *The Colony* (New York: Scribner, 2006).

54. Jack London visited the colony and attempted to debunk these tales of inhumane conditions. Jack London, "The Lepers of Molokai," *Women's Home Companion* 35(January 1908).

55. Arthur L. Dean and Richard Wrenshall, "Fractionation of Chaulmoogra Oil," *Journal of the American Chemical Society* 42 (1920): 2626–45.

56. John Parascandola, "Chaulmoogra Oil and the Treatment of Leprosy," *Pharmacological History* 45, no. 2 (2003): 47–57.

57. David Fairchild, *The World Was My Garden: Travels of a Plant Explorer* (New York: Scribner's, 1941), 372. The genus *Hydnocarpus* had the alternative name of *Tarakntogenos*.

58. Rock's letters to Fairchild are preserved at the Hunt Institute for Botanical Documentation, Pittsburg, PA.

59. Gong Yin, *Zhongguo tusi zhidu* (Kunming, Yunnan minzu chubanshe, 2000), 598–600; Volker Grabowsky and Andrew Turton, *The Gold and Silver Road of Trade and Friendship* (Chiang Mai: Silkworm Books, 2003), 173–93.

60. Rock, Diary, February 23, 1922, RBGE.

61. Rock to Fairchild, March 24, 1922, Hunt Institute.

62. William Wright-Smith to Forrest, September 9, 1924, RBGE.

63. Gong, *Zhongguo tusi zhidu*, 632. Rock to Fairchild, November 10, 1922, Hunt Institute.

64. David Arnold, *The Tropics and the Traveling Gaze: India, Landscape, and Science, 1800–1856, Culture, Place, and Nature* (Seattle: University of Washington Press, 2006), 185.

65. Rebecca Solnit, *River of Shadows: Eadweard Muybridge and the Technological Wild West* (New York: Viking, 2003), 45.

66. Francis Kingdon Ward, Diary, March 24, 1924, Royal Botanic Garden, Kew.

67. Timothy Mitchell, *Colonising Egypt* (Cambridge: Cambridge University Press, 1998).

68. Webb Keane, *Christian Moderns: Freedom and Fetish in the Missionary Encounter* (Berkeley: University of California Press, 2007).

CHAPTER 1

1. "Report by Mr. Litton (sent home by Mr. Ottewill) on Journey from Tengyueh to Upper Salwen," May 1906, Foreign Office 881/8682, PRO. In 1908, George Forrest presented an edited version of Litton's report, accompanied by Forrest's photographs, to the Royal Geographic Society: "Journey on the Upper Salwen, October–December, 1905," *The Geographical Journal* 32 (1908): 239–66.

2. Forrest, "Journey," 265. The great scientific traveler Xu Hongzu, writing as Xu Xiake, solved this puzzle for Chinese geographers in the seventeenth century. See Xu Hongzu, *Xu Xiake Youji*, ed. Ding Wenjiang (Beijing: Shang wu yin shu guan, 1986) and Joseph Needham, *Science and Civilization in China*, vol. 3 (Cambridge: Cambridge University Press, 1959), 524–25.

3. A copy of the manuscript is published in Joseph Rock, comp., *Na-khi Manuscripts*, ed. Klaus L. Janert (Wiesbaden: F. Steiner, 1965), 7:2:400–7.

4. Xie Benshu, "Cong Pianma shijian dao Banhong shijian: Zhong Mian bianjie lishi yange wenti," *Yunnan shehui kexue*, no. 4 (2000): 72–81. Chinese-language scholarship on the history of the Yunnan border remains very nationalistic. On the southern border, see, for example, Hong Congwen, "Cong Banhong shijian kan Yunnan bianjiang guanli jigou de yunzuo," *Zhongguo bianjiang shidi yanjiu*, no. 3 (1997): 69–79.

5. Quoted in Dorothy Woodman, *The Making of Burma* (London: Cresset Press, 1962), 282.

6. Litton, "Report by Mr. Litton."

7. Shi Hongshao, letters to the *Waiwubu*, 1905, Foreign Office 228/1604, PRO.

8. Litton, "Report by Mr. Litton."

9. Forrest to Clementina Trail, 1904 (typescript letter no. 3), RBGE.

10. See Bernard Cohn, "Introduction," in *Colonialism and its Forms of Knowledge: The British in India* (Princeton: Princeton University Press, 1996).

11. Litton, "Report by Mr. Litton."

12. Forrest, "Journey to the Upper Salwen," 275.

13. Litton, "Report by Mr. Litton," 11.

14. For an account of the "Pianma incident," see Xie, "Cong Pianma shijian."

15. Woodman, *The Making of Burma*, 494–517. The final border followed the Gaoligong range except for two locations: Pianma, which China successfully retrieved from Burma, and, in the far northwest, the Dulong (formerly Qiu; known to the British as the Tarong and to Tibetans as the Drong), a tributary of the Irrawaddy.

16. Litton wrote in Forrest's stead to Balfour. Litton to Balfour, December 13, 1905, RBGE.

17. Joseph Rock, *A Na-Khi-English Encyclopedic Dictionary*, vol. 1 (Rome: Istituto Italiano per il Medio ed Estremo Oriente, 1963), 234; Joseph Rock, *A Na-Khi-English Encyclopedic Dictionary*, vol. 2 (Rome: Istituto Italiano per il Medio ed Estremo Oriente, 1972), 564.

18. Quoted in Joseph Rock, *The Ancient Na-Khi Kingdom of Southwest China* (Cambridge: Harvard University Press, 1947), 334n46.

19. Joseph Rock, *The Na-Khi Naga Cult and Related Ceremonies* (Rome: Istituto Italiano per il Medio ed Estremo Oriente, 1952), 2:607, 2:613.

20. Forrest to Williams, August 8, 1921, RBGE.

21. Partha Chatterjee, *The Nation and Its Fragments* (Princeton: Princeton University Press, 1993), 14; see also George Steinmetz, "'The Devil's Handwriting': Precolonial Discourse, Ethnographic Acuity, and Cross-Identification in German Colonialism," *Comparative Studies in Society and History* 45, no. 1 (2003): 41–95.

22. Ann Laura Stoler, *Carnal Knowledge and Imperial Power: Race and the Intimate in Colonial Rule* (Berkeley: University of California Press, 2002).

23. Catherine Hall, *Civilising Subjects: Metropole and Colony in the English Imagination, 1830–1867* (Oxford: Polity, 2002), 65.

24. The quotations from Forrest in this and the next paragraphs are extracts from a letter to his mother, Mary Forrest, and family, July 3, 1904, in a private collection. The parts of the letter extracted here are quoted in McLean, *George Forrest,* 34–37.

25. Hall, *Civilising Subjects,* 27; see also John Tosh, *A Man's Place: Masculinity and the Middle-Class Home in Victorian England* (New Haven: Yale University Press, 1999).

26. Forrest to Mary Forrest, July 3, 1904, quoted in McLean, *George Forrest,* 35.

27. As an example, take Henry Charles Sirr's *China and the Chinese: Their Religion, Character, Customs and Manufactures,* 2 vols. (London: Wm S. Orr, 1849). Sirr endeavors to "depict national character as it is, without romantic colouring . . . or puerile deprecation." He praises filial piety, quotes Confucious at length, and describes the Chinese as vicious, revolting, sexually promiscuous, addicted to gambling, cheating, and fraud, avaricious, fatalistic, stoical, careless of life, and subject to disease (vol. 2, 415–25).

28. Steinmetz, "Devil's Handwriting," 79.

29. Forrest to Mary Forrest, July 3, 1904, quoted in McLean, *George Forrest,* 36.

30. Forrest calls the place "Namsa." It is most likely Nanzha, where in 1905 several companies of the Pu'er Border Pacification Battalion were stationed. Foreign Office 228/1598, PRO.

31. Forrest to Mary Forrest, July 21, 1904, RBGE.

32. Forrest to Clementina Trail, August 4–12, 1904, RBGE.

33. "Enc. no. 2 in Tengyueh no. 17 of 22 May 1907," Foreign Office 228/1671, PRO; Wilkinson to Foreign Office, October 18, 1905, Foreign Office 228/1598, 218–21, PRO.

34. Forrest to Mary Forrest, undated typescript (circa September 1904), "Letter No. 5," RBGE.

35. Forrest to Mary Forrest, undated typescript (circa September 1904), "Letter No. 6," RBGE.

36. Ibid.

37. Timothy Mitchell, *Colonising Egypt* (Cambridge: Cambridge University Press, 1988).

38. Anne Maxwell, *Colonial Photography and Exhibitions: Representations of the "Native" and the Making of European Identities* (London: Leicester University Press, 1999).

39. E. Ann Kaplan, *Looking for the Other: Feminism, Film and the Imperial Gaze* (London: Routledge, 1997).

40. Cohn, *Colonialism and its Forms of Knowledge.*

41. Deborah Poole, *Vision, Race, and Modernity: A Visual Economy of the Andean Image World* (Princeton: Princeton University Press, 1997), 15.

42. Mary Louise Pratt, *Imperial Eyes: Travel Writing and Transculturation* (London: Routledge, 1992).

43. Forrest to Mary Forrest, undated typescript (circa September 1904), "Letter No. 5," RBGE.

44. See Bin Yang, "Horses, Silver and Cowries: Yunnan in Global Perspective," *Journal of World History* 15, no. 3 (2004): 281–322.

45. Zhongdian Xian difangzhibianzuan weiyuanhui, *Zhongdian Xian zhi* (Kunming: Yunnan minzu chubanshe, 1997), 145.

46. G. Litton, "Report on a survey in Thibetan Yunnan with notes on the condition of the country in the autumn of 1904," Foreign Office 228/1562, PRO.

47. Ibid.

48. Ibid.

49. Forrest to Mary Forrest, "Letter No. 5."

50. Ibid.

51. Forrest to Mary Forrest, undated typescript (circa December 1904) "Letter No. 3," RBGE.

52. This would change dramatically in the 1920s, when warlords in Sichuan began selling rifles to Yi and Tibetan populations east and north of Zhongdian in exchange for opium revenue, and the British began to supply rifles to Lhasa.

53. Forrest to Mary Forrest, "Letter No. 3."

54. Paul Landau, "Empires of the Visual: Photography and Colonial Administration in Africa," in *Images and Empires: Visuality in Colonial and Postcolonial Africa*, eds. Paul Landau and Deborah Kaspin, 141–71 (Berkeley: University of California Press, 2002).

55. Francis Kingdon Ward, *The Land of the Blue Poppy* (Cambridge: Cambridge University Press, 1913), 170.

56. Francis Kingdon Ward, *The Mystery Rivers of Tibet* (London: Seeley, Service, 1923), 164.

57. Michael Taussig, *Mimesis and Alterity: A Particular History of the Senses* (New York: Routledge, 1993), 208.

58. Martin Jay, "Scopic Regimes of Modernity," in *Vision and Visuality*, ed. Hal Foster, 3–27 (Seattle: Bay Press, 1988).

59. David Arnold, *The Tropics and the Traveling Gaze: India, Landscape, and Science, 1800–1856, Culture, Place, and Nature* (Seattle: University of Washington Press, 2006), 209.

CHAPTER 2

1. George Forrest to J. C. Williams, June 6, 1913, RBGE.

2. Joseph Rock, *The Ancient Na-Khi Kingdom of Southwest China* (Cambridge: Harvard University Press, 1947), 216–19. In 1950, when a census was taken of the Lijiang valley, Nvlvk'ö contained 136 households and 816 people. See He Yaohua and Yang Fuyuan, eds., *Lijiang Yulong Shan quyu cunzhai fazhan yu shengtai diaocha* (Kunming: Yunnan renmin chubanshe, 1998).

3. Charles Fremont McKhann, "Fleshing Out the Bones: Kinship and Cosmology in Naqxi Religion" (PhD diss., University of Chicago, 1992), 248–51.

4. Rock, *Ancient Na-Khi Kingdom*, 218.

5. He and Yang, *Lijiang Yulong Shan*, 11.

6. Ibid., 13.

7. For photographs of Nvlvk'ö villagers performing a "sacrifice to heaven" ceremony in the 1930s, see "The Propitiation of Heaven, Photographs by Joseph Rock," in *Naxi and Moso Ethnography, Kin, Rites, Pictographs*, eds. Michael Oppitz and Elisabeth Hsu, 173–88 (Zurich: Völkerkundemuseum, 1998). See also Joseph Rock, "The Mùan-bpö Ceremony or the Sacrifice to Heaven as Practiced by the Na-khi," *Monumenta Serica* 13 (1948): 1–160; and McKhann, "Fleshing out the Bones."

8. He and Yang, *Lijiang Yulong Shan*, 13–15.

9. Ibid., 1–9.

10. Rock, *Ancient Na-Khi Kingdom*, 156.

11. Ibid., 156; He and Yang, *Lijiang Yulong Shan*, 2.

12. Rock, *Ancient Na-Khi Kingdom*, 149.

13. Xin Fachun, *Ming Mu shi yu Zhongguo Yunnan zhi kai fa* (Taibei Shi: Wenshi zhe chubanshe, 1985).

14. Mu Gong's first chronicle is reproduced in full in Rock, *Ancient Na-Khi Kingdom*. A copy of the second chronicle, with its preface by Yang Shen, was collected by Jaques Bacot and published by Eduoard Chavannes first in "Documents historiques et géographiques relatifs à Likiang," *T'oung Pao* 13 (1912): 564–653 and then in *Les Mo-so* (Leiden: Brill, 1913). For a list of additional sources on the Mu chiefs, see Chen Zidan, "Lijiang Mushi tusi dang'an wenxian pingshu," *Guzhai zhengli yajiu xuekan* 6 (Nov. 2004): 25–28.

15. Rock, *Ancient Na-Khi Kingdom*, 77.

16. Charles Backus, *The Nan-Chao Kingdom and T'ang China's Southwestern Frontier* (Cambridge: Cambridge University Press, 1981).

17. Mosozhao was the name of one of the six petty kingdoms in the Erhai lake area that had been unified three centuries before to create the Nanzhao. In Chinese texts of the Tang and later, Moso was also a name given to the peoples whom we presume were the ancestors of the present-day Naxi. Mu Gong was apparently attempting to appropriate the name of the famous kingdom to give prestige to the polity that, he claims, his ancestors ruled during the Song period. On the Nanzhao, see Backus, *The Nan-Chao Kingdom*.

18. Christine Mathieu, *A History and Anthropological Study of the Ancient*

Kingdoms of the Sino-Tibetan Borderland—Naxi and Mosuo (Lewiston, NY: Edwin Mellen Press, 2003), 73.

19. Ibid., 51–93.

20. "Geneaological Chronicles of the Mu Dynasty," in Rock, *Ancient Na-Khi Kingdom*, 97–99.

21. The Mu chiefs carefully preserved their letters from the emperor. When the first appointed regular prefectural magistrate *(liu zhifu)* of Lijiang burned the Mu records, Mu Xing's wife hid these letters in a locked box, saving them from destruction. They remain an important source on the history of Lijiang.

22. I have quoted Joseph Rock's translation, *Ancient Na-Khi Kingdom*, 282.

23. Quoted in Julian Ward, *Xu Xiake (1587–1641): The Art of Travel Writing* (Richmond, Surrey: Curzon, 2001), 139.

24. The foregoing is based upon the Mu chronicles, reproduced in Rock, *Ancient Na-Khi Kingdom*; Naxizu jianshi bianxiezu., *Naxizu jianshi* (Kunming: Yunnan renmin chubanshe, 1984); Mathieu, *History*; and Cai Hua, *A Society Without Fathers or Husbands: The Na of China* (New York: Zone Books, 2001).

25. James Lee, "The Legacy of Immigration in Southwest China, 1250–1850," *Annales de démographie historique* (1982): 279–304.

26. Charles McKhann, "Naxi, Rerkua, Moso, Meng: Kinship, Politics and Ritual on the Yunnan-Sichuan Frontier," in *Naxi and Moso Ethnography*, eds. Oppitz and Hsu, (Zurich: Völkerkundemuseum Zurich, 1998), 31.

27. Mathieu, *History*, 86.

28. McKhann, "Fleshing out the Bones." McKhann, "Naxi, Rerkua, Moso, Meng."

29. Claude Lévi-Strauss, *Elementary Structures of Kinship* (Boston: Beacon Press, 1969 [1949]), 544–66.

30. Forrest to Balfour, December 26, 1907; Balfour to Forrest, December 28, 1907, both at RBGE.

31. Forrest to Balfour, June 13, 1909, RBGE.

32. Forrest to Balfour, March 30, 1909; April 1, 1909; April 12, 1909; April 25, 1909, all at RBGE.

33. Forrest to Balfour, undated (circa December 1909), RBGE.

34. Forrest to Balfour, January 4, 1910, RBGE.

35. Forrest to Balfour, July 12, 1910, RBGE.

36. Forrest to Balfour, December 26, 1909, RBGE.

37. Forrest to Balfour, July 12, 1910, RBGE.

38. Forrest to Williams, October 25, 1912; Forrest, "Notes on Collecting," in "Field Notes, 1904, 1906, 1911," RBGE.

39. Forrest, "Notes on Collecting."

40. McLean, *George Forrest*, 103.

41. Several accounts of Li Genyuan's occupation of Tengyue are given in letters from British consuls to the British minister in Beijing in 1912, Foreign Office 228/1842, PRO.

42. Smith to Sir John Jordan, November 29, 1911; December 4, 1911; December 13, 1911, Foreign Office 228/1807, PRO.

43. Zhang Dezhi, Zhou Conglong, Wang Pei and Li Qiaofu, "Petition to the President," Foreign Office 228/1842, PRO.

44. On the Guomingdang invasion of the upper Salween, see Tang Rong and Li Zicheng, "Minguo chunian Nujiang zhibian shuping," *Chuxiong Shifan xuebao* 15, no. 4 (Oct. 2000): 81–85; Lisuzu jianshi bianxiezu, *Lisuzu jianshi* (Kunming: Yunnan renmin chubanshe, 1983); Nuzu jianshi bianxiezu, *Nuzu jianshi* (Kunming: Yunnan renmin chubanshe, 1987).

45. McLean, *George Forrest*, 105–8. Forrest to Balfour, April 24, 1911; August 30, 1911; November 1, 1911; December 6, 1911; Forrest to Williams, October 23, 1911, all at RBGE.

46. George Forrest, "Rhododendrons in China," *Gardeners' Chronicle* 51, no. 4 (May, 1912): 291–92.

47. Forrest to Williams, May 18, 1912, RBGE.

48. Forrest to Williams, July 12, 1912, RBGE.

49. Forrest to Balfour, August 31, 1912; Forrest to Williams, August 31, 1912; September 26, 1912; October 10, 1912, all at RBGE. "Tengyue Consulate Intelligence Report," September 30, 1912, Foreign Office 228/1842, PRO.

50. Forrest to Williams, August 31, 1912, RBGE.

51. Forrest to Williams, February 17, 1913, RBGE.

52. Ibid.

53. Forrest to Williams, June 16, 1913, RBGE.

54. This description of household space is from McKhann, "Fleshing Out the Bones," 248–60.

55. Forrest to Mary Forrest, "Letter No. 3," 1904, RBGE.

56. *Primula beesiana* and *Primula bulleyana* were among Forrest's discoveries on his first expedition, named after the seed firm that employed him and its owner, Arthur Bulley.

57. Forrest to Williams, June 26, 1913; Forrest to William Wright Smith, June 24, 1913, both at at RBGE.

58. Forrest to Williams, July 17, 1913, RBGE.

59. Forrest to Williams, June 16, 1913; Forrest to Smith, June 24, 1913, both at RBGE.

60. Forrest to Williams, August 1, 1913, RBGE.

61. For a history and ethnography of the three-river basin, see Xiaolin Guo, *State and Ethnicity in China's Southwest* (Leiden: Brill, 2008).

62. Cai Hua, *A Society Without Fathers or Husbands*, 45; Li Ruming, ed., *Lijiang Naxizu Zizhixian zhi* (Yunnan: Kunming renmin chubanshe, 2001), 52.

CHAPTER 3

1. *Mu shi, tusi zhuan,* quoted in Naxizu jianshi bianxiezu, *Naxizu jianshi* (Kunming: Yunnan renmin chubanshe, 1984).

2. For a summary, see Yu Haibo and Yu Jianhua, *Mu shi tusi yu Lijiang* (Kunming: Yunnan minzu chubanshe, 2002), 248–62.

3. Li Ruming, ed., *Lijiang Naxizu Zizhixian zhi* (Kunming: Yunnan renmin

chubanshe, 2000), 678; "Geneaological Chronicles of the Mu," reproduced in Joseph Rock, *The Ancient Na-Khi Kingdom of Southwest China* (Cambridge: Harvard University Press, 1947), 143.

4. Chen Hongmou's career has been described in detail by William T. Rowe in *Saving the World: Chen Hongmou and Elite Consciousness in Eighteenth-Century China* (Stanford: Stanford University Press, 2001).

5. Guan Xuexuan, and Wan Xianyan, *Lijiang Fu zhi lue*, quoted in Rock, *Ancient Na-Khi Kingdom*, 46.

6. William T. Rowe, "Education and Empire in Southwest China: Ch'en Hung-mou in Yunnan, 1733–1738," in *Education and Society in Late Imperial China, 1600–1900*, eds. Benjamin Elman and Alexander Woodside (Berkeley: University of California Press, 1994), 419.

7. Lijiang did, however, produce seven *jinshi* and sixty *juren* between 1723 and 1911, an unusual number of successful candidates for a sparsely populated peripheral region.

8. Rowe, "Education and Empire," 435.

9. Li Ruming, *Lijiang Naxizu Zizhixian zhi*, 676–85.

10. Ibid., 679. For descriptions of elementary-level education in the Qing, see Angela Leung, "Elementary Education in the Lower Yangtze Region in the Seventeenth and Eighteenth Centuries," in *Education and Society in Late Imperial China, 1600–1900*, eds. Elman and Woodside, 381–416; and Evelyn Rawski, *Education and Popular Literacy in China* (Ann Arbor: University of Michigan Press, 1979).

11. "Geneaological Chronicles of the Mu," reproduced in Rock, *Ancient Na-Khi Kingdom*, plate 12.

12. Joseph Rock, "Studies in Na-khi Literature, I: The Birth and Origin of Dto-Mba Shi-Lo, the Founder of Mo-so Shaminism According to Mo-so Manuscripts." *Bulletin de l'École Française d'Extrême-Oriente* 37 (1937): 1–39; Joseph Rock, *A Na-Khi-English Encyclopedic Dictionary*, vol. 1 (Rome: Istituto Italiano per il Medio ed Estremo Oriente), 9; Christine Mathieu, *A History and Anthropological Study of the Ancient Kingdoms of the Sino-Tibetan Borderland* (Lewiston, NY: Edwin Mellen Press, 2003), 144, 192.

13. Matheiu, *History*, 156. Partly on the basis of these late dates, Jackson and Pan speculate that dongba script was probably invented very late and that most of the books were first written in the nineteenth century. Anthony Jackson and Pan Anshi, "The Authors of Naxi Ritual Books, Index Books, and Books of Divination," in *Naxi and Moso Ethnography*, eds. Michael Oppitz and Elisabeth Hsu, 237–73 (Zürich: Völkerkundemuseum Zürich, 1998).

14. These are the authors whom Rock called the Dto-la brothers. Rock believed there to be three Dto-la brothers. Since their manuscripts were particularly beautiful, he endeavored to collect them all, and a large number have been preserved in Western libraries. Analyzing these books on the basis of their style, Jackson and Pan identified nine authors; one wrote in a style identical to that of a famous dongba from Baisha whom He Zhiwu and Guo Dalie identify as Jiu Zhilao. Jackson and Pan, "The Authors of Naxi Ritual Books."

15. Rock, who collected and translated the manuscript, gives the date according to the Chinese calendar as 1573. But as Jackson and Matheiu both argue, this is most likely not a Chinese date because if it were, it would identify the dynasty as well as the imperial cycle. See Matheiu, *History*, 160–62; Anthony Jackson, "Kinship, Suicide, and Pictographs among the Na-khi," *Ethnos* 36 (1971), 52–93.

16. Jackson and Pan, "The Authors of Naxi Ritual Books," 246.

17. Matheiu, *History*, 163–64.

18. Ibid., 165.

19. Indeed, Matheiu argues that the centering of dongba script in the Lijiang basin and its absence in Yongning, with its closely allied Daba religious tradition, is evidence that the script was invented after the Ming removed Yoning from the jurisdiction of the Mu chiefs in 1382. Matheiu, *History*, 165–66.

20. On the origins of Yi scripts, see David Bradley, *Proto-Loloish* (London: Curzon Press, 1979).

21. He Zhiwu and Guo Dalie, "Dongba jiao de paixi he xianzhuang," in *Dongba wenhua lunji*, eds. Guo Dalie and Yang Shiguang, 38–54 (Kunming: Yunnan renmin chubanshe, 1985).

22. Jackson and Pan, "The Authors of Naxi Ritual Books," 242.

23. He and Guo, "Dongba jiao," 38.

24. Matheiu, *History*, 147.

25. He and Guo, "Dongba jiao," 39. He and Guo estimate that there were around a thousand dongba. This, however, includes the various local traditions of ritual practitioners who were officially named Naxi in the 1950s but who did not use writing (Nari, Naheng, Lare, Ruanke, Laluo, and Tanglang).

26. Versions of this text have been translated separately by Joseph Rock and Li Lincan and commented on by Pan Anshi. See Joseph Rock, *The Zhimä Funeral Ceremony of the Na-Khi of Southwest China* (Vienna-Mödling: St. Gabriel's Mission Press, 1955), 51–52; Li Lincan, *Moxiezu de jingdian yanjiu* (Taibei: Dongfang wenhua shuju, 1971), 181; Pan Anshi, "The Translation of Naxi Religious Texts," in *Naxi and Moso Ethnography*, eds. Oppitz and Hsu, 277. Here, I follow Rock's translation and explanations. A photostat of another version of the text is in Joseph Rock, comp. *Na-Khi Manuscripts*, ed. Klaus Ludwig Janert, (Wiesbaden: F. Steiner, 1965), 7:2:351–56.

27. Rock, *Dictionary*, vol. 1, xxii.

28. Joseph Rock, *The Na-Khi Naga Cult and Related Ceremonies* (Rome: Istituto Italiano per il Medio ed Estremo Oriente, 1952), 1:12.

29. He Limin and He Sicheng, "The Dto-mbà Ceremony to Propitiate the Demons of Suicide," in *Naxi and Moso Ethnography*, eds. Oppitz and Hsu, 150. I have modified He and He's translation.

30. Michael Oppitz, "Ritual Drums of the Naxi in the Light of their Origin Stories," in *Naxi and Moso Ethnography*, eds. Oppitz and Hsu, 334.

31. Pan Anshi, "The Translation of Naxi Ritual Texts," in *Naxi and Moso Ethnography*, eds. Oppitz and Hsu.

32. Joseph Rock, *The Mùan Bpö Ceremony: Or, the Sacrifice to Heaven as Practiced by the Na-khi* (Peiping: Catholic University, 1948), 72, 182.

33. Jean-Jacques Lecercle, *Philosophy of Nonsense: The Intuitions of Victorian Nonsense Literature* (London: Routledge, 1994), 21n34.

34. Pan, "The Translation of Naxi Ritual Texts," in *Naxi and Moso Ethnography,* eds. Oppitz and Hsu, (Zurich: Völkerkundemuseum Zurich, 1998), 278.

35. Lecercle, *Philosophy of Nonsense,* 60.

36. Charles Fremont McKhann, "Fleshing Out the Bones: Kinship and Cosmology in Naqxi Religion," (PhD diss., University of Chicago, 1992), 272.

37. Rock, *Zhimä Funeral Ceremony,* xv.

38. Rock, *Zhimä Funeral Ceremony,* 61–66. Rock's translations vary enormously in quality. Or, rather, aspects of them vary; his full English renditions are reliably horrible. In his translation of this text, Rock took unusual care to have the entire pictographic text copied, to give the full vocal transcription of the entire text, and to give a full explanation of every single pictograph. This has allowed me to give a modified English translation that I think better reflects the original.

39. Rock, *Zhimä Funeral Ceremony,* 92–100.

40. Joseph Rock, "The D'a Nv Funeral Ceremony with Special Reference to the Origin of Na-khi Weapons," *Anthropos* 50 (1955): 21–22.

41. Rock, *Zhimä Funeral Ceremony,* 109. I have again taken some liberties with Rock's translation, aided by his detailed footnotes, which include many of the pictographs for this text. Unfortunately, however, Rock provides neither the full original text nor a transcription.

42. Ibid., 196.

43. Ibid., 201.

44. Several "gods' road" scrolls have been preserved and published. The Lijiang Naxi Culture Research Institute has published a fine example as color photographs in Li Xi, ed., *Jinshen zhi lu: Naxizu dongba shenlu tu* (Kunming: Yunnan meishu chubanshe, 2001). Rock published two scrolls as photographs in Joseph Rock, "Studies in Na-khi Literature, II: The Na-khi Ha zhi p'i, or the Road the Gods Decide," *Bulletin de l'École Française d'Extrême-Oriente* 37 (1937): 40–119. My account below is based on the latter.

45. The notion of species or kind in regard to trees is expressed in dongba manuscripts with the word *o,* symbolized with two circles joined by a line. *O* is a word for tree trunk, but also for bone, and, by extension, for agnatic relationships: paternal relatives are colloquially called *o-k'o,* "bone relatives." Examples are Rock, *Dictionary,* vol. 1, 365, 371; and He Wanbao and He Jiaxiu, eds., *Naxi dongba gu ji yi zhu quan ji* (Kunming, Yunnan renmin chubanshe, 1999), vol. 27, 110 and vol. 28, 46. The latter is a collection of Naxi dongba manuscripts, given as pictographs, as transcriptions, and as interlinear translations in Chinese, in one hundred volumes.

46. Rock, *Dictionary,* vol. 1, 46.

47. Ibid., 296.

48. He and He, eds., *Naxi Dongba gu ji,* vol. 26, 30.

49. Ibid., vol. 22, 245.

50. Ibid.

51. See for example He and He, eds., *Naxi dongba gu ji,* vol. 41, 206, and Rock, *Na-Khi Naga Cults,* 1:277n590.

52. Among these were puppets called Auntie Rhododendron Woman *(Mùnkhi äèe)* and Uncle Willow Man *(Zhvkhi äbpà),* or sometimes Uncle Rhododendron Man *(Mùnkhi äbpà),* drawn in texts as human figures, dressed in hemmed clothing, with *Rhododendron* or willow twigs in their hats. He and He, eds., *Naxi dongba gu ji,* vol. 27, 263; Rock, *Dictionary,* vol. 1, 294.

53. Rock, comp., *Na-khi Manuscripts,* 7:2:281–82; Rock, *Dictionary,* vol. 1, 329–30.

54. Rock, comp., *Na-khi Manuscripts,* 7:2:282.

55. Ibid., 279.

56. He and He, eds., *Naxi dongba gu ji,* vol. 27, 262.

57. Ibid.

58. Rock, comp., *Na-khi Manuscripts,* 7:2:281.

59. Rock, *Zhimä Funeral Ceremony,* 50; Rock, *Dictionary,* vol. 1, 324. Another suggestive graph in the "Origin of Sorrow" song is the verb *nv̀.* Meaning "to weep," and used when Ssussa finds no years for sale in Kunming, it is expressed as a man walking with enlarged eyes, dripping tears. Rock, *Zhimä Funeral Ceremony,* 72–73.

CHAPTER 4

1. Lorraine Daston and Peter Galison, "The Image of Objectivity," *Representations* 40 (1992): 83. See also Peter Louis Galison, "Judgement against Objectivity," in *Picturing Science, Producing Art,* eds. Caroline Jones and Peter Galison, 327–59 (New York: Routledge, 1998).

2. The phrase *"Rhododendron* whirlpool" is from Francis Kingdon Ward, *Rhododendrons* (London: Latimer House, 1949), 36.

3. Forrest to Williams, December 14, 1912, RBGE.

4. On the Kilmarnock Academy, see William Boyd, *Education in Ayrshire through Seven Centuries* (London: University of London Press, 1961).

5. Forrest to Williams, April 29, 1913, RBGE.

6. Forrest to Mary Forrest, undated typescript (circa September 1904) "Letter No. 5," RBGE.

7. Immanuel Kant, *Observations on the Feeling of the Beautiful and the Sublime,* trans. John T. Goldthwait (Berkeley: University of California Press, 1960 [1764]), 47.

8. Immanuel Kant, *Critique of Judgement,* trans. James Creed Meredith (Oxford: Oxford University Press, 1973 [1790]), 247, quoted in Gilles Deleuze, *Kant's Critical Philosophy: The Doctrine of the Faculties,* trans. Hugh Tomlinson and Barbara Habberjam (Minneapolis: University of Minnesota Press, 1984), 51.

9. Isaac Bailey Balfour to Reginald Farrer, February 6, 1915, RBGE.

10. J. C. Williams to E. A. Bowles, November 22, 1914, Royal Horticultural Institute.

11. Forrest to Balfour, June 6, 1915, RBGE.

12. Elwes had collected in Sikkim; he had authored a monograph on lilies, and he had coauthored a seven-volume study of the trees of Great Britain and Ireland. Henry John Elwes, *A Monograph of the Genius Lilium* (London: Taylor and Francis, 1877); Henry John Elwes and Augustine Henry, *The Trees of Great Britain and Ireland* (Edinburgh: privately printed, 1906).

13. Brenda McLean, *George Forrest, Plant Hunter* (Edinburgh: Antique Collectors' Club, 2004), 136. Most of the information in this paragraph and the preceding one is from McLean.

14. Sir Harry J. Veitch to Mr. Weeks, October 11, 1916, RBGE.

15. Forrest to Keeble, March 18, 1917, RBGE.

16. Forrest to Keeble, March 29, 1917, RBGE.

17. Forrest to Keeble, May 25, 1917, RBGE.

18. Forrest romanized this name as Tsedjrong.

19. Melvyn Goldstein, *A History of Modern Tibet, 1913–1951: The Demise of the Lamaist State* (Berkeley: University of California Press, 1989), 65–87. Warren Smith, *Tibetan Nation: A History of Tibetan Nationalism and Sino-Tibetan Relations* (Boulder: Westview Press, 1996), 168–230; Yang Xuezheng, ed., *Yunnan zongjiao shi* (Kunming: Yunnan renmin chubanshe, 1999), 374.

20. Forrest to Keeble, July 17, 1917, RBGE.

21. George Forrest, "Plant Collecting in China," *Gardener's Chronicle*, October 27, 1917, 165–66.

22. Forrest to Smith, undated (July 1917), RBGE.

23. Willis's three papers on the subject were "The Endemic Flora of Ceylon, with reference to Geographical Distribution and Evolution in General," *Philosophical Transactions of the Royal Society of London. Series B, Containing Papers of a Biological Character* 206 (1915): 307; "The Evolution of Species in Ceylon, with reference to the Dying Out of Species," *Annals of Botany* 30 (1916): 1; and "The Distribution of Species in New Zealand," *Annals of Botany* 30 (1916): 437.

24. On Noah's ark and Mt. Ararat, see Margaret Hogden, *Early Anthropology In the Sixteenth And Seventeenth Centuries* (Philadelphia: University of Pennsylvania Press, 1964); on Humboldt's South American Eden, see H. Walter Lack, *Alexander Von Humboldt and the Botanical Exploration of the Americas* (Munich: Prestel, 2009). The Russian botanist N. I. Vavilov would further develop ideas about centers of origin, particularly of cultivated plants, in the 1920s and 1930s. See N. I. Vavilov, *The Origin, Variation, Immunity and Breeding of Cultivated Plants: Selected Writings,* trans. K. Starr Chester (Waltham, MA: Chronica Botanica, 1951).

25. William Wright Smith, "Obituary of George Forrest, for the Rhododendron Society," 1932, RBGE.

26. Forrest to Balfour, November 1, 1917, RBGE.

27. Ibid.

28. Ibid.

29. Forrest to Smith, July 21, 1917, RBGE.

30. Francis Kingdon Ward, *Land of the Blue Poppy: Travels of a Naturalist in Eastern Tibet* (Cambridge: Cambridge University Press, 1913), 36.

31. Forrest to Williams, March 26, 1918, RBGE.

32. Forrest to Balfour, November 1, 1917, RBGE.

33. Forrest to Balfour, March 26, 1918, RBGE.

34. "Tengyueh Intelligence Report for the Quarter Ended September 30, 1918," Foreign Office 228/3275, PRO.

35. Smith, *Tibetan Nation*, 205–10; Goldstein, *History of Modern Tibet*, 82–84.

36. Yanjing in Chinese. Forrest romanized the Tibetan name as Jerkalo.

37. Forrest to Balfour, May 2, 1918; Forrest to Chittenden, June 20, 1918; Forrest to Balfour, July 7, 1918; Forrest to Chittenden, August 25, 1918, all at RBGE.

38. Forrest to Balfour, July 7, 1918, RBGE.

39. Ibid.

40. Forrest to Balfour, September 28, 1918, RBGE.

41. Forrest to Chittenden, March 26, 1919, RBGE.

42. Valentin to Forrest, May 20, 1919, RBGE.

43. Forrest to Williams, July 31, 1919; Forrest to Chittenden, August 23, 1919, both at RBGE.

44. George Forrest, "A Lecture by Mr. George Forrest on Recent Discoveries of Rhododendrons in China," *Rhododendron Society Notes* 2 (1920): 17, 20.

45. McLean, *George Forrest*, 157–58.

46. "Tengyueh Intelligence Report, June 15, 1921," Foreign Office 228/3275, PRO.

47. Forrest to Balfour, July 1921, RBGE.

48. Forrest to Cory, July 29, 1921, RBGE.

49. Forrest to Smith, August 19, 1921; Forrest to Balfour, August 30, 1921, both at RBGE.

50. Forrest to Balfour, September 3, 1921; Forrest to Cory, September 7, 1921, both at RBGE.

51. Ibid.

52. Forrest to Balfour, March 23, 1922, RBGE.

53. Ibid.; Forrest to Cory, August 4, 1922, RBGE.

54. "Burma North Eastern Frontier, China, Yunnan Province, Sheet no. 23," India Survey, 1899–1900.

55. G. Litton, "Report on the tribes on the Upper Salwen and Upper Irrawaddy with reference to the undelimited India-Yunnan frontier," 1904, Foreign Office 228/1562, PRO.

56. Stéphane Gros, "Centralization et intégration du système égalitaire Drung sous l'anfluence des pouvoirs voisins," *Péninsule* 35, no. 2 (1997): 95–

118, and Yunnan Sheng bianjizu, eds., *Dulongzu shehui lishi diaocha* (Kunming: Yunnan renmin chubanshe, 1984).

57. Li Quanming, ed., *Dulongzu wenhua daguan* (Kunming: Yunnan minzu chubanshe, 1999), 12; Liu Dacheng, ed., *Nuzu wenhua daguan* (Kunming: Yunnan minzu chubanshe, 1999), 6–8.

58. Another Wang lineage, with the grade of *tubazong*, also resided at Kangpu in the eighteenth century. For a time at least, these Wangs held jurisdiction over the Western Circuit, which included the southern portion of the upper Nu. The Wang *mugua (tuqianzong)* lineage, which later moved to Yezhi, appears to have established marriage alliances with the Wang *tubazong*. And it appears to have absorbed the Wang *tubazong*'s fiefdom into its own, as it did with the Nan *tubazong*'s territories. Gong Yin, *Zhongguo tusi zhidu* (Kunming: Yunnan minzu chubanshe, 2000), 564.

59. He Zhiwu, "Gudai Naxizu de 'mugua' zhidu," in *Naxi dongba wenhua*, ed. He Zhiwu (Changchun: Jilin jiaoyu chubanshe, 1989).

60. Joseph Rock, *The Ancient Na-Khi Kingdom of Southwest China* (Cambridge: Harvard University Press, 1947), 30.

61. Quoted in Joseph Rock, *Ancient Na-Khi Kingdom*, 334n46.

62. In the 1930s an administrator named Zhang Jiabing found a collection of such tallies in the *mugua*'s house, from the Jiaqing through the Tongzhi reigns (1796–1875), which he cited as proof that the inhabitants of the Qiu river had paid tribute to the *mugua* continuously from the eighteenth century to the early twentieth. Zhang Jiabing, "Dian Mian beiduan weiding bianjie nei zhi xiankuang" (1940), reprinted in *Yunnan Sheng Dulongzu lishi ziliao huibian*, ed. Zhongguo kexueyuan minzu yanjiusuo, Yunnan minzu diaochaozu (Beijing: Zhongguo kexueyuan minzu yanjiusuo, 1964).

63. Litton, "Report on the Tribes," Foreign Office 228/1562, PRO.

64. A. Desgodins, "De Yerkalo à Ts'é-Kou," *Bulletin de la Société de Géographie*, 6th ser., vol. 13 (1877): 176, quoted in Stéphane Gros, "Terres de confins, terres de colonisation," *Péninsule* 33, no. 2 (1996): 178.

65. Jacques Bacot, "Anthropologie du Tibet, les populations du Tibet sudoriental," *Bulletins et Mémoires de la Société d'Anthropologie de Paris*, ser. 5, vol. 9 (1909): 465. Quoted in Gros, "Terres de confins," 178. The Tibetan rulers of Tsarong also returned salt for the tribute from the Nu and Qiu.

66. The official was Xia Hu, whose report on his journey is *Nu Qiu bian'ai xiangqing* [1908], reprinted in *Yunnan Sheng Dulongzu lishi ziliao huibian*, ed. Zhongguo Kexueyuan Minzu Yanjiusuo.

67. Zhang, "Dian Mian beiduan"; Gros, "Centralization et intégration." During his visits to Yezhi in 1923, Joseph Rock found that some Nu and Drung were still paying tribute to the Wang lineage.

68. Forrest to Balfour, October 1, 1905, RBGE.

69. Litton, "Report on the tribes."

70. Joseph Rock, *A Na-Khi-English Encyclopedic Dictionary*, vol. 2 (Roma: Istituto Italiano per il Medio ed Estremo Oriente, 1972), 564, 572.

71. Rock, *Ancient Na-Khi Kingdom*, 78, 213.

72. Joseph Rock, *The Na-Khi Naga Cult and Related Ceremonies* (Rome: Istituto Italiano per il Medio ed Estremo Oriente, 1952), 2:608. I have modified Rock's translation.

73. My translation of this line is uncertain, as Rock's transliteration in *Na-Khi Naga Cult*, 613 does not match the manuscript in Joseph Rock, comp., *Na-Khi Manuscripts*, ed. Klaus L. Janert (Wiesbaden: F. Steiner, 1965), 5:4:1516.

74. Rock, *Na-Khi Naga Cult*, 607–19 and the numerous notes from other parts of these two volumes indexed to these pages.

75. Forrest to Balfour, September 3, 1921, RBGE.

76. Joseph Rock, *The Zhimä Funeral Ceremony of the Na-Khi of Southwest China* (Vienna: St. Gabriel's Mission Press, 1955), 92–100, my retranslation.

77. Forrest to Cory, August 14, 1923; Forrest to Cory, August 21, 1923, both at RBGE.

78. Forrest to Smith, May 7, 1925, RBGE.

79. Forrest made a film of the trunk section carried in procession through the main street of Tengyue; the film is in the RBGE.

80. "Inquest held at the house of George Forrest in Tengyueh, at 5.45 p.m., 6ᵗʰ January, 1932," Foreign Office 656/152, PRO.

81. Foreign Office to de Rothschild, March 3, 1932; H. Prideux-Brune to Foreign Office, January 13, 1932.; H. Prideaux-Brune to Mrs. Forrest, January 8, 1932, all at PRO.

82. While the species diversity of the *hymenanthes* subgenus of *Rhododendron* is greatest in the general region of the southeast Himalayas, recent genetic research on the section suggests that this is not its center of origin. *Hymenanthes* appears to have originated in northeast Asia. See Z. Xi et al., "Regional DNA Variation within *Rhododendron macrophyllum*," *Journal of the American Rhododendron Society* 60, no. 1 (2006): 37–41.

83. Alfred Gell, *Art and Agency: An Anthropological Theory* (Oxford: Oxford University Press, 1989), 37.

CHAPTER 5

1. Joseph Rock, Diary, March 25, 1934, RBGE.

2. Muir traveled a great deal and published very little. For a brief biography, see "Dr. Frederick Muir," *Nature* 127 (June 13, 1931): 900.

3. Joseph Rock, "A Herbarium," *Newsletter of the Hawaiian Botanical Society* 2, no. 1 (1963 [1913]): 14–15.

4. E. M. Bryan, "An Anecdote Concerning Joseph Rock," *Newsletter of the Hawaiian Botanical Society* 2, no. 1 (1963): 16. Though Bryan wrote his article forty-three years after his course with Rock, he worked from journal entries he had written at the time.

5. Todd Gunning, "Never Seen This Picture Before," in *Time Stands Still: Muybridge and the Instantaneous Photography Movement*, ed. Philip Prodger (New York: Oxford University Press, 2003), 247.

6. Rebecca Solnit makes this argument forcefully in *River of Shadows: Ead-*

weard Muybridge and the Technological Wild West (New York: Viking, 2003). But see also the essays in Prodger, ed., *Time Stands Still,* and Marta Braun, *Picturing Time: The Work of Etienne-Jules Marey, 1830–1904* (Chicago: University of Chicago Press, 1992).

7. See Rock's resume of the book, included with his 1913 report on the College of Hawaii herbarium to the Hawai'ian legislature, reprinted as Joseph Rock, "A Herbarium," *Newsletter of the Hawaiian Botanical Society* 2, no. 1 (1963 [1913]): 14–15.

8. Bryan, "An Anecdote," 16.

9. Rock, Diary, April 12, 1922, RBGE.

10. Rock, Diary, April 13, 1922, RBGE.

11. Joseph Rock to David Fairchild, April 20, 1922, Hunt Institute for Botanical Documentation, Pittsburg, PA (hereafter Hunt Institute).

12. George Forrest to J. C. Williams, May 21, 1922, RBGE.

13. Rock, Diary, April 19, 1922, RBGE.

14. Rock, Diary, February 14, 1922, RBGE; Rock, Diary, February 13, 1925, RBGE.

15. Rock, Diary, October 4, 1922, RBGE.

16. Rock, Diary, January 10, 1922, RBGE.

17. Rock, Diary, March 16, 1923, RBGE.

18. Maurice Merleau-Ponty, *The Visible and the Invisible: Followed by Working Notes,* ed. Claude Lefort (Evanston, Ill.: Northwestern University Press, 1968 [1964]), 137.

19. Ibid., 143.

20. Javier Sanjines, "Visceral Cholos: Desublimation and the Critique of Mestizaje in the Bolivian Andes," in *Impossible Presence: Surface and Screen in the Photogenic Era,* ed. Terry Smith (Chicago: University of Chicago Press, 2001), 223.

21. Gilles Deleuze and Francis Bacon, *Francis Bacon: The Logic of Sensation* (London: Continuum, 2003), 378.

22. Rock, Diary, December 18, 1924, RBGE.

23. Rock, Diary, February 13, 1931, RBGE.

24. Ibid.

25. Rock to David Fairchild, November 1, 1922, Hunt Institute.

26. Ibid.

27. See for example, Linxia Liang, *Delivering Justice in Qing China: Civil Trials in the Magistrate's Court* (Oxford: Oxford University Press, 2007).

28. Rock, Diary, May 9, 1922, RBGE.

29. Rock to Ralph E. Graves, May 17, 1925, Arnold Arboretum.

30. Rock, Diary, January 17, 1930, RBGE.

31. Rock to David Fairchild, September 22, 1922, Hunt Institute.

32. Joseph Rock, "Bandits—a Government Asset," Arnold Arboretum.

33. Rock, Diary, December 14, 1924, RBGE.

34. Rock, Diary, April 4 and 5, 1925, RBGE.

35. Rock, Diary, April 11, 1925, RBGE.

36. Rock, Diary, May 9, 1926, RBGE.

37. For a few of the many analyses of this evolution, see Christopher Pinney, "The Parallel Histories of Anthropology and Photography," in *Anthropology and Photography, 1860–1920*, ed. Elizabeth Edwards, 74–96 (New Haven: Yale University Press / Royal Anthropological Institute, London, 1992), and the other essays in this volume; Alison Griffiths, "Knowledge and Visuality in Turn of the Century Anthropology: The Early Ethnographic Cinema of Alfred Cort Haddon and Walter Baldwin Spencer," *Visual Anthropology Review* 12, no. 2 (1996): 18–43; Deborah Poole, *Vision, Race, and Modernity: A Visual Economy of the Andean Image World* (Princeton: Princeton University Press, 1997).

38. Pinney, "Parallel Histories," 80.

39. Roland Barthes, *Camera Lucida: Reflections on Photography*, trans. Richard Howard (New York: Hill and Wang, 1981 [1980]), 96.

40. Christopher Pinney, "Notes from the Surface of the Image: Photography, Postcolonialism and Vernacular Modernism," in *Photography's Other Histories*, eds. Christopher Pinney and Nicholas Peterson, 202–20 (Durham: Duke University Press, 2003).

41. Pinney, "Notes from the Surface"; Christian Metz, "Photography and Fetish," *October* 34 (1986): 82–83.

42. Ibid., 84.

43. Rock, Diary, May 13, 1926, RBGE.

44. "America's Advance Guard," *New York Times*, May 5, 1910.

45. "Thrills in Motion Picture. 'The Wild Heart of Africa' Shows Game at Close Range," *New York Times*, May 27, 1929, quoted in Robert J. Gordon, *Picturing Bushmen: The Denver African Expedition of 1925* (Athens, OH: Ohio University Press, 1997).

CHAPTER 6

1. The ruling lineage of Muli, the Bar lineage, was given the Chinese family name of Xiang in 1781; after this, the Chinese version of their names is most consistently available in written records, even though the part of the name following Xiang had a Tibetan origin. The Tibetan names are not always consistently rendered in writing. For this reason, I have chosen to use the Chinese version of names for the Bar lineage.

2. Forrest to Balfour, July 7, 1918, RBGE.

3. Winifred Kingdon-Ward, "The Flower Chief," unpublished manuscript, Royal Botanic Garden, Kew, 62. Forrest "lent" Ward the men who took him to Muli. Forrest to Reginald Cory, July 29, 1921, RBGE.

4. Muli Zangzu Zizhixian zhi bianzuan weiyuan hui, *Muli Zangzu Zizhixian zhi* (Chengdu: Sichuan renmin chubanshe, 1995), 201.

5. For Qiogyi Sangyi's name I am using the romanization commonly used

in Chinese documents, which are, other than the *Muli chos 'byung,* the only information about pre-1949 Muli.

6. *Muli chos 'byung [Muli zheng jiao shi]* (Chengdu: Sichuan minzu chubanshe, 1992).

7. This according to *Muli Zangzu Zizhxian zhi,* 81–83. Muli Zangzu Zizhixian gaikuang bianxiezu, *Muli Zangzu Zizhixian gaikuang* (Chengdu: Sichuan minzu chubanshe, 1985) lists twenty-one generations, inserting two after the fifth generation. After 1726, the two accounts match.

8 Quoted in Dai Xu, "Muli Zangzu Zizhixian zui zao de dang'an," in *Muli chos-'byung: Muli zheng jiao shi, 1580–1735.* Translated by Lobsang Gedun, compiled by Ngawang Khenrab. (Chengdu: Sichan minzu chubanshe, 1993), 4.

9. Ibid., 6.

10. Rock reads the character for *xiang* as *hang.* Joseph Rock, *The Ancient Na-Khi Kingdom of Southwest China* (Cambridge: Harvard University Press, 1947), 357n6. It is likely that he was recording its local pronunciation.

11. In fact, however, between 1726 and 1944 the succession was solely to younger brothers or brothers' sons. The Eighteenth Great Lama, Xiang Songdian Chunpin, who reigned from 1944 to 1950, was the Seventeenth's maternal cousin; the Nineteenth Great Lama, Xiang Peichu Zhaba, whose reign was from 1950 to 1953, was Songdian Chunpin's maternal cousin. *Muli Zangzu Zhizhixian zhi,* 544.

12. Joseph Rock, Diary, December 27, 1928, RBGE.

13. Geoffrey Samuel, *Civilized Shamans: Buddhism in Tibetan Societies* (Washington DC: Smithsonian Institution Press, 1993), 284.

14. In Chinese transliteration, the treasurer was the greater *suban,* the Tibetan-language secretary was the greater *zhongyi,* his assistant the lesser *zhongyi,* and the chief of the bodyguards the lesser *suban.* The Chinese-language secretary had the Chinese title of *shifu.*

15. Dai Xu, "Muli Zangzu Zizhixian zui zao de dang'an," 2.

16. *Lding-dpon* in the Wyle system.

17. *Muli Zangzu Zizhixian gaikuang,* 57.

18. Ibid., 66.

19. The best description of Yongning's political structure is in Chuan-Kang Shih, *Quest for Harmony: The Moso Traditions of Sexual Union And Family Life* (Stanford: Stanford University Press, 2010).

20. Rock, Diary, August 27, 1928, RBGE.

21. Lu Hui, "Preferential Bilateral-Cross-Cousin Marriage among the Nuosu in Liangshan," in *Perspectives on the Yi of Southwest China,* ed. Stevan Harrell (Berkeley: University of California Press, 2001), 68–80; Ann Maxwell Hill and Eric Diehl, "A Comparative Approach to Lineages among the Xiao Lianshan Nuosu and Han," in *Perspectives on the Yi,* ed. Harrell, 51–67; Ma Erzi, "Names and Geneaologies among the Nuosu of Liangshan," in *Perspectives on the Yi,* ed. Harrell, 81–93.

22. *Muli Zangzu Zizhixian gaikuang,* 57–58.

23. Rock, Diary, December 13, 1928, RBGE.

24. Rock, Diary, December 14, 1928, RBGE.

25. Joseph Rock, "The Murder of the Mu-li King," unpublished manuscript, Arnold Arboretum; Joseph Rock, "The End of Mu-li," unpublished manuscript, Arnold Arboretum.

26. Rock, Diary, January 28, 1924, RBGE. In 1921 Francis Kingdon Ward had given the Great Lama a cheap Chinese-made camera, now nowhere in evidence.

27. Ibid.

28. Rock, Diary, February 2, 1924, RBGE.

29. Clifford Geertz, *Negara: The Theatre State in Nineteenth-Century Bali* (Princeton: Princeton University Press, 1980), 122.

30. Geertz, *Negara*, 13.

31. Stanley Jeyaraja Tambiah, *The Buddhist Saints of the Forest and the Cult of Amulets: A Study in Charisma, Hagiography, Sectarianism, and Millennial Buddhism* (Cambridge: Cambridge University Press, 1984).

32. Samuel, *Civilized Shamans*, 32.

33. Ibid., 171, 220.

34. Rock, Diary, January 10, 1929, RBGE.

35. Rock, Diary, June 2, 1928, RBGE.

36. Rock, Diary, August 27, 1929, RBGE.

37. Clare Harris, "The Politics and Personhood of Tibetan Buddhist Icons," in *Beyond Aesthetics: Art and the Technologies of Enchantment*, eds. Christopher Pinney and Nicholas Thomas, (Oxford: Berg, 2001), 188; Lawrence Babb, "Glancing: Visual Interaction in Hinduism," *Journal of Anthropological Research* 37, no. 4 (1981): 398.

38. Harris, "Politics and Personhood," 185; Anjan Chakraverty, *Sacred Buddhist Painting* (New Delhi: Lustre Press, 1998).

39. Christopher Pinney, "Notes from the Surface of the Image: Photography, Postcolonialism and Vernacular Modernism," in *Photography's Other Histories*, eds. Christopher Pinney and Nicholas Peterson (Durham: Duke University Press, 2003), 2; Bernard Faure, "The Buddhist Icon and the Modern Gaze," *Critical Inquiry* 24:3 (1998): 804.

40. Harris, "Politics and Personhood," 189.

41. I am indebted to Donald Lopez (personal communication) for this point.

42. Rock, Dairy, May 18, 1926, RBGE.

43. Joseph Rock, "Inane Lamaist Practices at Ragya Monastery," unpublished manuscript, Arnold Arboretum.

44. Harris, "Politics and Personhood," 191.

45. Ibid.

46. "A King from Muli, Sikang, China," Photoprint from Scenes in Tibet, made on the Ernest Schäfer expeditions to Tibet, 1930–1939, LC-USZ62–75973, Library of Congress, Washington, D.C. See also Isrun Engelhardt, *Tibet in 1938–1939: Photographs from the Ernst Schäfer Expedition to Tibet* Chicago: Serindia, 2007).

47. Rock was commenting not on the portrait itself but on the costume. He

had shipped it to the National Geographic Society, where it surfaced at a ball, worn by Melville Grosnevor, son of the society's director, who was photographed as the "Muli King." Fisher to Rock, February 18, 1935; Rock to Fisher, March 26, 1935, both at Arnold Arboretum.

48. Rock, Diary, August 15, 1928, RBGE.

49. Rock, Diary, August 16, 1928, RBGE.

50. Rock, Diary, August 23, 1928, RBGE.

51. Rock, Diary, August 24, 1928, RBGE; September 2, 1928, RBGE.

52. On this trip, Rock had an American companion, a young man from Kansas named Hagen who worked for the American consulate in Kunming.

53. Rock, Diary, December 18, 1928, RBGE; December 19, 1928, RBGE; December 20, 1928, RBGE.

54. Rock, Diary, January 6, 1929, RBGE.

55. Rock, Diary, December 31, 1928, RBGE.

56. Rock to Ralph Graves, February 5, 1929, Arnold Arboretum.

57. Rock, Diary, February 10, 1929, RBGE.

58. Rock, Diary, January 7, 1929, RBGE.

59. Rock, Diary, August 5, 1929, RBGE.

60. Ibid.

61. Rock, Diary, August 7, 1929, RBGE.

62. Rock, Diary, August 7, 1929, RBGE; August 8, 1929, RBGE.

63. Pinney, "Notes from the Surface of the Image," 203.

CHAPTER 7

1. Albert Bumstead to John Oliver LaGorce, September 22, 1927, Hunt Institute.

2. Joseph Rock, Diary, June 13, 1926, RBGE.

3. Joseph Rock, *The Amnye Ma-chhen Range and Adjacent Regions: A Monographic Study* (Rome: Istituto Italiano per il Medio ed Estremo Oriente, 1956), 116.

4. Rock, *Amnye Ma-chhen*, 117–22.

5. Rock to David Fairchild, October 10, 1923, Hunt Institute.

6. S. B. Sutton, *In China's Border Provinces: The Turbulent Career of Joseph Rock, Botanist-Explorer* (New York: Hastings House, 1974), 82–83.

7. Rock to Charles Sprague Sargent, March 21, 1925, Arnold Arboretum; Rock, Diary, December 16, 1924, RBGE. Rock, Diary, December 16, 1924, RBGE.

8. Rock, Diary, December 13, 1924 to December 26, 1924, RBGE; Rock to Sargent, January 5, 1925, Arnold Arboretum.

9. Rock, Diary, December 27, 1924 to January 11, 1925, RBGE.

10. Rock, Diary, January 11, 1925, RBGE.

11. Rock to Sargent, January 5, 1925, Arnold Arboretum.

12. Rock to Sargent, March 21, 1925, Arnold Arboretum.

13. Rock, Diary, December 16, 1924, RBGE.

14. Rock, Diary, December 18, 1924, RBGE.

15. Rock, Diary, January 11, 1925, RBGE.

16. Rock, Diary, April 6, 1925, RBGE.

17. For an account of the *hu fa* movement, see Leslie H. Dingyan Chen, *Chen Jiongming and the Federalist Movement: Regional Leadership and Nation Building in Early Republican China* (Ann Arbor: University of Michigan Press, 1999).

18. Robert A. Kapp, *Szechwan and the Chinese Republic: Provincial Militarism and Central Power, 1911–1938* (New Haven: Yale University Press, 1973).

19. Kristin Stapleton, *Civilizing Chengdu: Chinese Urban Reform* (Cambridge, MA: Harvard University Press, 2000).

20. S. A. M. Adshead, "Salt and Warlordism in Szechwan 1914–1922," *Modern Asian Studies* 24, no. 4 (1990): 729–43.

21. Ma Xuanwei, *Sichuan junfa Yang Sen* (Chengdu: Sichuan renmin chubanshe, 1983).

22. Rock, Diary, February 12, 1925, RBGE.

23. Rock, Diary, January 1, 1925, RBGE.

24. Rock, Diary, February 16 to March 16, 1925, RBGE; Rock to Sargent, February 15, 1925; Rock to Sargent, February 18, 1925; Rock to Sargent, March 21, 1925, all letters at Arnold Arboretum.

25. Rock, Diary, March 17, 1925 to March 21, 1925, RBGE; Rock to Sargent, March 21, 1925, Arnold Arboretum.

26. Rock, Diary, March 20, 1924, RBGE.

27. Jonathan Lipman, *Familiar Strangers: A History of Muslims in Northwest China* (Seattle: University of Washington Press, 1997), 173; Jonathan Lipman, "The Border World of Gansu, 1895–1935" (PhD diss., Stanford University, 1981), 251.

28. Rock, Diary, March 14, 1925, RBGE; March 15, 1925, RBGE.

29. Rock, Diary, March 28, 1925, RBGE.

30. Lipman, *Familiar Strangers*, 18.

31. Ibid., 71.

32. Mongours were a large group of agriculturalists of uncertain origin, called by the Chinese *turen,* meaning "people of the earth." They inhabited enclaves in the areas occupied mainly by non-Muslim Chinese, in the Huangzhong district around Xining. Some were Muslim; more were adherants of the Gelugpa order. See Louis M. J. Schram's study, originally published in three parts in 1954, 1957, and 1961 in *Transactions of the American Philosophical Society,* and reprinted in a single volume as Louis M. J. Schram, *The Monguors of the Kansu-Tibetan Frontier,* ed. Kevin Stuart (Xining: Plateau Publications, 2006).

33. Schram, *Monguors*, 288.

34. Ibid., 89.

35. Ibid., 338.

36. Ibid., 349.

37. Peter Perdue, *China Marches West: The Qing Conquest of Central Eurasia* (Cambridge, MA: Belknap Press of Harvard University Press, 2005), 244–48.

38. Fernanda Pirie, "Segmentation within the State: The Reconfiguration of Tibetan Tribes in China's Reform Period," *Nomadic Peoples* 9, nos. 1–2 (2005): 83–102.

39. Robert Brainerd Ekvall, *Cultural Relations on the Kansu-Tibetan Border* (Chicago: University of Chicago Press, 1939), 69; Matthias Hermanns, *Die Nomaden von Tibet* (Vienna: Verlag Herold, 1949), 231, quoted in Geoffrey Samuel, *Civilized Shamans: Buddhism in Tibetan Societies* (Washington DC: Smithsonian Institution Press, 1993), 93.

40. See Charlene Makley, *The Violence of Liberation: Gender and Tibetan Buddhist Revival in Post-Mao China* (Berkeley: University of California Press, 2007).

41. Gong Yin, *Zhongguo tusi zhidu* (Kunming: Yunnan minzu chubanshe, 2000), 1300; Bao Yongchang and Zhang Yandu, *Taozhou Ting zhi* [18 juan] (Taibei: Chengwen chubanshe, 1970). Yang Jiqing, *tusi* from 1902 to 1937, told Rock that his ancestors had come from central Tibet, making their way across Sichuan to Chone, pacifying the tribes as they went. They settled in Chone and intermarried with the lineage of the Mongol prince of Alashan, a territory near Ningxia. Tibetan and Chinese language records, however, state that the lineage originated in Chone during the Yuan dynasty.

42. Yang Shihong, ed., *Zhuoni tusi lishi wenhua* (Lanzhou: Gansu minzu chubanshe, 2007), 71–77.

43. Ibid., 94. Thus, the grandmother of the fourteenth-generation *tusi* Yang Sheng served as his regent, "guarding the seal" from 1754 until her death in 1760. Yang Shihong, ed. *Zhuoni tusi*, 94; Gong, *Zhongguo tusi zhidu*, 1300.

44. Yang Shihong, ed., *Zhuoni tusi*, 94.

45. Yang Yong, "Taolun Zhuoni Dasi zai Anduo Zangqu de lishi diwei he yingxiang," *Journal of the Northwest University for Nationalities (Philosophy and Science)*, no. 4 (2004): 93–99.

46. Yang Maosen, "Zhuoni yinjing yuan gaisong," *Xizang yishu yanjiu*, no. 3 (2003): 69–71.

47. Qi Dianchen and Zhuoni Xian zhi bianzuan weiyuanhui, *Zhuoni Xian zhi* (Lanzhou: Gansu minzu chubanshe, 1994), 97.

48. Rock, Diary, April 22, 1925 and notes following March 3, 1927, RBGE.

49. Ekvall, *Cultural Relations*, 31.

50. Rock to Sargent, April 26, 1925, Arnold Arboretum.

51. Rock, Diary, November 22, 1925, RBGE.

52. Rock to Sargent, May 10, 1925, Arnold Arboretum.

53. Rock, Diary, April 28, 1925, RBGE. Rock to Sargent, May 10, 1925, Arnold Arboretum, makes it 270,000 ounces.

54. Rock, Diary, March 4, 1925, RBGE.

55. Rock, Diary, May 20, 1925, RBGE. Most of Rock's information about Drokwa affairs came from young Will Simpson, a missionary who, having grown up near Labrang, lived in Drokwa camps, spoke the Drokwa dialect, and had many Drokwa informants. ("He smelled so strong of butter and other unspeakable

odors that I made him sleep outside" [Rock, Diary, May 20, 1925, RBGE].) See also the autobiography of the Jamyang Shepa's military minister and brother, Apo Alo (Huang Zhengqing), who played a crucial role in the events and their aftermath. Huang Zhengqing, *Huang Zhengqing yu wu shi Jiamuyang* (Lanzhou: Gansu renmin chubanshe, 1989).

56. Rock, Diary, May 23, 1925, RBGE.

57. Rock, Diary, May 26, 1925, RBGE.

58. In his diaries and letters, Rock calls this gompa Ankur and Drakur; on his map published in *National Geographic* 57 (1930), 131–185, "Seeking the Mountains of Mystery," he calls it Dokar.

59. Huang Zhengqing, *Huang Zhengqing yu wu shi Jiamuyang,* 16–23. In this autobiography, Apa Alo gives a detailed account of the events leading up to and following the "war to expel Ma," from the point of view of Labrang's protectors. Rock, Diary, May 28, 1925, RBGE.

60. Rock, Diary, July 8, 1925, RBGE. See also Rock, "Seeking the Mountains of Mystery" for Rock's account of atrocities, possibly exaggerated, since they do not appear in his diaries.

61. After a formal protest by Ma Qi that Lu had encouraged the Tibetans to rebel, the Beijing government demoted Lu.

62. Rock, Diary, September 12, 1925, RBGE; Rock to Sargent, September 12, 1925, Arnold Arboreum.

63. Rock, Diary, November 29, 1925, RBGE. The gompa of Heicuo, between Old Taozhou and Labrang, was not so fortunate. The Muslim troops had carried off everything: cattle, clothing, grain, butter and tea, the doorknobs, window frames and yak-hair curtains. Sixteen monks had been killed, including an incarnate lama, and the highest incarnation had fled.

64. Liu Jihua, "Mingguo shiqi Gansu tusi zhidu bianqian yanjiu," *Lanzhou Jiaoyu Xueyuan xuebao,* no. 2 (2003).

65. Ma Haotian, *Gan Qing Zang bianqu kaochaji,* 3 vols. (1942–1947 [1936]), 46, quoted in Charlene E. Makley, *The Violence of Liberation: Gender and Tibetan Buddhist Revival in Post-Mao China* (Berkeley: University of California Press, 2007), 74.

66. James E. Sheridan, *Chinese Warlord: The Career of Feng Yü-hsiang* (Stanford: Stanford University Press, 1966), 193–97.

67. Lipman, "Border World of Gansu," 257.

68. Qi Dianchen, ed., *Zhuoni Xian zhi,* 21.

69. Robert Ekvall, "Revolt of the Crescent in Western China," *Asia and the Americas* 29 (1929): 946–47; Lipman, "Border World of Gansu," 266–67.

70. Lipman, "Border World of Gansu," 268; Sheridan, *Chinese Warlord,* 250–52.

71. Qi Dianchen, ed., *Zhuoni Xian zhi,* 22. Rock, however, citing missionary sources, insists that Yang Jiqing's Tibetan troops fought against the Red Army troops at Wancang, in Chone. Joseph Rock, "Rebellion in Kan-su," unpublished manuscript, Arnold Arboretum.

72. Qi Dianchen ed., *Zhuoni Xian zhi,* 22; Liu, "Mingguo shiqi Gansu tusi zhidu," 11. Rock wrote a hearsay account of the murder: Joseph Rock, "The End of the Chone Prince," unpublished manuscript, Arnold Arboretum.

73. Joseph Rock, "The Principality of Cho-ni," unpublished manuscript, Arnold Arboretum.

74. Rock learned the ethnonym "Tebbu" from American missionaries; in Chone, the people of Diebu were known generically as T'epa or more specifically by the names of their banners.

75. Rock, Diary, June 13, 1925, RBGE.

76. Rock, Diary, August 1, 1925 to August 9, 1925, RBGE; Rock to Sargent, August 9, 1925, Arnold Arboretum.

77. Rock, Diary, August 14, 1925 to August 28, 1925, RBGE. On the politics of the Sixth Panchen Lama's sojourn in Gansu and subsequent journey to Beijing, see Fabienne Jagou, "The Sixth Panchen Lama's Chinese Titles," in *Kamps pa Histories: Visions of People, Place and Authority,* ed. Lawrence Epstein, 85–102 (Boston: Brill, 2002).

78. Rock, Diary, September 5, 1925, RBGE.

79. Rock, Diary, September 11, 1925 to October 25, 1925, RBGE; Rock to Sargent, October 9, 1925; Rock to Sargent, October 25, 1925, both letters at Arnold Arboretum.

80. Rock, Diary, November 28, 1925, RBGE.

81. Rock, Diary, December 4 to December 10, 1925, RBGE.

82. Rock, Diary, January 8, 1926, RBGE.

83. Rock, "Rebellion in Kan-su."

84. Rock, Diary, July 20, 1926, RBGE.

85. Rock to Sargent, August 10, 1926, Arnold Arboretum.

86. Rock to Sargent, June 8, 1926, Arnold Arboretum.

87. Rock, "Seeking the Mountains of Mystery," 185.

88. Rock, *Amnye Ma-chhen,* 113.

CHAPTER 8

1. I have chosen to transcribe these names, as with all the Chinese names in the text, into *hanyu pinyin.* Rock's transliterations of the names were idiosyncratic and inconsistent, and he almost never wrote them in characters, so I am uncertain about my romanizations of a few. It is possible that some of these men did work for Forrest for short periods. At least one man, whom Rock hired earlier, but who did not take part in the Gansu expedition, had worked for Forrest. In a thorough comparison of faces in Rock's and Forrest's photographs, however, I find no faces that I can identify as the same.

2. Joseph Rock, Diary, November 8, 1928, RBGE. "Now my English tongue is silent and I must henceforth speak Chinese to my Nashi boys a number of whom do not speak that tongue. " The men with Rock on this trip, to Yongning, Muli, and Kangding (Dajianlu) had all participated in the Gansu expedition.

3. Joseph Rock, "Work Accomplished," typescript, Arnold Arboretum.

4. Rock to Sargent, August 17, 1924, Arnold Arboretum.

5. Rock, Diary, December 25, 1924, RBGE.

6. Rock to Sargent, December 20, 1925, Arnold Arboretum.

7. Rock to Sargent, September 22, 1926, Arnold Arboretum.

8. He Jiangyu, He Laoyu, and Joseph Rock, *Gudu zhi lu: Zhiwuxuejia, ren-leixuejia Yuesefu Luoke he ta zai Yunnan de tanxian jingli* (Kunming: Yunnan jiaoyu chubanshe, 2000), 168.

9. This was the rate Zhao Chengzhang paid his muleteers during those years.

10. Sydney White, "Fame and Sacrifice in the Gendered Construction of Naxi Identities," *Modern China* 23 no. 3 (1997): 298–327.

11. Rock, Diary, February 10, 1925, RBGE.

12. Rock, Diary, October 22, 1925, RBGE.

13. Rock, Diary, December 25, 1924, RBGE.

14. Rock, Diary, November 9, 1925, RBGE.

15. Rock, Diary, November 10, 1925, RBGE.

16. Joseph Rock, "Affidavit sworn before Robert Yoder, Vice Consul to the United States of America," August 8, 1949, Hunt Institute.

17. Rock, Diary March 13, 1926, RBGE.

18. Edgar Snow, *Journey to the Beginning* (New York: Random House, 1958), 56, quoted in S. B. Sutton, *In China's Borderlands: The Turbulent Career of Joseph Rock, Botanist-Explorer* (New York: Hastings House, 1974), 212.

19. Rock, Diary, April 8, 1927, RBGE.

20. Rock, "Affadavit," Hunt Institute.

21. The species collected are described in Outram Bangs and James L. Peters, "Birds Collected by Dr. Joseph Rock in Western Kansu and Eastern Tibet," *Bulletin of the Museum of Comparative Zoology* 68, no. 7 (1928): 313–81.

22. Rock to Outram Bangs, January 1, 1926; Bangs to Sargent, March 17, 1927, both at Arnold Arboretum.

23. Rock, Diary, February 3, 1936, RBGE.

24. Rock, Diary, August 5, 1925, RBGE. Three of the Nvlvk'ö men had stayed behind in Chone to collect there.

25. Rock, Diary, August 5, 1925, RBGE; August 7, 1925, RBGE.

26. Rock to Sargent, November 5, 1924, Arnold Arboretum.

27. James E. Sheridan, *Chinese Warlord: The Career of Feng Yü-Hsiang* (Stanford: Stanford University Press, 1966), 196.

28. Rock to Sargent, August 9, 1926, Arnold Arboretum.

29. Rock to Sargent, November 18, 1926; Rock to Sargent, November 24, 1926, both at Arnold Arboretum.

30. Rock, Diary, December 3, 1926, RBGE.

31. See for instance Aram Yengoyan, "Simmel and Frazer: The Adventure and the Adventurer," in *Tarzan Was an Eco-Tourist and Other Tales in the Anthropology of Adventure,* eds. Luis Antonio Vivanco and Robert Gordon, 27–42 (New York: Berhahn Books, 2006).

32. Georg Simmel, "The Adventure," in *Simmel on Culture: Selected Writings,* eds. David Frisby, and Mike Featherstone (London: Sage, 1997), 225.

33. Ibid., 226.
34. Ibid., 225.
35. Ibid.
36. Rock, Diary, January 22 to January 27, 1927, RBGE.
37. Rock, Diary, March 14 to March 15, 1927, RBGE; Joseph Rock, "Escape from China in 1927," unpublished manuscript, Arnold Arboretum.
38. Rock, Diary, March 16 to April 16, 1927, RBGE.
39. Rock, Diary, April 17, 1927, RBGE.
40. Rock to Graves, November 25, 1927, Hunt Institute.
41. Rock to Graves, May 24, 1927, Hunt Institute.
42. Rock, Diary, December 31, 1927, RBGE.
43. Rock to Graves, November 25, 1927, Hunt Institute.
44. Rock, Diary, April 23, 1928, RBGE.
45. Joseph Rock, "The Romance of K'a-Mä-Gyu-Mi-Gkyi," *Bulletin de l'École française d'Extrême-Orient* 39 (1939): 149, 151. I have simplified Rock's transcription system slightly, as described in the Note on Transcription," and indented text in order to highlight parallelism.
46. Rock, "Romance," plates 10, 11.
47. Rock, Diary, May 13, 1928, RBGE.
48. He, He and Rock, *Gudu zhi lu*, 167. Rock did not record his version of this incident: he did not write his diaries during those sedentary weeks.
49. Rock, Diary, January 29, 1927, RBGE.
50. Rock, Diary, September 2, 1928, RBGE.
51. Rock, Diary, September 6, 1928, RBGE.
52. Rock, Diary, October 12, 1928, RBGE.
53. Rock, Diary, November 12, 1928, RBGE.
54. Rock, Diary, December 19, 1929, RBGE.
55. Paul Fussell, *Abroad: British Literary Traveling between the Wars* (New York: Oxford University Press, 1980).

CHAPTER 9

1. Joseph Rock, Diary, November 30, 1929, RBGE.
2. Rock, Diary, December 1, 1929, RBGE.
3. Rock, Diary, December 14, 1929, RBGE; Joseph Rock, *A Na-Khi-English Encyclopedic Dictionary*, vol. 2 (Roma: Istituto Italiano per il Medio ed Estremo Oriente, 1972), 286, 497; Joseph Rock, *The Na-Khi Naga Cult and Related Ceremonies* (Rome: Istituto Italiano per il Medio ed Estremo Oriente, 1952), 1:215n375, 2:694–98, 2:729–34.
4. Rock, Diary, December 8, 1929, RBGE.
5. Rock, Diary, December 15, 1929, RBGE. Later, Rock would suggest that the unusually high incidence of suicide in the Lijiang valley, especially among young women, was the fault of dongba. Youths who killed themselves were honored with a long and elaborate special funeral called *harlallü*, for which dongba were hired to dance and read texts. Though many dongba rituals had died away

in the Lijiang valley, this one remained extremely popular in the 1920s and 1930s. Rock published a detailed translation of a text central to this rite that described the death of K'amägyumigyki, a young woman who made a suicide pact with her lover, rather than marry the man her parents chose for her. Though written in the recondite ancient language of dongba texts, the story was also told and retold as a folk tale throughout the Lijiang valley. Rock believed that dongba encouraged young lovers to imitate K'amägyumigyki by describing a heavenly land for suicide spirits in which they could live in bliss forever. See Joseph Rock, "The Romance of K'a-Mä-Gyu-Mi-Gkyi," *Bulletin de l'École Française d'Extrême-Oriente* 39, no. 1 (1939): 1–155. This has been a point of contention in Naxi studies ever since.

6. Joseph Rock, "Southwest China Expedition, Dr. Joseph F. Rock, Director, Statement of Receipts and Disembursements from March 17, 1928 through May 20, 1930," Hunt Institute; Graves to LaGorce, April 16, 1930, Hunt Institute; "Statement of appropriations to Dr. Joseph F. Rock for National Geographic Society, Southwest China Expedition, 1927–1930 Inclusive," Hunt Institute; Rock to La Gorce, May 23, 1930, Hunt Institute.

7. Joseph Rock, "Introduction," in *Na-Khi Manuscripts,* comp. Joseph Rock, ed. Claus L. Janert (Wiesbaden: Franz Steiner, 1965) 7:1:xv.

8. Rock, Diary, April 14, 1931, RBGE.

9. Rock, Diary, April 30, 1931, RBGE.

10. Rock, Diary, April 19, 1931, RBGE.

11. Jacques Bacot and Edouard Chavannes, *Les Mo-so: Ethnographie des Mo-so, Leurs Religions, Leur Langue et Leur écriture* (Leiden: E. J. Brill, 1913).

12. See Christine Matheiu, *A History and Anthropological Study of the Ancient Kingdoms of the Sino-Tibetan Borderland* (Lewiston, NY: Edwin Mellen Press, 2003) for a thorough and clever comparison of the two chronicles.

13. Joseph Rock, *The Ancient Na-Khi Kingdom of Southwest China* (Cambridge: Harvard University Press, 1947), 87n2.

14. Rock, Diary, December 31, 1932, RBGE.

15. Rock, Diary, June 30, 1932, RBGE.

16. Walter Benjamin, "The Task of the Translator," in Walter Benjamin, *Illuminations,* ed. Hannah Arendt, trans. Harry Zorn (New York: Harcourt, Brace & World, 1968) 69.

17. Ibid., 82.

18. Ibid.

19. Ibid.

20. Joseph Rock, *Zhimä Funeral Ceremony of the Na-Khi of Southwest China* (Vienna: St. Gabriel's Mission Press, 1955), 99.

21. Ibid., 158n10.

22. Ibid., 166n7.

23. Ibid., 225. I have been liberal in trying to create a more readable English version out of Rock's translation.

24. Ibid., 115. Again, I have taken the liberty of attempting to clean up Rock's version.

25. Joseph Rock, "The D'a Nv Funeral Ceremony with Special Reference to the Origin of Na-khi Weapons," *Anthropos* 50 (1955): 2. Later, Rock's speculations about Naxi origins proved attractive to Chinese scholars who theorized that most Tibeto-Burman peoples of the southwest might be descendants of ancient Qiang peoples of eastern Tibet. More recent scholarship, however, suggests that Naxi originated as an alliance of peoples living in and around the Lijiang valley.

26. Rock, "D'a Nv Funeral Ceremony," 4–5.

27. Rock, Diary, April 16, 1936, RBGE. Rock to Collector of Internal Revenue, Baltimore MD, March 20, 1955, Hunt Institute.

28. Rock was very clear, however, that no one would ever use the text as a guide, for by the time of publication, a new motor road had reached Dali and beyond, almost to Lijiang. Rock, *Ancient Na-Khi Kingdom,* 11.

29. Rock, *Ancient Na-Khi Kingdom,* 15.

30. Simpich to LaGorce, May 23, 1930, Arnold Arboretum.

31. Anonymous to Grosevenor, February 16, 1931, Arnold Arboretum.

32. Rock, *Ancient Na-Khi Kingdom,* 213.

33. Rock, Diary, April 24, 1937, RBGE.

34. Rock, Diary, July 5, 1938, RBGE.

35. S. B. Sutton, *In China's Border Provinces: The Turbulent Career of Joseph Rock, Botanist-Explorer* (New York: Hastings House, 1974), 269. Rock to Collector of Internal Revenue, Baltimore, MD, March 20, 1955, Hunt Institute.

36. Alvin Chock, "J. F. Rock, 1884–1962," *Newsletter of the Hawaiian Botanical Society* 2, no. 1 (1963): 6.

37. Rock to Elisseeff, June 15, 1944, Hunt Institute.

38. Sutton, *China's Border Provinces,* 284–85. Merill to Rock, June 30, 1944, Hunt Institute.

39. Rock, Diary, April 12, 1931, RBGE.

40. He Zhihui to Joseph Rock, 1944–1946 (twenty-one letters), Hunt Institute.

41. Sutton describes these difficulties. Sutton, *China's Border Provinces,* 286–87.

42. Joseph Rock, comp., *Na-Khi Manuscripts,* ed. Klaus L. Janert (Wiesbaden: F. Steiner, 1965), 7:2:260–61.

43. Ibid., 279.

44. Rock to Merrill, December 4, 1950; Rock to Merrill, December 12, 1950; Rock to Merrill, January 18, 1951, all at Arnold Arboretum.

45. Rock to Collector of Internal Revenue, Baltimore MD, March 20, 1955, Hunt Institute.

46. Chock, "J. F. Rock," 7.

47. Rock to Giuseppe Tucci, undated, Hunt Institute.

48. The first volume was published in 1963. The second volume was delayed for five years because of the great difficulty of completing the work without Rock's aid and a lack of funding.

49. Rock to H. R. Fletcher, January 30, 1960; Fletcher to Rock, February 11,

1960; Rock to Fletcher, April 3, 1960, all at RBGE. At the time, Fletcher was the regius keeper at the Royal Botanic Garden, Edinburgh.

50. In fact, Rock had copied these sketches from a far more skilled cartographer, the young missionary Will Simpson who had accompanied him on that journey. The sketches were simple route maps with distances measured by time, directions taken with a handheld compass from a horse's back, and no attempt to determine latitude, longitude, or compass declination. A cartographer for the National Geographic Society, Albert Bumstead, had worked them into a single, line-drawn map, published in Rock's *National Geographic* article about his trip, as "Map of the Route to the Amnyi Machen Mountains compiled from sketches made in the field by Joseph F. Rock," *National Geographic Magazine* 57 (1930): 138–39. After seeing the article, William Simpson, still in Amdo, wrote the editor to say that all the sketches had actually been his own. Several he had made when traveling alone, of areas to which Rock had never been. Rock had paid him well, he wrote, and he didn't dispute his right to publish the map, but it would have been more honorable to say by whom the original data was collected. William Simpson to Ralph Graves, May 31, 1930, Hunt Institute. Simpson did not complain about not being given credit for the maps in the 1956 Amnye Machen monograph, having been killed in Gansu by discharged soldiers in 1932. On the final set of ten maps, prepared during the last year of Rock's life and never published, Rock and Sherman finally, and without explanation, gave Simpson credit as a coauthor for the charts of northern Amdo.

51. John A. Sherman to Joseph Rock, April 16, 1961; July 13, 1961; January 12, 1962, all in Hunt Institute. Fifteen maps were planned, ten completed. The maps are now owned by the Arnold Arboretum and can be seen in digitalized form on their website: http://www.arboretum.harvard.edu/library/tibet/map.html.

52. Rock, Diary, February 11 to February, 14, 1922, RBGE.

53. Rock, small, undated notebook, RBGE.

Bibliography

ARCHIVES

Arnold Arboretum, Boston, MA
Hunt Institute for Botanical Documentation, Pittsburg, PA
Public Records Office, Kew (PRO)
Royal Botanic Garden, Edinburgh (RBGE)
Royal Botanic Garden, Kew
Royal Geographical Society, London
Royal Horticultural Institute, London

PRIMARY AND SECONDARY SOURCES

Adamson, William. *The Life of the Rev. James Morison*. London: Hodder and Stoughton, 1898.
Adshead, S. A. M. "Salt and Warlordism in Szechwan 1914–1922." *Modern Asian Studies* 24 (1990): 729–43.
Agrawal, Arun. "Dismantling the Divide between Indigenous and Scientific Knowledge." *Development and Change* 26, no. 3 (1995): 413–39.
"America's Advance Guard." *New York Times*. May 5, 1910.
Aris, Michael. *Lamas, Princes, and Brigands: Joseph Rock's Photographs of the Tibetan Borderlands of China*. Assisted by Patrick Booz. Contributions by S. B. Sutton and Jeffrey Wagner. New York: China Institute in America, 1992.
Arnold, David. *The Tropics and the Traveling Gaze: India, Landscape, and Science, 1800–1856, Culture, Place, and Nature*. Seattle: University of Washington Press, 2006.
Babb, Lawrence. "Glancing: Visual Interaction in Hinduism." *Journal of Anthropological Research* 37, no. 4 (1981): 387–401.
Backus, Charles. *The Nan-Chao Kingdom and T'ang China's Southwestern Frontier*. Cambridge: Cambridge University Press, 1981.
Bacot, Jacques, and Edouard Chavannes. *Les Mo-So: Ethnographie Des Mo-So, Leurs Religions, Leur Langue et Leur Écriture*. Leiden: E. J. Brill, 1913.

Bangs, Outram and James L. Peters. "Birds Collected by Dr. Joseph Rock in Western Kansu and Eastern Tibet." *Bulletin of the Museum of Comparative Zoology* 68, no. 7 (1928): 313–81.

Bao Yongchang and Zhang Yandu. *Taozhou Ting zhi* [18 juan]. Taibei: Chengwen chubanshe, 1970.

Barthes, Roland. *Camera Lucida: Reflections on Photography*. Translated by Richard Howard. New York: Hill and Wang, 1981 [1980].

Beale, Lionel S. *How to Work with the Microscope: Fifth Edition Revised Throughout and Much Enlarged, with One Hundred Plates, Comprising More Than Six Hundred Engravings, Some Printed in Colours*. London: Harrison, 1880.

Becaari, Odoardo, and Joseph Francis Charles Rock. *A Monographic Study of the Genus Pritchardia*. Honolulu: Bishop Museum Press, 1921.

Bello, David A. "To Go Where No Han Could Go for Long: Malaria and the Qing Construction of Ethnic Administrative Space in Frontier Yunnan." *Modern China* 31, no. 3 (2005): 283–317.

Benjamin, Walter. "The Task of the Translator." In *Illuminations*, edited by Hannah Arendt, 69–82. New York: Schocken Books, 1968.

Bower, F. O. *A Course of Practical Instruction in Botany*. London: Macmillan, 1888.

———. *Sixty Years of Botany in Britain (1875–1935): Impressions of an Eyewitness*. London: Macmillan, 1938.

Boyd, William. *Education in Ayrshire through Seven Centuries*. London: University of London Press, 1961.

Bradley, David. *Proto-Loloish*. London: Curzon Press, 1979.

Braun, Marta. *Picturing Time: The Work of Etienne-Jules Marey (1830–1904)*. Chicago: University of Chicago Press, 1992.

Bretschneider, E. *History of European Botanical Discoveries in China*. London: S. Low, Marston and Co., 1898.

Brockway, Lucile L. *Science and Colonial Expansion: The Role of the British Royal Botanic Gardens*. New York: Academic Press, 1979.

Brook, Timothy. *Geographical Sources of Ming-Qing History*. Ann Arbor: Center for Chinese Studies, University of Michigan Press, 1988.

Bryan, E. H. "An Anecdote Concerning Joseph F. Rock." *Newsletter of the Hawaiian Botanical Society* 2, no. 1 (1963): 14–16.

Carter, Paul. *The Road to Botany Bay: An Exploration of Landscape and History*. New York: Knopf, 1988.

Chakrabarty, Dipesh. *Provincializing Europe: Postcolonial Thought and Historical Difference*. Princeton: Princeton University Press, 2000.

Chakraverty, Anjan. *Sacred Buddhist Painting*. New Delhi: Lustre Press, 1998.

Chatterjee, Partha. *The Nation and Its Fragments*. Princeton: Princeton University Press, 1993.

Chavannes, Edouard. "Documents historiques et géographiques relatifs à Likiang." *T'oung Pao* 13 (1912): 564–653.

Chen, Leslie H. Dingyan. *Chen Jiongming and the Federalist Movement: Re-*

gional Leadership and Nation Building in Early Republican China. Ann Arbor: Center for Chinese Studies, University of Michigan Press, 1999.

Chen Zidan. "Lijiang Mushi tusi dang'an wenxian pingshu." *Guzhai zhengli yajiu xuekan* 6 (Nov. 2004).

Chen Zonghai. *Tengyue Ting zhi*. 1887.

Chock, Alvin. "J. F. Rock, 1884–1962." *Newsletter of the Hawaiian Botanical Society* 2, no. 1 (1963): 1–13.

Christensen, Carl Frederik Albert, and Joseph Francis Charles Rock. *Asiatic Pteridophyta Collected by Joseph F. Rock, 1920–1924*. Washington, DC: U.S. Government Printing Office, 1931.

Coates, P. D. *The China Consuls: British Consular Officers, 1843–1943*. Hong Kong: Oxford University Press, 1988.

Cohn, Bernard S. *Colonialism and Its Forms of Knowledge: The British in India*. Princeton: Princeton University Press, 1996.

Coleman, William M. "The Uprising at Batang: Khams and its Significance in Chinese and Tibetan History." In *Kamps pa Histories: Visions of People, Place and Authority*, edited by Lawrence Epstein, 31–56. Boston: Brill, 2002.

Cosgrove, Denis. *Social Formation and Symbolic Landscape*. London: Croom Helm, 1984.

———. *Geography and Vision: Seeing, Imagining and Representing the World*. London: I.B. Tauris, 2008.

Cosgrove, Denis, and Stephen Daniels, eds. *The Iconography of Landscape: Essays on the Symbolic Representation, Design, and Use of Past Environments*. Cambridge: Cambridge University Press, 1988.

Cowan, John Macqueen. *The Journeys and Plant Introductions of George Forrest, V.M.H.* London: Royal Horticultural Society / Oxford University Press, 1952.

Cox, E. H. M. *Plant-Hunting in China*. Hong Kong: Oxford University Press, 1986.

Cox, E. H. M., and Helen T. Maxwell. *Farrer's Last Journey*. London: Dulau, 1926.

Cox, E. H. M., and William T. Stearn. *The Plant Introductions of Reginald Farrer*. London: New Flora and Silva, 1930.

Cresswell, Tim. *Mobility in the Western World*. New York: Routledge, 2006.

Das, Veena, and Deborah Poole, eds. *Anthropology in the Margins of the State*. Santa Fe: School of American Research Press, 2004.

Daston, Lorraine, and Peter Galison. "The Image of Objectivity." *Representations* 40 (1992): 81–128.

David, Armand. *Journal de mon troisième voyage d'exploration dans L'Empire Chinois*. Paris: Hachette, 1875.

———. *Abbé David's Diary: Being an Account of the French Naturalist's Journeys and Observations in China in the Years 1866 to 1869*. Translated and edited by Helen M. Fox. Cambridge, MA: Harvard University Press, 1949.

David, Armand, and Émile Oustalet. *Les Oiseaux de la Chine*. Paris: G. Masson, 1877.

Davidian, H. H. *The Rhododendron Species*, 4 vols. Portland: Timber Press, 1982.

Davies, H. R. *Yunnan: The Link Between India and the Yangtze.* Cambridge: Cambridge University Press, 1909.

Dayrat, Benoit. *Les Botanistes et la flore de France: Trois siècles de découvertes.* Paris: Muséum national d'Histoire naturelle, 2003.

Dean, Arthur L., and Richard Wrenshall. "Fractionation of Chaulmoogra Oil." *Journal of the American Chemical Society* 42 (1920): 2626–45.

———. "The Treatment of Leprosy with Especial Reference to Some New Chaulmoogra Oil Derivatives." *Public Health Report* 35 (1920): 1959–74.

———. "Preparation of Chaulmoogra Oil Derivatives for the Treatment of Leprosy." *Public Health Report* 37 (1922): 1395–99.

Deleuze, Gilles. *Kant's Critical Philosophy: The Doctrine of the Faculties.* Translated by Hugh Tomlinson and Barbara Habberjam. Minneapolis: University of Minnesota Press, 1984.

———. *Francis Bacon: The Logic of Sensation.* Translated by Daniel W. Smith. London: Continuum, 2003.

Desmond, Ray. *The European Discovery of the Indian Flora.* Oxford: Clarendon Press, 1992.

Edwards, Elizabeth, Chris Gosden, and Ruth B. Phillips. *Sensible Objects: Colonialism, Museums, and Material Culture.* Oxford: Berg, 2006.

Ekvall, Robert Brainerd. "Revolt of the Crescent in Western China." *Asia and the Americas* 29 (1929): 946–47.

———. *Cultural Relations on the Kansu-Tibetan Border.* Chicago: University of Chicago Press, 1939.

———. *Fields on the Hoof: Nexus of Tibetan Nomadic Pastoralism.* New York: Holt, 1968.

Elman, Benjamin A., and Alexander Woodside, eds. *Education and Society in Late Imperial China, 1600–1900.* Berkeley: University of California Press, 1994.

Elwes, Henry John, and W. H. Fitch. *A Monograph of the Genius Lilium.* London: Taylor and Francis, 1877.

Elwes, Henry John, and Augustine Henry. *The Trees of Great Britain and Ireland.* Edinburgh: privately printed, 1906.

Engelhardt, Isrun. *Tibet in 1938–1939: Photographs from the Ernst Schäfer Expedition to Tibet.* Chicago: Serindia, 2007.

Fairchild, David. *The World Was My Garden: Travels of a Plant Explorer.* New York: Scribner's, 1941.

Fan, Fa-ti. *British Naturalists in Qing China: Science, Empire, and Cultural Encounter.* Cambridge, MA: Harvard University Press, 2003.

Farrer, Reginald. *On the Eaves of the World.* London: E. Arnold, 1917.

———. *The Rainbow Bridge.* London: E. Arnold, 1926.

Faure, Bernard. "The Buddhist Icon and the Modern Gaze." *Critical Inquiry* 24, no. 3 (1998): 768–813.

Fletcher, Harold Roy. *A Quest of Flowers: The Plant Explorations of Frank Ludlow And George Sherriff Told From Their Diaries And Other Occasional Writings.* Edinburgh: Edinburgh University Press, 1975.

Fletcher, Harold Roy, and William H. Brown. *The Royal Botanic Garden, Edinburgh, 1670–1970.* Edinburgh: Her Majesty's Stationary Office, 1970.

Forrest, George. "Journey on the Upper Salwen, October–December, 1905." *The Geographical Journal* 32 (1908): 239–66.

———. "A Lecture by Mr. George Forrest on Recent Discoveries of Rhododendrons in China." *Rhododendron Society Notes* 2 (1920): 3–23.

———. "Exploration of N.W. Yunnan and S.E. Tibet, 1921–1922." *Journal of the Royal Horticultural Society of London* 49 (1924): 25–36.

———. "The Perils of Plant Collecting." *Gardeners' Chronicle,* May 21, 1924, 325–26; May 28, 1924, 344.

———. "The Explorations and Work of George Forrest" and "Exploration for Rhododendron, 1917–22." In *Rhododendrons and the Various Hybrids,* edited by John Guille Millais, 16–26. London: Longmans, Green and Co., 1924.

Franchet, Adrien. *Plantae Davidianae Ex Sinarum Imperio,* 2 vols. Paris: G. Masson, 1884–1888.

———. *Plantæ Delavayanæ.* Paris: P. Klincksieck, 1890.

Fussell, Paul. *Abroad: British Literary Traveling between the Wars.* New York: Oxford University Press, 1980.

Fyfe, Gordon, and John Law. *Picturing Power: Visual Depiction and Social Relations.* London: Routledge, 1988.

Galison, Peter Louis. "Judgement against Objectivity." In *Picturing Science, Producing Art,* edited by Caroline Jones and Peter Galison, 327–59. New York: Routledge, 1998.

Gao Fayuan, ed. *Yunnan minzu cunzhai diaocha, Dulongzu.* Kunming: Yunnan Daxue chubanshe, 2001.

Gascoigne, John. *Science in the Service of Empire: Joseph Banks, the British State, and the Uses of Science in the Age of Revolution.* Cambridge: Cambridge University Press, 1998.

Ge Agan. *Dongba gubu wenhua.* Kunming: Yunnan renmin chubanshe, 1999.

———., ed. *Dongba wenhua zhenji.* Kunming: Yunnan meishu chubanshe, 2001.

Geertz, Clifford. *Negara: The Theatre State in Nineteenth-Century Bali.* Princeton: Princeton University Press, 1980.

Geismar, Haidy. "Malakua: A Photographic Collection." *Comparative Studies in Society and History* 48, no. 3 (2006): 520–63.

Gell, Alfred. *Art and Agency: An Anthropological Theory.* Oxford: Oxford University Press, 1989.

Giersch, C. Pat. "A Motley Throng: Social Change on Southwest China's Early Modern Frontier, 1700–1880." *The Journal of Asian Studies* 60, no. 1 (2001): 67–94.

Goh, Daniel P. "States of Ethnography: Colonialism, Resistance, and Cultural Transcription in Malaya and the Philippines, 1890s—1930s." *Comparative Studies in Society and History* 49, no. 1 (2007): 109–42.

Goldstein, Melvyn C. *A History of Modern Tibet, 1913–1951: The Demise of the Lamaist State.* Berkeley: University of California Press, 1991.

Gong Yin. *Zhongguo tusi zhidu.* Kunming: Yunnan minzu chubanshe, 2000.

Gordon, Robert J. *Picturing Bushmen: The Denver African Expedition of 1925.* Athens, OH: Ohio University Press, 1997.

Grabowsky, Volker, and Andrew Turton. *The Gold and Silver Road of Trade and Friendship: The McLeod and Richardson Diplomatic Missions to Tai States in 1837.* Chiang Mai: Silkworm Books, 2003.

Griffiths, Alison. "Knowledge and Visuality in Turn of the Century Anthropology: The Early Ethnographic Cinema of Alfred Cort Haddon and Walter Baldwin Spencer." *Visual Anthropology Review* 12, no. 2 (1996): 18–43.

Gros, Stéphane. "Terres de confins, terres de colonisation: Essai sur les Marches sino-tibétaines du Yunnan à travers l'implantation de la Mission du Tibet." *Péninsule* 33, no. 2 (1996): 147–211.

———. "Centralization et intégration du système égalitaire Drung." *Péninsule* 35, no. 2 (1997): 95–114.

Groves, Colin P., Yingxiang Wang, and Peter Grubb. "Taxonomy of Musk Deer, Genus Moschus (Moschidae, Mammalia)." *Acta Theriologica Sinica* 15, no. 3 (1995): 181–97.

Guan Xuexuan and Wan Xianyan. *Lijiang Fu zhi lue.* 1743.

Gunning, Todd. "Never Seen This Picture Before." In *Time Stands Still: Muybridge and the Instantaneous Photography Movement*, edited by Phillip Prodger. New York: Oxford University Press, 2003.

Guo Dalie. *Zhongguo shaoshu minzu da cidian: Naxizu juan.* Nanning: Guangxi minzu chubanshe, 2002.

Guo, Xiaolin. *State and Ethnicity in China's Southwest.* Leiden: Brill, 2008.

Hall, Catherine. *Civilising Subjects: Metropole and Colony in the English Imagination, 1830–1867.* Oxford: Polity, 2002.

Han, Lianxian. "Looking for the Giant Azalea." *The Garden* 115, no. 5 (1990): 261–64.

Handel-Mazzetti, Heinrich. *A Botanical Pioneer in South West China: Experiences And Impressions of an Austrian Botanist During the First World War; With 48 Photographs Taken by the Author And Seven Maps.* Translated by David Winstanley. Chippenham: Antony Rowe, 1996 [1927].

Harrell, Stevan, ed. *Perspectives on the Yi of Southwest China.* Berkeley: University of California Press, 2001.

Harris, Clare. "The Politics and Personhood of Tibetan Buddhist Icons." In *Beyond Aesthetics: Art and the Technologies of Enchantment*, edited by Christopher Pinney and Nicholas Thomas, 181–99. Oxford: Berg, 2001.

Harris, Clare, and Tsering Shakya. *Seeing Lhasa: British Depictions of the Tibetan Capital, 1936–1947.* Chicago: Serinda Publications, 2003.

Harris, Jonathan. *Writing Back to Modern Art: After Greenberg, Fried and Clark.* New York: Routledge, 2005.

He Hong, Guo Dalie, and Yang Yihong. *Naxi xiangxingzi dongba wen ying yong.* Kunming: Yunnan daxue chubanshe, 2006.

He Jiangyu, He Laoyu, and Joseph Rock. *Gudu zhi lu: Zhiwuxuejia, renleixuejia Yuesefu Luoke he ta zai Yunnan de tanxian jingli.* Kunming: Yunnan jiaoyu chubanshe, 2000.

He Jiquan, He Baolin, Guo Dalie, and Yang Yihong. *Naxizu dongba gu ji xuan du*. Kunming: Yunnan daxue chubanshe, 2006.

He Wanbao and He Jiaxiu, eds. *Naxi dongba gu ji yi zhu quan ji*. 100 vols. Kunming: Yunnan renmin chubanshe, 1999.

He Yaohua and Yang Fuyuan, eds. *Lijiang Yulong Shan quyu cunzhai fazhan yu shengtai diaocha*. Kunming: Yunnan renmin chubanshe, 1998.

He Zhiwu. "Gudai Naxizu de 'mugua' zhidu." In *Naxi dongba wenhua*, edited by He Zhiwu. Changchun: Jilin jiaoyu chubanshe, 1989.

He Zhiwu and Guo Dalie. "Dongba jiao de paixi he xianzhuang." In *Dongba wenhua lunji*, edited by Guo Dalie and Yang Shiguang, 38–54. Kunming: Yunnan renmin chubanshe, 1985.

Henry, Augustine. *Notes on the Economic Botany of China*. Shanghai: Presbyterian Mission Press, 1893.

Hogden, Margaret. *Early Anthropology in the Sixteenth and Seventeenth Centuries*. Philadelphia: University of Pennsylvania Press, 1964.

Hong Congwen. "Cong Banhong shijian kan Yunnan bianjiang guanli jigou de yunzuo." *Zhongguo bianjiang shidi yanjiu* 3 (1997): 69–79.

Hooker, Joseph Dalton. *The Rhododendrons of Sikkim-Himalaya: Being an Account, Botanical And Geographical, of the Rhododendrons Recently Discovered In the Mountains of Eastern Himalaya; From Drawings And Descriptions Made On the Spot, During a Government Botanical Mission to That Country*. London: Reeve, 1849.

———. *Himalayan Journals: Notes of a Naturalist In Bengal, the Sikkim And Nepal Himalayas, the Knasia Mountains*. London: J. Murray, 1855.

Hosie, Alexander. *Three Years in Western China: A Narrative of Three Journeys in Ssu-Chuan, Kuei-Chow, and Yün-Nan*. London: G. Philip and Son, 1890.

———. *On the Trail of the Opium Poppy: A Narrative of Travel in the Chief Opium-Producing Provinces of China*. London: G. Philip and Son, 1914.

Hua, Cai. *A Society without Fathers or Husbands: The Na of China*. Cambridge, MA: Zone Books, 2001.

Huang Zhengqing. *Huang Zhengqing yu wu shi Jiamuyang*. Lanzhou: Gansu renmin chubanshe, 1989.

Jackson, Anthony. *Elementary Structures of Na-Khi Ritual*. Göteborg, 1970.

———. "Kinship, Suicide, and Pictographs among the Na-khi." *Ethnos* 36 (1971): 52–93.

———. *Na-Khi Religion: An Analytical Appraisal of the Na-Khi Ritual Texts*. The Hague: Mouton, 1979.

Jackson, Anthony, and Anshi Pan. "The Authors of Naxi Ritual Books, Index Books and Books of Divination." In *Naxi and Moso Ethnography: Kin, Rites, Pictographs*, edited by Michael Oppitz and Elisabeth Hsu, 237–73. Zürich: Völkerkundemuseum Zürich, 1998.

Jacobs, Nancy J. "The Intimate Politics of Ornithology in Colonial Africa." *Comparative Studies in Society and History* 48, no. 3 (2006): 564–603.

Jagou, Fabienne. "The Sixth Panchen Lama's Chinese Titles." In *Kamps pa His-*

tories: Visions of People, Place and Authority, edited by Lawrence Epstein, 85–102 (Boston: Brill, 2002).

Jardine, N., J. A. Secord, and E. C. Spary, eds. *Cultures of Natural History,* Cambridge: Cambridge University Press, 1996.

Jay, Martin. "Scopic Regimes of Modernity." In *Vision and Visuality,* edited by Hal Foster, 3–27. Seattle: Bay Press, 1988.

Jones, Caroline A. *Eyesight Alone: Clement Greenberg's Modernism and the Bureaucratization of the Senses.* Chicago: University of Chicago Press, 2005.

Jones, Caroline A., and Peter Louis Galison, eds. *Picturing Science, Producing Art.* New York: Routledge, 1998.

Kant, Immanuel. *Observations on the Feeling of the Beautiful and the Sublime.* Translated by John T. Goldthwait. Berkeley: University of California Press, 1960 [1764].

———. *Critique of Judgement.* Translated by James Creed Meredith. Oxford: Oxford University Press, 1973 [1790].

Kaplan, E. Ann. *Looking for the Other: Feminism, Film, and the Imperial Gaze.* New York: Routledge, 1997.

Kapp, Robert A. *Szechwan and the Chinese Republic: Provincial Militarism and Central Power, 1911–1938.* New Haven: Yale University Press, 1973.

———. "Chunking: A Center of Warlord Power, 1926–1937." In *The Chinese City Between Two Worlds,* edited by Mark Elvin and G. William Skinner, 143–70. Stanford: Stanford University Press, 1974.

Keane, Webb. *Christian Moderns: Freedom and Fetish in the Missionary Encounter.* Berkeley: University of California Press, 2007.

Kingdon Ward, Francis. *The Land of the Blue Poppy: Travels of a Naturalist in Eastern Tibet.* Cambridge: Cambridge University Press, 1913.

———. *The Mystery Rivers of Tibet: A Description of the Little-Known Land Where Asia's Mightiest Rivers Gallop in Harness Through the Narrow Gateway of Tibet; Its Peoples, Fauna, and Flora.* London: Seeley, Service and Co., 1923.

———. *Rhododendrons.* London: Latimer House, 1949.

Kristeva, Julia. *Powers of Horror: An Essay on Abjection, European Perspectives.* New York: Columbia University Press, 1982.

Lack, H. Walter. *Alexander Von Humboldt and the Botanical Exploration of the Americas.* Munich: Prestel, 2009.

Landau, Paul, and Deborah Kaspin, eds. *Images and Empires: Visuality in Colonial and Postcolonial Africa.* Berkeley: University of California Press, 2002.

Leach, Edmund Ronald. *Political Systems of Highland Burma: A Study of Kachin Social Structure.* Cambridge, MA: Harvard University Press, 1954.

Lecercle, Jean-Jacques. *Philosophy of Nonsense: The Intuitions of Victorian Nonsense Literature.* London: Routledge, 1994.

Lee, James. "The Legacy of Immigration in Southwest China, 1250–1850." *Annales de démographie historique* (1982): 279–304.

Lempert, Michael P. "Denotational Textuality and Demeanor Indexicality in Tibetan Buddhist Debate." *Journal of Linguistic Anthropology* 15, no. 2 (2005): 171–93.

Lévi-Strauss, Claude. *Elementary Structures of Kinship*. Boston: Beacon Press, 1969 [1949].

Li Guowen. *Dongba wenhua cidian*. Kunming Shi: Yunnan jiaoyu chubanshe, 1997.

Li Jinming. *Dulongzu wenhua daguan*. Kunming Shi: Yunnan minzu chubanshe, 1999.

Li Lincan. *Moxiezu de jingdian yanjiu*. Taibei: Dongfang wenhua shuju, 1971.

Li Quanming, ed. *Dulongzu wenhua daguan*. Kunming: Yunnan minzu chubanshe, 1999.

Li Ruming, ed. *Lijiang Naxizu Zizhixian zhi*. Kunming: Yunnan renmin chubanshe, 2001.

Li Xi, ed. *Jinshen zhi lu: Naxizu dongba shenlu tu*. Kunming: Yunnan meishu chubanshe, 2001.

Liang, Linxia. *Delivering Justice in Qing China: Civil Trials in the Magistrate's Court*. Oxford: Oxford University Press, 2007.

Lipman, Jonathan Neaman. "The Border World of Gansu, 1895–1935." PhD diss., Stanford University, 1981.

———. *Familiar Strangers: A History of Muslims in Northwest China*. Seattle: University of Washington Press, 1997.

Lisuzu jianshi bianxiezu, ed. *Lisuzu jianshi*. Kunming: Yunnan renmin chubanshe, 1983.

Liu Dacheng, ed. *Nuzu wenhua daguan*. Kunming: Yunnan minzu chubanshe, 1999.

Liu Jihua. "Mingguo shiqi Gansu tusi zhidu bianqian yanjiu." *Lanzhou Jiaoyu Xueyuan xuebao*, no 2 (2003): 9–13.

London, Jack. "The Lepers of Molokai." *Women's Home Journal*, January 1908.

Lyte, Charles. *Frank Kingdon-Ward: The Last of the Great Plant Hunters*. London: J. Murray, 1989.

Ma Xuanwei. *Sichuan junfa Yang Sen*. Chengdu: Sichuan renmin chubanshe, 1983.

Makley, Charlene E. *The Violence of Liberation: Gender and Tibetan Buddhist Revival in Post-Mao China*. Berkeley: University of California Press, 2007.

Martland, S. P. "Caruso's First Recordings: Myth and Reality." *ARSC Journal* 25, no. 2 (1994): 192–201.

Mathieu, Christine. *A History and Anthropological Study of the Ancient Kingdoms of the Sino-Tibetan Borderland—Naxi and Mosuo*. Lewiston, NY: Edwin Mellen Press, 2003.

Maxwell, Anne. *Colonial Photography and Exhibitions: Representations of the "Native" and the Making of European Identities*. London: Leicester University Press, 1999.

McKhann, Charles Fremont. "Fleshing Out the Bones: Kinship and Cosmology in Naqxi Religion." PhD diss., University of Chicago, 1992.

———. "Naxi, Rerkua, Moso, Meng: Kinship, Politics and Ritual on the Yunnan-Sichuan Frontier." In *Naxi and Moso Ethnography: Kin, Rites, Pictographs,* edited by Michael Oppitz and Elizabeth Hsu, 23–45. Zürich: Völkerkundemuseum Zürich, 1998.

McLean, Brenda. *A Pioneering Plantsman: A. K. Bulley and the Great Plant Hunters.* London: Stationary Office, 1997.

———. *George Forrest, Plant Hunter.* Edinburgh: Antique Collectors' Club, 2004.

Merleau-Ponty, Maurice. *The Visible and the Invisible: Followed by Working Notes.* Edited by Claude Lefort. Translated by Alphonso Lingis. Evanston: Northwestern University Press, 1968 [1964].

Metz, Christian. *The Imaginary Signifier: Psychoanalysis and the Cinema.* Translated by Celia Britton, Annwyl Williams, Ben Brewster, and Alfred Guzzetti. Bloomington, IN: Indiana University Press, 1982.

———. "Photography and Fetish." *October* 34 (1986): 81–90.

Mitchell, Timothy. *Colonising Egypt.* Cambridge: Cambridge University Press, 1988.

Mueggler, Erik. *The Age of Wild Ghosts: Memory, Violence, and Place in Southwest China.* University of California Press, 2001.

———. "The Lapponicum Sea: Matter, Sense and Affect in the Botanical Exploration of Southwest China and Tibet." *Comparative Studies in Society and History* 47, no. 3 (2005): 442–79.

———. "Reading, Glaciers, and Love in the Botanical Exploration of Southwest China and Tibet." *Michigan Quarterly Review* 44, no. 4 (2005): 722–53.

Mu-li chos-'byung: Muli zheng jiao shi, 1580–1735. Translated by Lobsang Gedun. Compiled by Ngawang Khenrab. Chengdu: Sichan minzu chubanshe, 1993.

Muli Zangzu Zizhiian gaikuang bianxiezu. *Muli Zangzu Zizhixian gai kuang.* Chengdu: Sichuan minzu chubanshe, 1985.

Muli Zangzu Zizhixian zhi bianzuan weiyuan hui. *Muli Zangzu Zizhixian zhi.* Chengdu: Sichuan renmin chubanshe, 1995.

Myint-U, Thant. *The Making of Modern Burma.* Cambridge: Cambridge University Press, 2001.

Naxizu jianshi bianxiezu. *Naxizu jianshi.* Kunming: Yunnan renmin chubanshe, 1984.

Naxizu wenxue shi bianxiezu. *Naxizu wenxue shi.* Chengdu: Sichuan minzu chubanshe, 1992.

Needham, Joseph. *Science and Civilization in China,*. Vol. 3. Cambridge: Cambridge University Press, 1959.

Nuzu jianshi bianxiezu. *Nuzu jianshi.* Kunming: Yunnan renmin chubanshe, 1987.

Olalquiaga, Celeste. *The Artificial Kingdom: A Treasury of the Kitsch Experience.* New York: Pantheon Books, 1998.

Oppitz, Michael, and Elisabeth Hsu, eds. *Naxi and Moso Ethnography: Kin, Rites, Pictographs.* Zürich: Völkerkundemuseum Zürich, 1998.

Parascandola, John. "Chaulmoogra Oil and the Treatment of Leprosy." *Pharmacological History* 45, no. 2 (2003): 47–57.

Perdue, Peter. *China Marches West: The Qing Conquest of Central Eurasia.* Cambridge, MA: Belknap, 2005.

———. "Sustainable Development on China's Frontiers." In *Natures Past,* edited by Paolo Squatriti, 172–79. Ann Arbor: University of Michigan Press, 2006.

Pinney, Christopher. "The Parallel Histories of Anthropology and Photography." In *Anthropology and Photography, 1860–1920,* edited by Elizabeth Edwards, 74–96. New Haven: Yale University Press, 1992.

———. "Underneath the Banyan Tree: William Crooke and Photographic Depictions of Caste." In *Anthropology and Photography, 1860–1920,* edited by Elizabeth Edwards, 165–73. New Haven: Yale University Press, 1992.

Pinney, Christopher, and Nicolas Peterson. *Photography's Other Histories.* Objects/Histories. Durham: Duke University Press, 2003.

Pinney, Christopher, and Nicholas Thomas. *Beyond Aesthetics: Art and the Technologies of Enchantment.* Oxford: Berg, 2001.

Pirie, Fernanda. "Segmentation within the State: The Reconfiguration of Tibetan Tribes in China's Reform Period." *Nomadic Peoples* 9, nos. 1–2 (2005): 83–102.

Poole, Deborah. *Unruly Order: Violence, Power, and Cultural Identity in the High Provinces of Southern Peru.* Boulder: Westview Press, 1994.

———. *Vision, Race, and Modernity: A Visual Economy of the Andean Image World.* Princeton: Princeton University Press, 1997.

Prakash, Gyan. "Science 'Gone Native' in Colonial India." *Representations* 40 (1992): 153–78.

Pratt, Mary Louise. *Imperial Eyes: Travel Writing and Transculturation.* London: Routledge, 1992.

Prodger, Phillip. *Time Stands Still: Muybridge and the Instantaneous Photography Movement.* New York: Oxford University Press, 2003.

Qi Dianchen and Zhuoni Xian zhi bianzuan weiyuanhui. *Zhuoni Xian zhi.* Lanzhou: Gansu minzu chubanshe, 1994.

Radlkoffer, L., and Joseph Francis Charles Rock. "New and Noteworthy Hawaiian Plants." *Hawaii Board of Agriculture and Forestry Botanical Bulletin* 1 (1911): 1–14.

Raffles, Hugh. *In Amazonia: A Natural History.* Princeton: Princeton University Press, 2002.

Rawski, Evelyn. *Education and Popular Literacy in China.* Ann Arbor: University of Michigan Press, 1979.

Reingold, Edward M., and Nachum Dershowitz. *Calendrical Tabulations, 1900–2200.* Cambridge: Cambridge University Press, 2002.

Richards, Thomas. *The Commodity Culture of Victorian England: Advertising and Spectacle, 1851–1914.* Stanford: Stanford University Press, 1990.

Rock, Joseph Francis Charles. "A New Hawaiian Scaevola (S. Swezayana)." *Bulletin of the Torrey Botanical Club* 36 (1909): 645–50.

———. *Notes Upon Hawaiian Plants, with Descriptions of New Species and Varieties.* Honolulu: College of Hawaii, 1911.

———. "A Herbarium." *Newsletter of the Hawaiian Botanical Society* 2, no. 1 (1963 [1913]): 14–15.

———. *The Indigenous Trees of the Hawaiian Islands.* Honolulu: privately printed, 1913.

———."List of Hawaiian Names of Plants." *Hawaii Board of Agriculture and Forestry Botanical Bulletin* 2 (1913): 2–20.

———. "The Sandalwoods of Hawaii: A Revision of the Hawaiian Species of the Genus Santalum." *Hawaii Board of Agriculture and Forestry Botanical Bulletin* 3 (1916): 1–43.

———. "The Ohia Lehua Trees of Hawaii: A Revision of the Hawaiian Species of the genus Metrosideros Banks, with Special Reference to the Varieties and Forms of Metrosideros Collina (Forster) A. Gray Subspecies Polymorpha (Gaud.) Rock." *Hawaii Board of Agriculture and Forestry Botanical Bulletin* 4 (1917): 1–76.

———. *The Ornamental Trees of Hawaii.* Honolulu: privately printed, 1917.

———. "The Arborescent Indigenous Legumes of Hawaii." *Hawaii Board of Agriculture and Forestry Botanical Bulletin* 5 (1919): 1–53.

———. "The Hawaiian genus Kokia: A Relative of the Cotton." *Hawaii Board of Agriculture and Forestry Botanical Bulletin* 6 (1919): 1–22.

———. "A Monographic Study of the Hawaiian Species of the Tribe Lobelioideae, Family Campanulaceae." *Bernice P. Bishop Museum Memorial* 7, no. 2 (1919): i–xvi, 1–395.

———. *The Leguminous Plants of Hawaii: Being an Account of the Native, Introduced and Naturalized Trees, Shrubs, Vines and Herbs, Belonging to the Family Leguminosae.* Honolulu: Hawaiian Sugar Planter's Association Experiment Station, 1920.

———. "Hunting the Chaulmoogra Tree." *National Geographic Magazine* 41 (1922): 243–76.

———. "Banishing the Devil of Disease among the Nashi." *National Geographic Magazine* 46 (1924): 473–99.

———. "Experiences of a Lone Geographer." *National Geographic Magazine* 48 (1925): 331–47.

———. "Life among the Lamas of Choni." *National Geographic Magazine* 54 (1928): 569–619.

———. "Seeking the Mountains of Mystery." *National Geographic Magazine* 57 (1930): 131–85.

———. "Glories of the Minya Konka." *National Geographic Magazine* 58 (1930): 385–437.

———. "Konka Risumgongba, Holy Mountain of the Outlaws." *National Geographic Magazine* 60 (1931): 1–65.

———. "Sungmas, the Living Oracles of the Tibetan Church." *National Geographic Magazine* 68 (1935): 475–86,

———. "Studies in Na-khi Literature, I: The Birth and Origin of Dto-Mba Shi-

Lo, the Founder of Mo-So Shaminism, According to Mo-so Manuscripts." *Bulletin de l'École Française d'Extrême-Oriente* 37 (1937): 1–39.

———. "Studies in Na-khi Literature, II: The Na-khi Ha zhi p'i; Or the Road the Gods Decide." *Bulletin de l'École Française d'Extrême-Oriente* 37 (1937): 40–119.

———. "The Romance of K'a-Mä-Gyu-Mi-Gkyi: A Na-khi Tribal Love Story Translated from Na-khi Pictographic Manuscripts." *Bulletin de l'École Française d'Extrême-Oriente* 39, no. 1 (1939): 1–155.

———. *The Ancient Na-Khi Kingdom of Southwest China.* Cambridge, MA: Harvard University Press, 1947.

———. "The Mùan-bpö Ceremony or the Sacrifice to Heaven as Practiced by the Na-khi," *Monumenta Serica* 13 (1948): 1–160.

———. *The Na-Khi Naga Cult and Related Ceremonies.* 2 vols. Rome: Istituto Italiano per il Medio ed Estremo Oriente, 1952.

———. *The Zhimä Funeral Ceremony of the Na-Khi of Southwest China: Described and Translated from Na-Khi Manuscripts.* Vienna: St. Gabriel's Mission Press, 1955.

———. "The D'a Nv Funeral Ceremony with Special Reference to the Origin of Na-Khi Weapons." *Anthropos* 50 (1955): 1–31.

———. *The Amnye Ma-Chhen Range and Adjacent Regions: A Monographic Study.* Rome: Istituto Italiano per il Medio ed Estremo Oriente, 1956.

———. *A Na-Khi-English Encyclopedic Dictionary.* Rome: Istituto Italiano per il Medio ed Estremo Oriente, 1963 (vol. 1), 1972 (vol. 2).

———., comp. *Na-Khi Manuscripts.* Edited by Klaus L. Janert. Wiesbaden: F. Steiner, 1965. Vol. 5, Part 4; Vol. 7, Parts 1 and 2.

Rowe, William T. "Education and Empire in Southwest China: Ch'en Hung-Mou in Yunnan, 1733–1738." In *Education and Society in Late Imperial China, 1600–1900,* edited by Benjamin A. Elman and Alexander Woodside, 417–57. Berkeley: University of California Press, 1994.

———. *Saving the World: Chen Hongmou and Elite Consciousness in Eighteenth-Century China.* Stanford: Stanford University Press, 2001.

Ryan, James R. *Picturing Empire: Photography and the Visualization of the British Empire.* London: Reaktion Books, 1997.

Samuel, Geoffrey. *Civilized Shamans: Buddhism in Tibetan Societies.* Washington, DC: Smithsonian Institution Press, 1993.

Samuel, Geoffrey, Hamish Gregor, and Elisabeth Stutchbury. *Tantra and Popular Religion in Tibet.* New Delhi: International Academy of Indian Culture / Aditya Prakashan, 1994.

Schram, Louis M. J. *The Monguors of the Kansu-Tibetan Frontier.* Edited by Kevin Stuart. Xining: Plateau Publications, 2006 [1957].

Schweinfurth, Ulrich, and Heidrun Marby Schweinfurth. *Exploration in the Eastern Himalayas and the River Gorge Country of Southeastern Tibet: Francis (Frank) Kingdon Ward (1885–1958): An Annotated Bibliography with a Map of the Area of His Expeditions.* Wiesbaden: Steiner, 1975.

Scottish Rock Garden Club. *George Forrest, V.M.H., Explorer and Botanist, Who by His Discoveries and Plants Successfully Introduced Has Greatly Enriched Our Gardens. 1873–1932.* Edinburgh: Stoddart and Malcolm, 1935.

Sheridan, James E. *Chinese Warlord: The Career of Feng Yü-Hsiang.* Stanford: Stanford University Press, 1966.

Shih, Chuan-kang. *Quest for Harmony: The Moso Traditions of Sexual Union and Family Life.* Stanford: Stanford University Press, 2010.

Shulman, Nicola. *A Rage for Rock Gardening: The Story of Reginald Farrer.* London: Short Books, 2002.

Sichuan Sheng bianjizu, ed. *Sichuan Sheng Naxizu shehui lishi diaocha.* Chengdu: Sichuan Sheng shehui kexueyuan chubanshe, 1987.

Simmel, George. "The Adventure." In *Simmel on Culture: Selected Writings,* edited by David Frisby and Mike Featherstone, 221–32 (London: Sage, 1997).

Sirr, Henry Charles. *China and the Chinese: Their Religion, Character, Customs and Manufactures.* London: Wm S. Orr, 1849.

Smith, Warren W. *Tibetan Nation: A History of Tibetan Nationalism and Sino-Tibetan Relations.* Boulder: Westview Press, 1996.

Snow, Edgar. *Journey to the Beginning.* New York: Random House, 1958.

Solnit, Rebecca. *River of Shadows: Eadweard Muybridge and the Technological Wild West.* New York: Viking, 2003.

Song Zhaolin. *Eya Da Cun: Yi kuai juda de shehui huo huashi.* Chengdu: Sichuan renmin chubanshe, 2003.

Stapleton, Kristin. *Civilizing Chengdu: Chinese Urban Reform.* Cambridge, MA: Harvard University Press, 2000.

Stein, Rolf Alfred. *The World in Miniature: Container Gardens and Dwellings in Far Eastern Religious Thought.* Stanford: Stanford University Press, 1990.

Steinmetz, George. "The Devil's Handwriting: Precolonial Discourse, Ethnographic Acuity, and Cross-Identification in German Colonialism." *Comparative Studies in Society and History* 45, no. 1 (2003): 41–95.

Stevenson, J. B., J. Hutchinson, H. V. Tagg, and A. Rehder. *The Species of Rhododendron.* Edinburgh: The Rhododendron Society, 1930.

Stoler, Ann Laura. *Carnal Knowledge and Imperial Power: Race and the Intimate in Colonial Rule.* Berkeley: University of California Press, 2002.

Sutton, S. B. *In China's Border Provinces: The Turbulent Career of Joseph Rock, Botanist-Explorer.* New York: Hastings House, 1974.

Tagliacozzo, Eric. "Ambiguous Commodities, Unstable Frontiers: The Case of Burma, Siam, and Imperial Britain, 1800–1900." *Comparative Studies in Society and History* 46, no. 2 (2004): 354–77.

Tambiah, Stanley Jeyaraja. *The Buddhist Saints of the Forest and the Cult of Amulets: A Study in Charisma, Hagiography, Sectarianism, and Millennial Buddhism.* Cambridge: Cambridge University Press, 1984.

Tang Rong and Li Zicheng. "Minguo chunian Nujiang zhibian shuping." *Chuxiong Shifan xuebao* 15, no. 4 (Oct. 2000): 81–85.

Taussig, Michael. *Mimesis and Alterity: A Particular History of the Senses.* New York: Routledge, 1993.

Tayman, John. *The Colony.* New York: Scribner, 2006.

Thornton, Thomas. *Being and Place among the Tlingit.* Seattle: University of Washington Press, 2008.

T'ien, Ju-K'ang. *Religious Cults of the Pai-I Along the Burma-Yunnan Border.* Ithaca, NY: Cornell Southeast Asia Program, 1986.

Tosh, John. *A Man's Place: Masculinity and the Middle-Class Home in Victorian England.* New Haven: Yale University Press, 1999.

Turrill, William Bertram. *Pioneer Plant Geography: The Phytogeographical Researches of Sir Joseph Dalton Hooker.* The Hague: M. Nijhoff, 1953.

———. *The Royal Botanic Gardens: Kew, Past and Present.* London: H. Jenkins, 1959.

———. *Joseph Dalton Hooker: Botanist, Explorer, and Administrator.* London: Thomas Nelson and Sons, 1963.

Vavilov, N. I. *The Origin, Variation, Immunity and Breeding of Cultivated Plants: Selected Writings.* Translated by K. Starr Chester. Waltham, MA: Chronica Botanica, 1951.

Veitch, James H. *Hortus Veitchii: A History of the Rise of the Nurseries of Messrs. James Veitch and Sons, Together with an Account of the Botanical Collectors and Hybridists Employed by them and a list of the most Remarkable of their Introductions.* London: James Veitch and Sons, 1906.

Vivanco, Luis Antonio, and Robert J. Gordon. *Tarzan Was an Eco-Tourist and Other Tales in the Anthropology of Adventure.* New York: Berhahn Books, 2006.

Ward, Julian. *Xu Xiake (1587–1641): The Art of Travel Writing.* Richmond, Surrey: Curzon, 2001.

White, Sydney. "Fame and Sacrifice in the Gendered Construction of Naxi Identities." *Modern China* 23, no. 3 (1997): 298–327.

Whittle, Tyler. *The Plant Hunters.* Philadelphia: Chilton Book Company, 1970.

Wile, Raymond R. "The Gramophone Becomes a Success in America, 1896–1898." *ARSC Journal* 27, no. 2 (1996): 139–70.

Willis, J. C. "The Endemic Flora of Ceylon, with Reference to Geographical Distribution and Evolution in General." *Philosophical Transactions of the Royal Society of London. Series B, Containing Papers of a Biological Character* 206 (1915): 307–42.

———. "The Distribution of Species in New Zealand." *Annals of Botany* 30 (1916): 437–57.

———. "The Evolution of Species in Ceylon, with reference to the Dying Out of Species." *Annals of Botany* 30 (1916): 1–23.

Wilson, Ernest H. *A Naturalist in Western China with Vasculum Camera and Gun, Being Some Account of Eleven Years' Travel, Exploration, and Observation in the more Remote Parts of the Flowery Kingdom.* New York: Doubleday, Page and Co., 1913.

Woodman, Dorothy. *The Making of Burma.* London: Cresset Press, 1962.

Wright, Stanley Fowler. *Hart and the Chinese Customs.* Belfast: W. Mullan, 1950.

Xi, Z., et al. "Regional DNA Variation within *Rhododendron macrophyllum.*" *Journal of the American Rhododendron Society* 60, no. 1 (2006): 37–41.

Xia Hu. *Nu Qiu bian'ai xiangqing* (1908), reprinted in *Yunnansheng Dulongzu lishi ziliao huibian*, edited by Zhongguo kexueyuan minzu yanjiusuo Yunnan minzu diaochazu, 11–15. Beijing: Zhongguo kexueyuan minzu yanjiusuo, 1964.

Xie Benshu. "Cong Pianma shijian dao Banhong shijian: Zhong Mian bianjie lishi yange wenti." *Yunnan shehui kexue*, no. 4 (2000): 72–81.

Xin Fachun. *Ming Mu shi yu Zhongguo Yunnan zhi kai fa*. Taibei Shi: Wenshi zhe chubanshe, 1985.

Xu Hongzu. *Xu Xiake Youji*. Edited by Ding Wenjiang. Beijing: Shang wu yin shu guan, 1986.

Yang, Bin. "Horses, Silver and Cowries: Yunnan in Global Perspective." *Journal of World History* 15, no. 3 (2004).

———. *Between Winds and Clouds: The Making of Yunnan (Second Century BCE to Twentieth Century CE)*. New York: Columbia University Press, 2009.

Yang Maosen. "Zhuoni yinjing yuan gaishu." *Xizang yishu yanjiu*, no. 3 (2003): 69–71.

———. "Zhuoni Shendingsi fawu yishu." *Xizang yishu yanjiu*, no. 2 (2006): 7–9.

Yang Shihong. *Zhuoni tusi lishi wenhua*. Lanzhou: Gansu minzu chubanshe, 2007.

Yang Xuezheng, ed. *Yunnan zongjiao shi*. Kunming: Yunnan renmin chubanshe, 1999.

Yong Yong. "Taolun Zhuoni dasi zai Anduo Zangqu de lishi diwei he yingxiang." *Journal of the Northwest University for Nationalities (Philosophy and Science)*, no. 4 (2004): 92–99.

Younghusband, Francis Edward, ed. *Peking to Llasa: The Narrative of Journeys in the Chinese Empire Made by the Late Brigadier General George Pereira, Jr.* Boston: Houghton Mifflin, 1926.

Yu Haibo and Yu Jianhua. *Mu shi tusi yu Lijiang*. Kunming: Yunnan minzu chubanshe, 2002.

Yunnan Sheng bianjizu, ed. *Dulongzu shehui lishi diaocha*. Kunming: Yunnan renmin chubanshe, 1984.

Zhang Jiabing. "Dian Mian beiduan weiding bianjie nei zhi xiankuang." In *Yunnan Sheng Dulongzu lishi ziliao huibian*, edited by Zhongguo kexueyuan minzu yanjiusuo, Yunnan minzu diaochaozu. Beijing: Zhongguo kexueyuan minzu yanjiusuo, 1964.

Zheng Feizhou. *Naxi dongba wenzi zi su yanjiu*. Beijing: Minzu chubanshe, 2005.

Zhongdian Xian difangzhibianzuan weiyuanhui. *Zhongdian Xian zhi*. Kunming: Yunnan minzu chubanshe, 1997.

Zhou Bin. *Dongba wen yitizi yan jiu*. Shanghai: Huadong shifan daxue chubanshe, 2005.

Index

George Forrest is identified as GF. Joseph Rock is identified as JR. Page numbers in *italics* denote images.

Text:	10/13 Aldus
Display:	Aldus
Compositor:	Integrated Composition Systems
Indexer:	Victoria Baker
Printer and binder:	IBT Global

Milton Keynes UK
Ingram Content Group UK Ltd.
UKHW030243030224
437175UK00002B/261

9 780520 269033